# Encyclopedia of Scientific Principles, Laws, and Theories

# Encyclopedia of Scientific Principles, Laws, and Theories

## Volume 2: L–Z

Robert E. Krebs

*Illustrations by Rae Déjur*

GREENWOOD PRESS
Westport, Connecticut • London

**Library of Congress Cataloging-in-Publication Data**

Krebs, Robert E., 1922–
Encyclopedia of scientific principles, laws, and theories / Robert E. Krebs ; illustrations by Rae Déjur.
p. cm.
Includes bibliographical references and index.
ISBN: 978-0-313-34005-5 (set : alk. paper)
ISBN: 978-0-313-34006-2 (vol. 1 : alk. paper)
ISBN: 978-0-313-34007-9 (vol. 2 : alk. paper)
1. Science—Encyclopedias. 2. Science—History—Encyclopedias. 3. Physical laws—Encyclopedias. I. Title.
Q121.K74 2008
503—dc22 2008002345

British Library Cataloguing in Publication Data is available.



Library of Congress Catalog Card Number: 2008002345
ISBN: 978-0-313-34005-5 (set)
978-0-313-34006-2 (vol. 1)
978-0-313-34007-9 (vol. 2)

First published in 2008

Greenwood Press, 88 Post Road West, Westport, CT 06881
An imprint of Greenwood Publishing Group, Inc.
www.greenwood.com

Printed in the United States of America

The paper used in this book complies with the Permanent Paper Standard issued by the National Information Standards Organization (Z39.48–1984).

10 9 8 7 6 5 4 3 2 1

To Carolyn, who contributed much to this encyclopedia as my researcher, pre-editor, constructive critic, proofreader, supportive wife, and friend.

# Contents


List of Entries ix

Preface xxi

Introduction xxiii

THE ENCYCLOPEDIA 1

Glossary 587

Appendix A: Alphabetical Listing of Entries by Scientific Discipline 607

Appendix B: Nobel Laureates in Chemistry (1901–2007) 613

Appendix C: Nobel Laureates in Physics (1901–2007) 619

Appendix D: Nobel Laureates in Physiology or Medicine (1901–2007) 627

Selected Bibliography 635

Index 645

# List of Entries


Abbe's Theory for Correcting Lens Distortions
Abegg's Rule and Valence Theory
Abel's Theory of Groups (Abelian Groups)
Adams' Concept of Hydrogenation
Adhemar's Ice Age Theory
Agassiz's Geological Theories
Agricola's Theories of Earthquakes and Volcanoes
Airy's Concepts of Geologic Equilibrium
Al-Battani's Theories
Alvarez's Hypotheses of Subatomic Collisions
Ambartsumian's Theory of Stellar Associations
Amdahl's Law
Ampère's Theories of Electrodynamics
Anaximander's Concepts and Ideas
Anderson's Positron Theory
Anderson's Theories and Model
Ångström's Principle of Spectrum Analysis and Related Theories
Arago's Wave Theory of Light and Arago's Disk
Arber's Concept of the Structure of DNA
Archimedes' Theories
Aristotle's Theories
Arrhenius' Theories, Principles, and Concepts
Aston's Whole Number Rule
Atomism Theories
The Auger Effect
Avogadro's Law, Hypotheses, and Number

Baade's Theories of Stellar Phenomena
Babbage's Theory of Computing
Babinet's Principle
Babo's Law
Bacon's Concept of Inductive Reasoning
Baekeland's Concept of Synthetic Polymerization
Baer's Laws of Embryonic Development
Baeyer's Strain Theory for Compound Stability
Bahcall's Theory for the Solar Neutrino Model
Bakker's Dinosaur Theory
Balmer Series
Baltimore's Hypothesis for the Reverse Transfer of RNA to DNA
Banach's Theory of Topological Vector Spaces

Banting's Theory for Isolating Pancreatic Insulin
Bardeen's Theory of Superconductivity
Barringer's Impact Theory of Craters
Beaumont's Theory for the Origin of Mountains
Becquerel's Hypothesis of X-Ray Fluorescence
Beer's Law
Behring's Theory of Immunology
Bell's Law (also known as the Bell–Magendie Law)
Bergeron's Theory of Cloud Processes
Bernoulli's Law of Large Numbers
Bernoulli's Principle
Berzelius' Chemical Theories
Bessel's Astronomical Theories
Bethe's Theory of Thermonuclear Energy
Biot–Savart Law
Birkeland's Theory of the Aurora Borealis
Bjerknes' Theory of Air Masses
Black's Theories of Heat
Bode's Law for Planetary Orbits
Bohm's Interpretation of the Uncertainty Theory for Electrons
Bohr's Quantum Theory of Atomic Structure
Bok's Globules Theory of Star Formation
Boltzmann's Laws, Hypotheses, and Constant
Bonnet's Theories of Parthenogenesis and Catastrophism
Boole's Theory of Symbolic Logic
Born–Haber Theory of Cycle Reactions
Boyle's Law
Bradley's Theory of a Moving Earth
Brahe's Theory of the Changing Heavens
Buffon's Theories of Nature
Bunsen's Theory of the Spectrochemistry of Elements
The $B_2FH$ (Burbidge–Burbidge–Fowler–Hoyle) Theory

Cagniard De La Tour's Concept of "Critical State"
Cailletet's Concept for Liquefying Gases
Calvin's Carbon Cycle
Candolle's Concept of Plant Classification
Cannizzaro's Theory of Atomic and Molecular Weights
Cantor's Mathematical Theories
Cardano's Cubic Equation
Carnot's Theories of Thermodynamics
Casimir Force (Effect)
Caspersson's Theory of Protein Synthesis
Cassini's Hypothesis for the Size of the Solar System
Cavendish's Theories and Hypothesis
Celsius Temperature Scale
Chadwick's Neutron Hypothesis
Chambers' Theory of the Origin of Life
Chandrasekhar Limit
Chang's Theories and Concepts
Chang's Theory of Capacitation
Chapman–Enskog Kinetic Theory of Gases
Chargaff's Hypothesis for the Composition of DNA
Charles' Law
Charney's Theoretical Meteorology
Charpak's Concept of Tracking Particles
Charpentier's Glacier Theory
Chevreul's Theory of Fatty Acids
Chu's Hypothesis for "High Temperature" Superconductivity
Clarke's Supergene Theory
Claude's Concept for Producing Liquid Air
Clausius' Laws and Theory of Thermodynamics
Cockcroft–Walton Artificial Nuclear Reaction
Cohn's Bacteria and Cell Theories
Compton's Wave–Particle Hypotheses
Conway's Game of Life Theory
Copernicus' Cosmology Theories
Corey's Theory of Retrosynthetic Analysis
Coriolis' Theory of Forces Acting on Rotating Surfaces
The Cori Theory of Catalytic Conversion of Glycogen

Coulomb's Laws
Couper's Theory for the Structure of Carbon Compounds
Crick–Watson Theory of DNA
Crookes' Radiation Theories
Crutzen's Theory of Ozone Depletion
Curies' Radiation Theories and Hypotheses
Curl's Hypothesis for a New Form of Carbon
Cuvier's Theories of Anatomy and Taxonomy

Daguerre's Concept of How to "Freeze" Images Made by the Camera Obscura
Dale's Theory of Vagus Nerve Stimuli
D'Alembert's Principle of Fluid Dynamics
Dalton's Law and Theories
Dana's Theory of Geosyncline
Daniell's Concept of the Electro-Chemical Cell
Darlington's Theory of Cell Nuclear Divisions
Darwin's Theory of Evolution by Natural Selection
Davisson's Theory of Diffraction of Electrons
Davy's Concept That Electric Current Can Be Used to Separate Elements
Dawkins' Theory of Evolution
De Beer's Germ-Layer Theory
De Broglie's Wave Theory of Matter
Debye–Hückel Theory of Electrolytes
Dehmelt's Electron Trap
Delbrück's and Luria's Phage Theory
Democritus' Atomic Theory of Matter
Descartes' Theories and Philosophy
De Vries' "Pangenes" Theory of Evolution
Dewar's Concept of Liquefying Gases
D'Herelle's Bacteriolytic Theory
Dicke's Theory of the Big Bang
Diesel's Concept of an Internal Combustion Engine
Dirac's Relativistic Theories
Djerassi's Theory for Synthetic Oral Contraception
Döbereiner's Law of Triads
Dobzhansky's Theory of Genetic Diversity
Domagk's Concept of Dyes as an Antibiotic
Doppler's Principle
Douglass' Theory of Dendrochronology
The Drake Equation
Draper's Ray Theory
Dulbecco's Theory of Cancer Cell Transformation
Dumas' Substitution Theory
Dyson's Theory of Quantum Electrodynamics

Eddington's Theories and Concepts
Edison's Theory of Thermionic Effect
Ehrlich's "Designer" Drug Hypothesis
Eigen's Theory of Fast Ionic Reactions
Einstein's Theories, Hypotheses, and Concepts
Einthoven's Theory that the Heart Generates an Electric Current
Ekman's Hypothesis of the Coriolis Effect on Ocean Currents
Eldredge–Gould Theory of Punctuated Evolution
Elion's Theory for Cell Differences
Elton's Theory of Animal Ecology
Enders' Theory for Cultivation of Viruses
Eötvös' Rule (Law)
Erasistratus' Theory of Anatomy and Physiology
Eratosthenes' Mathematical Concepts
Ernst's Theory of the Magnetic Moment of Atomic Nuclei
Esaki's Theory of Tunnel Diodes
Euclid's Paradigm for All Bodies of Knowledge
Eudoxus' Theory of Planetary Motion
Euler's Contributions in Mathematics
Everett's Multiple-Universe Theory of Reality
Ewing's Hypothesis for Undersea Mountain Ridges
Eyring's Quantum Theory of Chemical Reaction Rates

Fabricius' Theory of Embryology
Fahrenheit's Concept of a Thermometer
Fairbank's Quark Theory
Fajans' Rules for Chemical Bonding
Fallopius' Theories of Anatomy
Faraday's Laws and Principles
Fermat's Principles and Theories
Fermi's Nuclear Theories
Fessenden's Concept of the Thermionic Diode
Feynman's Theory of Quantum Electrodynamics (QED)
Fibonacci's Numbering System
Fick's Laws of Diffusion
Fischer's Projection Formulas
Fitzgerald's Concept of Electromagnetic Contraction
Fizeau's Theory of the Nature of Light as a Wave
Fleischmann's Theory for Cold Fusion
Fleming's Bactericide Hypothesis
Fleming's Rules for Determining Direction of Vectors
Flerov's Theory of Spontaneous Fission
Florey's Theory of Mucus Secretions
Flory's Theory of Nonlinear Polymers
Foucault's Theories of Light and Earth's Rotation
Fourier's Theories of Heat Conduction and Harmonic Wave Motion
Fowler's Theory of Stellar Nucleosynthesis
Fox's Theory of Proteinoid Microspheres
Fracastoro's Theory of Disease
Franck's Theory of Discrete Absorption of Electrons
Frankland's Theory of Valence
Franklin's Concept of DNA Structure
Franklin's Theories of Electricity
Fraunhofer's Theory of White Light
Fresnel's Theory for Multiple Prisms
Friedman's Theory of the Quark Structure of Nucleons
Friedmann's Theory of an Expanding Universe
Frisch's Theory of a Chain Reaction

Gabor's Theory of Reproducing Three-Dimensional Images
Galen's Theories of Anatomy and Physiology
Galileo's Theories
Gallo's HIV-AIDS Theory
Galton's Theory of Eugenics
Galvani's Theories of Galvanization and Animal Tissue Electricity
Gamow's Theories of the Universe and DNA
Garrod's Theory of Congenital Metabolic Disorders
Gassendi's Theories
Gauss' Mathematics and Electromagnetism Theorems
Gay-Lussac's Law of Combining Volumes
Geiger–Nutter Law (Rule) for Decay of Radioactive Isotopes
Geller's Theory of a Nonhomogeneous Universe
Gell-Mann's Theories for Subatomic Particles
Gerhardt's Type Theory for Classifying Organic Compounds
Giauque's Theory of Adiabatic Demagnetization
Gibbs' Theory of Chemical Thermodynamics
Gilbert's Theory for DNA Sequencing
Gilbert's Theory of Magnetism
Glaser's Concept of a Bubble Chamber for Detecting Subnuclear Particles
Glashow's Unifying Theory of the Weak Forces
Godel's Incompleteness Theorem
Gold's Cosmological Theories
Goldstein's Theory for the Metabolism of Cholesterol, Fats, and Lipids
Gould's Hypothesis of "Punctuated Equilibrium"
Graham's Laws of Diffusion and Effusion
Guth's Theory of an Inflationary Universe

Haber's Theories
Hadamard's Theory of Prime Numbers

Hadley's Hypothesis for the Cause of the Trade Winds
Haeckel's Biological Theories
Hahn's Theories of Nuclear Transmutations
Haldane's Theories of Genetics, Evolution, and Origins of Life
Hale's Solar Theories
Halley's Theories for Comets and Stars
Hall Effect of Electrical Flow
Hamilton's Mathematical Theories
Hardy's Mathematical Theories
Harkins' Nuclear Theories
Harvey's Theory for the Circulation of the Blood
Haüy's Geometric Law of Crystallization
Hawking's Theories of the Cosmos
Haworth's Formula
Heisenberg's Uncertainty Principle and Theory of Nucleons
Helmholtz's Theories and Concepts
Helmont's Theory of Matter and Growth
Henry's Principles of Electromagnetism
Herschel's Stellar Theories and Discoveries
Hertzsprung's Theory of Star Luminosity
Hertz's Theory for Electromagnetic Waves
Hess' Sea-Floor Spreading Hypothesis
Hess' Theory for the Ionization of Gases
Hewish's Theory of Pulsars
Higgins' Law of Definite Composition
Higgs' Field and Boson Theories
Hodgkin's Theory of Organic Molecular Structure
Hoffmann's Theory of Orbital Symmetry
Hooke's Laws, Theories, and Ideas
Hoyle's Theories of the Universe
Hubble's Law and Constant
Hückel's MO Theory or Rule and the Debye–Hückel Theory
Huggins' Theory of Spectrosopic Astronomy
Huygens' Theories of Light and Gravity

Ideal Gas Law
I-Hsing's Concepts of Astronomy
Ingenhousz's Theory of Photosynthesis
Ingold's Theory for the Structure of Organic Molecules
Ingram's Sickle Cell Theory
Ipatieff's Theory of High-Pressure Catalytic Reactions
Isaacs' Theory of Proteins Attacking Viruses

Jacob–Monod Theory of Regulator Genes
Jansky's Theory of Stellar Radio Interference
Janssen's Theory of Spectral Lines of Sunlight
Jeans' Tidal Hypothesis for the Origin of the Planets
Jeffreys' Theory of Genetic (DNA) Profiling
Jeffreys' Seismological Theories
Jenner's Inoculation Hypothesis
Jerne's Theory of Clonal Selection of Antibodies
Johanson's Theory for the Evolution of Humans
Joliot-Curies' Theory of Artificial Radioactivity
Josephson's Theory of Semiconductors
Joule's Law and Theories

Kamerlingh-Onnes' Theory of Matter at Low Temperatures
Kapitsa's Theory of Superfluid Flow
Kapteyn's Theory of Galactic Rotation
Karle's Theory for Determining Molecular Structure
Kekule's Theory of Carbon Compounds
Kelvin's Concept of Energy
Kendall's Theory for Isolating Adrenal Steroids
Kepler's Three Laws of Planetary Motion
Kerr's Theory of Quadratic Electro-Optic Effect (i.e., Kerr Effect)
Kerst's Theory for Accelerating Nuclear Particles
Khorana's Theory of Artificial Genes
Kimura's Neo-Darwinian Theory for Mutations

Kimura's Theory for Variations in Earth's Latitudes
Kipping's Theory of Inorganic-Organic Chemistry
Kirchhoff's Laws and Theories
Kirkwood's Asteroid Gap Theory
Klitzing's Theory for the Quantization of the Hall Effect
Koch's Germ-Disease Postulate
Kohlrausch's Law for the Independent Migration of Ions
Krebs Cycle
Kroto's "Buckyballs"
Kuiper's Theory for the Origin of the Planets
Kusch's Theory for the Magnetic Moment of the Electron

Lagrange's Mathematical Theorems
Lamarck's Theories of Evolution
Lambert's Theories
Lamb's Theory for the Quantum States of the Hydrogen Atom
Landau's "Two-Fluid Model" for Helium
Landauer's Principle for Very-Large-Scale Integration
Landsteiner's Theories of Blood Groups
Langevin's Concept for Use of Ultrasound
Langley's Theories of the Nervous System
Langmuir's Theories of Chemical Bonding and Adsorption of Surface Chemistry
Laplace's Theories and Nebular Hypothesis
Larmor's Theories of Matter
Laurent's Theories for Chemical "Equivalents" and "Types"
Lavoisier's Theories of Combustion, Respiration, and Conservation of Mass
Lawrence's Theory for the Acceleration of Charged Particles
Leakeys' Anthropological Theories
Leavitt's Theory for the Periodicity/Luminosity Cycle of Cepheid Variable Stars
Le Bel's Theory of Isomers
Le Chatelier's Principle
Lederberg's Hypothesis for Genetic Engineering
Lederman's Two-Neutrino Hypothesis
Lee's Theories of Weak Nuclear Interaction
Leeuwenhoek's Theory of Microscopic Life
Leibniz's Theory for "The Calculus"
Leishman's Hypothesis for Parasitic Diseases
Lemaître's Theory for the Origin of the Universe
Lenard's Theory for Electron Emission
Lenz's Law of Electromagnetics
Levene's Tetra-Nucleotide Hypothesis
Levi-Montalcini Cell Growth Theory
Lewis' Theory of Covalent Bonds
Liebig's Theory of Isomers and Organic Compound Radicals
Lindemann's Theory of Pi
Linnaeus' Theories for the Classification of Plants and Animals
Lister's Hypothesis of Antisepsis
Lockyer's Solar Atmosphere Theories
Lorentz's Physical Theories of Matter
Lorenz's Theory for Complex/Chaotic Systems
Lovell's Theory of Radio Astronomy
Lowell's Theory of Life on Mars
Lyell's' Theory of Uniformitarianism
Lysenko's Theory of the Inheritance of Acquired Characteristics

Mach's Number
Maiman's Theory for Converting the Maser to the Laser
Malpighi's Theory for the Detailed Structure of Animals and Plants
Malthusian Population Catastrophe Theory
Malus' Law for the Polarization of Light
Mansfield's Theory of Magnetic Resonance
Marconi's Theory of Radio Telegraphy
Margulis' Endosymbiotic Cell Theory

Martin's Theory of Chromatography
Matthias' Theory of Superconductivity
Maunder's Theory for Sunspots' Effects on Weather
Maupertuis' Principle of Least Action
Maxwell's Theories
Maynard Smith's Theory of Evolution
McClintock's Theory of Cytogenetics
McMillan's Concept of "Phase Stability"
Meissner Effect
Meitner's Theory of Nuclear Fission
Mendeleev's Theory for the Periodicity of the Elements
Mendel's Law of Inheritance
Merrifield's Theory of Solid-Phase Peptide Synthesis
Meselson–Stahl Theory of DNA Replication
Mesmer's Theory of Animal Magnetism
Metcalfe's Law
Meyer's Theory for the Periodicity of the Elements
Michelson's Theory for the "Ether"
Miescher's Nuclein Theory
Miller's Theory for the Origin of Life
Millikan's Theory for the Charge of Electrons
Minkowski's Space-Time Theory
Minsky's Theory of Artificial Intelligence (AI)
Misner's Theory for the Origin of the Universe
Mitscherlich's Law of Isomorphism
Mohorovicic's Theory of the Earth's Interior Structure
Montagnier's Theory for the HIV Virus
Moore's Law
Moseley's Law
Muller's Theory of Mutation
Mulliken's Theory of Chemical Bonding
Mullis' Theory for Enzymatic Replication of DNA

Nambu's Theory for the "Standard Model"
Nash's Embedding Theorems
Nathans' Theory for Restriction Enzymes
Natta's Theory for High Polymers
Néel's Theories of Ferrimagnetism and Antiferromagnetism
Neher's "Patch Clamp" to Record Small Ionic Currents
Nernst's Heat Theorem
Newcomb's Theory for the Speed of Light
Newlands' Law of Octaves
Newton's Laws and Principles
Nicholas' Theory of an Incomplete Universe
Nicolle's Theory for the Cause of Typhus
Noddack's Hypothesis for Producing Artificial Elements
Noether's Theorem
Norrish's Theory of Very Fast Reactions
Northrop's Hypothesis for the Protein Nature of Enzymes
Noyce's Concept for the Integrated Circuit

Ochoa's Theory for the Synthesis of RNA
Odling's Valence Theory
Oersted's Theory of Electromagnetism
Ohm's Law
Oken's Cell Theory
Olbers' Paradox
Oliphant's Concepts of Isotopes for Light Elements
Oort's Galaxy and Comet Cloud Theories
Oparin's Theory for the Origin of Life
Oppenheimer's Contributions to Theoretical Physics
Ostwald's Theories and Concept of Chemistry

Paracelsus' Concepts of Medicine
Pardee's Theory for Cell Enzyme Synthesis
Parkes' Theory for Separating Metals from Ores
Pascal's Concepts, Laws, and Theorems
Pasteur's Germ and Vaccination Theories

Pauling's Theory of Chemical Bonding
Pauli's Exclusion Principle
Pavlov's Theory of Associative Learning by Respondent Conditioning
Peano's Axioms and Curve Theorem
Pearson's Statistical Theories
Peierls' Concept for Separating U-235 from U-238
Penrose's Theories for Black Holes, "Twistors," and "Tiling"
Penzias' Theory for the Big Bang
Perl's Theory for a New Lepton
Perrin's Theory of Molecular Motion
Perutz's Theory of Molecular Structure of Hemoglobin
Pfeiffer's Phenomenon: The Theory of Bacteriolysis
Planck's Formula and Quantum Theory
Pogson's Theory for Star Brightness
Ponnamperuma's Chemical Theory for the Origin of Life
Porter's Theory for the Structure of Human Gamma Globulin
Poseidonius' Concept of Earth's Circumference
Poynting's Theories
Prévost's Theory for the Exchange of Heat Radiation
Priestley's Theories of Electrical Force and Dephlogisticated Air
Prigogine's Theories of Dissipative Structures and Complex Systems
Proust's Law of Definite Proportions
Ptolemy's Theory of a Geocentric Universe
Purcell's Theory of Nuclear Magnetic Resonance (NMR)
Pythagoras' Theorem

Quantum Theories: From 1900 to 2008

Rabi's Theory of Magnetic Moment of Particles
Raman's Theory of Light Scattering
Ramón y Cajal's Neuron Theory
Ramsay's Hypothesis for Inert Gases
Ramsey's Chemical Shift Theory for Improved MRI
Raoult's Law
Raup's Theory of Cyclic Extinction of Animals
Rayleigh's Light Scattering Law
Ray's Theories of Fossils and Plant Classification
Redi's Theory of Spontaneous Generation
Reed's Theory of the Transmission of Yellow Fever
Regiomontanus' Theory for Trigonometry
Reichenbach's Theory of Probability Based on Logical Empiricism (aka Logical Positivism)
Reichstein's Theory of the Chemical Role of the Adrenal Gland
Reines' Theory of Natural Neutrinos
Revelle's Theory of Global Warming
Ricciolo's Theory of Falling Bodies
Richardson's Law of Thermionic Emission
Richter's Theory of Earthquake Magnitude
Riemann's Theory for Differential Geometry
Robbins' Theory for the Poliovirus
Roberts' Theory of Split Genes
Roche's "Limit" Theory
Röentgen's Theory of X-Rays
Romer's Theory for the Speed of Light
Rossi's Theory for Cosmic Radiation
Rowland's Theory of Chlorofluorocarbons' Effects on the Ozone
Rubbia's Theory of Intermediate Vector Bosons
Rubin's Theory of Dark Matter
Rumford's Theory of Relating Work to Heat
Russell's Theory of Stellar Evolution
Rutherford's Theories of Radioactivity/ Transmutation and Atomic Structure
Rydberg's Theory of Periodicity for Atomic Structure
Ryle's Theory of Using Radio Astronomy for Observing Distant Galaxies

Sabin's Theory for Attenuated Live Polio Vaccine
Sachs' Theory of Photosynthesis
Sagan's Theories of Nuclear Winter and the Cosmos
Saha's Theory of Thermal Ionization
Sakharov's Nuclear Fusion Theory
Salam's Theory for the Properties of Elementary Particles
Sandage's Theories of Quasars and the Age of the Universe
Sanger's Theories of the Structure of Proteins and Gene Splitting
Sarich's Theory of Utilizing Protein to Genetically Date Man/Ape Divergence
Scheele's Theory of the Chemical Composition of Air
Schiaparelli's Theory of Regularity in the Solar System
Schleiden's Cell Theory for Plants
Schmidt's Theory of the Evolution and Distribution of Quasars
Schneider's Theory of Biological Systems and Climate Change
Schrödinger's Theory of Wave Mechanics
Schwann's Theory of Animal Cells
Schwarzschild's "Black Hole" Theory
Schwinger's Theory for Renormalization
Seaborg's Hypothesis for Transuranium Elements
Seebeck's Theory of Thermoelectricity
Segrè's Hypothesis for the Antiproton
Shapley's Theory of Globular Clusters
Sharp's Theory for the "Splicing" of DNA
Shepard's Theory of Submarine Canyon Formation
Shockley's Theory of Semiconductors
Sidgwick's Theory of Coordinate Bonds
Siemens' Theory for Regenerating Heat
Simon's Third Law of Thermodynamics
Slipher's Theories of Interstellar Gases and Andromeda
Smoot's Theory of a Nonuniform Universe
Snell's Law
Soddy's Displacement Law for Radioactive Decay and Theory of Isotopes
Sorensen's Negative Logarithms Representing Hydrogen Ion Concentration
Spallanzani's Theory Refuting Spontaneous Generation
Spedding's Theories
Spencer-Jones' Concept for Measuring Solar Parallax
Stahl's Phlogiston Theory
Stark's Theories
Stefan's Theory of Black Box Radiation
Steinberger's Two-Neutrino Theory
Steno's Theory for Fossil Formation
Stern's Theory for the Magnetic Moment of the Proton
Stokes' Laws of Hydrodynamics and Fluorescence
Stoney's Theory of the Electron
Strasburger's Law of Cytology
Struve's Theory of Interstellar Matter
Suess' Theory of Continental Drift
Swammerdam's Theory of Preformation
Szilard's Theory of Neutrons Sustaining a Chain Reaction

Tamm's Theory of the Cherenkov Effect
Tartaglia's Mathematical Solution to Cubic Equations
Tatum's Theory of Gene-Controlling Enzymes
Taylor's Theory of Gravitational Waves
Teller's Theory for the Hydrogen Bomb
Temin's Theory for Transcribing RNA Information into DNA
Tesla's Concept of High-Voltage Alternating Current
Thales' Theory That Water Is the Basis for All Things
Theophrastus' Concepts for Plant Classification
Theorell's Theory of Enzyme Action
Thomson's Electron Theory
Ting's Theory for a New Photon-Like Particle

Tiselius' Hypothesis for Protein Analysis
Todd's Theory for the Structure and Synthesis of Nucleotides, Nucleosides, and Nucleotide Co-enzymes
Tomonaga's Theory of Relativistic Quantum Electrodynamics
Tonegawa's Theory of Antibodies and the Immune System
Torricelli's Vacuum and Theorem
Townes' Theory for Amplifying Electromagnetic Waves
Townsend's Theory of Collision Ionization
Turing's Theory for Testing Computer Intelligence
Turner's Theory for Measuring Outer Energy Levels of Molecules
Tyndall's Theory for the Transmission of Light through Gases

Uhlenbeck's Theory of Electron Spin
Ulam's "Monte Carlo" System
Urey's Gaseous Diffusion and Origin-of-Life Theories

Van Allen Radiation Belts
Van de Graaff's Concept of Producing High Voltage
Van der Meer's Theory of Particles to Confirm the "Weak Force"
Van der Waals' Equation for Gas Molecules
Van't Hoff's Theory of Three-Dimensional Organic Compounds
Van Vleck's Theory of Paramagnetism
Vesalius' Theories of Anatomy and Physiology
Virchow's Cell Pathology Theory
Volta's Concept of an Electric Current
Von Laue's Theory for the Diffraction of X-Rays in Crystals
Von Neumann's Theory of Automata

Waddington's Theory of Genetic Assimilation
Waldeyer-Hartz Neuron Theory
Wallace's Theory of Evolution by Natural Selection
Wallach's Theory for the Molecular Structure of Organic Compounds
Walton's Concept for Transmuting Atomic Particles
Watson–Crick Theory of DNA
Watson's Theory of Electricity as a Fluid
Watson-Watt's Concept of Radar
Weber's Theory of Gravitational Waves
Wegener's Theory of Continental Drift
Weinberg's Grand Unification Theories
Weismann's Germ Plasm Theory
Weizsäcker's Theories of Star and Planet Formation
Werner's Coordination Theory of Chemistry
Werner's Neptunian Theory (Neptunism)
Wheeler's "Geon" Theory
Whipple's "Dirty Snowball" Theory of Comets
Whitehead's "Action-at-a-Distance" Theory of Relativity
Wien's Displacement Law
Wigner's Concept of Parity/Symmetry in Nuclear Reactions
Wilkinson's Concept of "Sandwich Compounds"
Williamson's Theory of Reversible Chemical Reactions
Wilson's Hypothesis of Cloud Condensation
Wilson's "Out-of-Africa" Theory
Wilson's Theory of Dynamic Equilibrium of Island Populations
Witten's Superstring Theory
Wohler's Theory for Nonliving Substances Transforming into Living Substances
Wolfram's Theory of Complex Systems
Wolf's Theory of the Dark Regions of the Milky Way
Woodward's Theory of Organic Molecular Synthesis

Wright's Theory of Genetic Drift (Sewall Wright Effect)
Wrinch's Cyclol Theory of Protein Structure
Wu's Theory of Beta Decay
Wurtz's Theory for Synthesizing Hydrocarbons
Wynne-Edwards' Theory of Group Selection

Yalow's Theory of Radioimmunoassay
Yang's Theory of Nonconservation of Parity in Weak Interactions
Yanofsky's Theory for Colinearity of DNA and Protein
Young's Wave Theory of Light
Yukawa's Meson Theory for the "Strong Interaction"

Zeeman's Theory of the Magnetic Effect on Light
Zeno's Paradoxes
Ziegler's Theory of Stereospecific Polymers
Zinn's Concept of a "Breeder Reactor"
Zuckerandl's Theory for Measuring the Rate of Evolution
Zwicky's Theory for Supernovas and Neutron Stars

# L

**LAGRANGE'S MATHEMATICAL THEOREMS:** Mathematics: *Comte Joseph-Louis Lagrange* (1736–1813), France.

***Lagrange's theory of algebraic equations:*** *Cubic and quartic equations can be solved algebraically without using geometry.*

Lagrange was able to solve cubic and quartic (fourth power) equations without the aid of geometry, but not fifth-degree (quintic) equations. Fifth-degree equations were studied for the next few decades before they were proved insoluble by algebraic means. Lagrange's work led to the theory of permutations and the concept that algebraic solutions for equations were related to group permutations (group theory). Lagrange's theory of equations provided the information that Niels Abel and others used to develop group theory (*see also* Abel; Euler; Fermat).

***Lagrange's mechanical theory of solids and fluids:*** *Problems related to mechanics can be solved by nongeometric means.*

Before Lagrange, Newtonian mechanics were used to explain the way things worked, as well as to solve problems dealing with moving bodies and forces. By applying mathematical analyses to classical mechanics, Joseph-Louis Lagrange developed an analytical method for solving mechanical problems that used equations having a different form from Newton's law ($F = ma$), by which acceleration is proportional to the applied force to accelerate the mass. Lagrange's equations, which can be shown to be equivalent to Newton's law and can be derived from Hamilton's formulation, are, like Hamilton's formulation, very convenient for studying celestial mechanics. In fact, Lagrange himself applied his equations to the mechanical problems of the moon's **librations** (oscillating rotational movement), as well as those dealing with celestial mechanics. For one example, he solved the three-body problem when he demonstrated by mechanical analysis that asteroids tend to oscillate around a central point—now referred to as the *Lagrangian point* (*see also* Einstein; Newton).

Ancient people used the natural motions and cycles of the sun and moon, the seasons, and other natural observable phenomena to determine some of their measurements of time. Historically, many countries had their own system of weights and measurements that were arbitrarily based on someone's idea of how much or how long something should be. Movement of people from region to region made communication and trade difficult when different systems of measurements as well as languages meshed. The introduction of the metric system is an example of the need for some standardization of units of weights and measurement. For instance, the metric system grew out of the Age of Reason in Europe and was spread widely across nations as the advances of Napoleon's army introduced it. For example, this was the first time that kilometers rather than miles were used throughout Europe. It was natural for the United Sates to adopt the English systems of weights and measures since we were an English colony. Even so many enlightened leaders, such as Thomas Jefferson, Benjamin Franklin, John Quincy Adams, and others recognized the utility of the metric system (e.g., it is easy to convert weight to volume because 1 gram of water equals 1 milliliter or cubic centimeter of water). Jefferson developed his own decimal system that was somewhat like the metric system except he used his own terminology and units. For example, he based his system on a decimal system that did not equate different units. He declared that the foot was just 10 inches (somewhat shorter than the English foot); each inch was divided into ten lines, and each line into 10 points. Ten feet equaled a decade, 100 feet equaled a rod, 1,000 feet a furlong, and 10,000 feet equaled a mile (the present English mile is 5,280 feet long). But his decimal system of weights and volume was not based on some natural phenomena, as was the metric meter that was based on a fraction of the distance of the meridian that extended from the North Pole to the equator through a particular point in Paris, which was divided by 1/10,000,000. This distance was named *meter* after the Greek word for "measure." Today the meter is defined as the length a path of light travels in one 299,792,458th on a second and is based on the speed of an electromagnetic light wave in a vacuum.

The history of the acceptance of the metric system in the United States is not pretty.

1. 1800—was one the first times that the metric system was used in the United States when the U.S. Coast Guard used the French standard of meters and kilograms in its Geodetic Survey.
2. 1866—Congress authorized the use of the metric system and supplied each state with weights and measures standards.
3. 1875—The Bureau of Weights and Measures was established and signed the Treaty of the Meter to use this standard.
4. 1893—The United States adopted the metric standards for length, mass, foot, pound, quart, as well as other metric units.
5. 1960—The Treaty of the Meter of 1875 was modernized and called the International System of Units (SI) as the metric system is known today.
6. 1965—Great Britain begins conversion to the metric system so they could become a member of the European Common Market.
7. 1968—U.S. Congress passes the Metric Study Act of 1968 to determine the feasibility of adopting the SI system.
8. 1975—U.S. Congress passes the "Metric Conversion Act" to plan the voluntary conversion to the SI system.
9. 1981—The Metric Board reports to Congress that it lacks authority to require a national conversion.
10. 1982—The Metric Board is abolished due to doubts about the commitments of the United States to convert.
11. 1988—U.S. Congress has introduced "carrot" incentives to U.S. industries to convert, and by the end of 1992 all federal agencies were required to use the SI system for procurements of grants, and so forth.

(*Continued*)

Today, there are both metric and English systems placed on commercial products (e.g., ounces and grams), but there is much opposition to changing transportation (road) signs to kilometers from miles. It seems the American public, despite years of learning the metric system in schools, still does not recognize or accept the utility of the SI metric system, and stubbornly adheres to the use of the archaic English system of weights and measures.

***Lagrange's concept for the metric system:*** *A base ten system will standardize all measurements and further communications among nations.*

Historically, all nations devised and used their own system for measuring the size, weight, temperature, distance, and so forth of objects. As the countries of Europe developed and commerce among them became more common, it was obvious that the jumble of different measuring systems was not only annoying but limited prosperity. At about the time of the French Revolution, a commission was established to solve this problem. Lagrange, Lavoisier, and others were determined to find a natural, constant unit on which to base the system. They selected the distance from the North Pole to the equator as a line running through Paris. This distance was divided into equal lengths of 1/10,000,000, which they called a *meter* ("measure" in Greek). A platinum metal bar of this length was preserved in France as the standard unit of length. Today, a meter is defined as the length of the path light travels in one 299,792,458th of a second and is based on the speed of electromagnetic waves (light) in a vacuum. Units for other measurements besides length were devised, using the base of ten to multiply or divide the selected unit. For instance, a unit of mass is defined as the mass accelerated one meter per second by a one-kilogram force. After several years of resistance, other countries recognized the utility of the metric system, which has since been adopted by all countries, except the United States, Liberia, and Myanmar (formerly Burma). Even so, international trade and commerce have forced the United States to use the metric system along with the archaic English system of measures. Despite several attempts to convert the United States to the metric system, the general public has refused to accept it.

**LAMARCK'S THEORIES OF EVOLUTION:** Biology: *Jean Baptiste Pierre Antoine de Monet, Chevalier de Lamarck* (1744–1829), France.

**Theory 1:** *New or changed organs of an animal are the result of changes in its environmental factors.*

Lamarck proposed that the first requirements for modifying the form or structure of an organism were changes in environmental circumstances. This was the basis for his view that there was a natural tendency for greater complexity and that a change in the environment was responsible for the changes in functions and forms of the organs of animals. In other words, the occurrence of new organs in an animal's body is the result of some new need that became "felt" by the animal.

***Theory 2:*** *Those parts of an animal not used will either not develop or will degenerate over time, and those parts of an animal that are used will continue to develop and change over time. The "need" responses by animals over many generations are acquired changes in functions and structures that will be inherited in future generations.*

Lamarck believed these changes resulted from environmental factors, which led to changes in the animals' behavior as well as structure, and, in time, this acquired behavior also become habitual. One of his examples was the behavior of antelopes fleeing from predators. As they ran faster, their leg muscles developed, thus passing this escaping behavior and fleetness to offspring. In other words, as the environment changes, so does an animal's behavior, as well as its organs' functions, and structure. The behavior becomes an active agent for the species' evolutionary development, and when this behavior becomes habitual behavior, it determines the extent and nature of the animal's structure. This is usually referred to as the *inheritance of acquired characteristics*, which includes an interpretation of the pre-Darwinian concept of natural selection. But this natural selection, according to Lamarck, resulted in either habitual use or disuse of a particular body part, which carries over from generation to generation. One of the classical examples of the inheritance of acquired characteristics is the theory of how the giraffe acquired its long neck. The giraffe had to stretch higher and higher to obtain tree leaves for food after the lower leaves were consumed. Thus, over time, the giraffes "acquired" longer and longer necks, a characteristic passed on to the next generations. Today, however, it is usually considered that possibly due to genetic mutations, some giraffes with slightly longer necks were able to secure more food, be healthier, live longer, and thus reproduce more giraffes with the altered genes.

Lamarck's ideas were not well received by other scientists, including Darwin. Even so, at one time Darwin accepted some aspects of the concept of acquired characteristics caused by environmental changes and incorporated this into his theory of natural selection.

*See also* Buffon; Cuvier; Darwin; Lysenko

**LAMBERT'S THEORIES:** Mathematics, Physics, and Astronomy: *Johann Heinrich Lambert* (1728–1777), Germany.

Johann Lambert was a multitalented scientist who made contributions to many branches of science and added to the knowledge of his day. Lambert's many theories and hypotheses in areas of physics and astronomy derived from his contributions to mathematics. Following is a short synopsis of some of his contributions to these fields.

In the field of optics the Lambert–Beer Law is also known as the *Beer–Lambert–Bouguer law* because three men contributed to this field (the law is named after Lambert, the German mathematician August Beer [1825–1863], and the French mathematician and astronomer Pierre Bouguer [1698–1758]). The law is based on an observable relationship between the absorption of light to the nature of the material through which the light is traveling. The law is a mathematical means of expressing the factors related to the absorption of light and is based on three physical phenomena:

1. The concentration of light through an absorbing medium is known as its "pathlength."
2. The optical "pathlength" (OPL) is the length to which the light must travel.
3. The *absorption coefficient* is the probability that a photon of a specific wavelength of light will be absorbed.

The connection between these factors can be expressed by the following equation: A = $\epsilon$dc, where A = absorption; $\epsilon$ = a coefficient; $d$ = is the pathlength in centimeters; and $c$ = the molar concentration.

For the *transmittance* of a beam of light passing through an absorbing medium, the amount of light absorbed is proportional to the light's intensity times the coefficient of the absorption medium. This relationship is expressed by one of Lambert's law of optics as expressed in the mathematical equation as follows:

> T = 10 − $\epsilon$cd, where $T$ = transmittance of the light; $\epsilon$ = the molar extinction coefficient; $c$ = the molar concentration; and $d$ = the pathlength in centimeters. Note: this equation can also be written as T = 10 − A, where -A is the same as ($\epsilon cd$) the absorption of the light.

The *absorption* of the transmitted light can also be expressed as the strength of the incident radiation that can be plotted against the concentration of the light. Because this relationship is not linear, it is expressed as a negative log 10. It can be expressed as follows: A = −log 10 (T).

The Lambert–Beer law is also related to the atmosphere. It can be used to describe the diminishing of the radiation of sunlight as it travels through Earth's atmosphere. There is a scattering of the sun's radiation as well as absorption by the extent of aerosols (tiny particles and gases) in the atmosphere (*see* the Tyndall Effect).

Lambert made several contributions to the field of mathematics. In his studies of light intensities and absorption he introduced the hyperbolic functions into the field of trigonometry. He also coined the word "albedo" which is the reflection factor of light off a surface. He proved that pi ($\pi$) is an irrational number by the continued use of fractions. By using non-Euclidean geometry he devised theories related to conic sections as a means to calculate the orbits that comets follow. He theorized that the sun and its planets travel together through the Milky Way, and that there are many galaxies in the universe, such as the Milky Way, with many sun-like stars that have their own planetary systems.

Lambert was familiar with Kant's nebular theory that stated that the planets in the solar system originated from a gassy cloud, which was evidence of God's existence and wisdom. Sometime later, Lambert published his own version of the origin of the solar system that was not dependent on God's wisdom. He theorized that there were many galaxies beyond the Milky Way, and they all had planetary systems revolving around the many suns in each galaxy. Lambert was also a successful inventor and was given credit for inventing the first *hygrometer* and the first *photometer*, both in 1760. A hygrometer measures the relative atmospheric humidity, and the photometer measures the brightness of light also known as flux.

*See also* Kant; Newton

**LAMB'S THEORY FOR THE QUANTUM STATES OF THE HYDROGEN ATOM:** Physics: *Willis Eugene Lamb, Jr.* (1913–), United States. Willis Lamb shared the 1955 Nobel Prize for Physics with Polykarp Kusch.

> *Each of the known states of hydrogen is actually two states having the same energy in the absence of a magnetic field. However, the two states exhibit slightly different energies in the presence of a weak magnetic field.*

Willis Lamb's theory on the quantum states of the hydrogen spectrum required a slight revision of Paul Dirac's electron theory, which stated, according to quantum mechanics, that the hydrogen spectrum should exhibit two different but equal states of energy. Lamb's research demonstrated that the spectrum of the hydrogen atom was split into two parts, but there was a small shift of the energy level of the hydrogen spectrum from that which Dirac predicted. This discrepancy for the predicted quantum electrodynamics of the hydrogen atom is now known as the *Lamb shift.* Willis Lamb first demonstrated it by splitting the spectrum for hydrogen into two distinct parts, each with slightly different energy states. His research revealed how electrons act within the influence of electromagnetic fields and is considered important for the electronics and computer industries in the development of new products.

*See also* Dirac; Kusch; Lorentz; Zeeman

**LANDAU'S "TWO-FLUID MODEL" FOR HELIUM:** Physics: *Lev Davidovich Landau* (1908–1968), Russia. Lev Landau was awarded the 1962 Nobel Prize for Physics for his work on condensed matter.

*Depending on its temperature the gas helium exists in two different liquid states, "phonon" and "roton."*

Landau founded a major field of theoretical physics known as "condensed matter" in the mid-twentieth century. This is matter that exists as either a liquid or solid at a very low temperature even though it may be a gas at room temperatures. He used this theory to explain the unique behavior of helium gas at temperatures below about 2.17 kelvin (K) at which point it becomes a liquid. At near absolute zero temperatures below 2.17 K helium exhibits superconductivity and superfluidity where it will flow up the sides of a beaker and over the edges. He and other scientists named this form of helium below 2.17 K as "helium II" and named its particle a "phonon" which is a "quantum of thermal energy." For particles of helium above the critical 2.17 K temperatures they called it "helium I," which is the elementary quantum measurement of the motion of a vortex formed in a liquid. The point at which helium becomes superfluid is called the critical point (or Lambda point) which, when graphed, exhibits a jump in specific heat, and exhibits a discontinuity in its density. This is one area of theoretical physics that has been proven experimentally. This area of super cold and superconductivity research in physics was speeded up when room temperature helium-3 could be produced in fairly large amounts by nuclear reactors. Although helium-4 which is normal helium with two protons and two neutrons in its nuclei works well when supercooled in establishing helium's critical point, helium-3 with two protons and one neutron in its nuclei showed promise for studying vortex motions below the critical point. Helium-3 could be cooled down to just 2 mK, which is just a tiny fraction of a degree above absolute zero (K) or −273 C. This is one thousand times cooler than the critical point of 2.2 K for helium-4. 2 mK is the point at which helium-3 becomes a *superfluid.* At this temperature the helium atoms "paired" up to form very slow moving boson particles that exhibited Bose–Einstein qualities of superfluids. (Bose–Einstein refers to the gas-like qualities of electromagnetic radiation. It is named after Albert Einstein and the Indian mathematician and physicist Satyendra Nath Bose [1894–1974] who collaborated on this theory.)

After graduating from universities in Russia in the 1920s, Landau visited centers in Europe where problems in theoretical physics were being explored. One of these centers was in Copenhagen, Denmark, where he became lifelong friends with Niels Bohr. Upon returning to Russia he became head of several university physics departments. Landau is responsible for developing several well-known schools of theoretical physics. His accomplishments ranged in many fields of physics including quantum electrodynamics (QED), atomic and nuclear theories, particle physics, astrophysics, thermodynamics, electrodynamics, quantum mechanics, and low-temperature physics.

**LANDAUER'S PRINCIPLE FOR VERY-LARGE-SCALE INTEGRATION:** Physics: *Rolf Landauer* (1927–1999), Germany and United States.

***Landauer's principle states, in essence:*** *Any irreversible computing of logical information must be accompanied by a corresponding increase in entropy as a dissipation of energy; thus there is an increase in noninformation.*

This principle is based on a consequence of the second law of thermodynamics that states that entropy in a closed environment will always increase and never decrease. This related to the development and advances in the number of transistors, etc., that can be integrated in a semiconductor chip. About sixty-five to seventy-five years ago the first practical computers used vacuum tubes to process data. These early room-sized computers required many heat-producing vacuum tubes that required giant air conditioning systems to keep them cool. Their computing power was also less than today's high-powered modern home computers. Single semiconductor chips soon replaced inefficient vacuum tubes. Several individual purpose circuits that were integrated into these single chips eventually followed. These were the first *small scale integration* (SSI) chips that contained several devices, such as diodes, transistors, resistors, and capacitors, on a single chip that made it possible to form more than one logic-gate on a single chip. The next generation of integrated chips was called *large-scale integration* (LSI) that contained about a thousand logic gates. *Very large-scale integration* (VLSI) chips that contained many thousands of logic functions on a single chip followed this generation. This nomenclature no longer makes sense because today there are semiconductor chips that provide many hundreds of millions of gates on a single chip. In a few years we will see the production of billion-transistor processors on a single chip that operate at just a few nanometer processes or at the molecular level of gates.

Rolf Landauer was born in 1927 in Stuttgart, Germany, and immigrated in 1938 to the United States at age eleven with his parents. He received his undergraduate degree at eighteen years of age from Harvard University and then served in the U.S. Navy. He earned his PhD from Harvard in 1950, after which he worked at IBM in Westchester County, New York. This is where he arrived at his principle that each bit of information lost in a computer circuit will result in the release of a specific amount of heat. However, it has since been determined that if there is no erasure of information it may be possible to reverse thermodynamics by not releasing heat. This has led computer scientists to develop the concept of reversible computing—which has yet to be realized.

*See also* Bardeen; Brattain; Shockley; Turing; von Neumann.

**LANDSTEINER'S THEORIES OF BLOOD GROUPS:** Biology: *Karl Landsteiner* (1868–1943), United States. Karl Landsteiner won the 1930 Nobel Prize for Medicine or Physiology.

> *Individuals within a species exhibit different proteins in their blood serum (plasma), just as different species also exhibit different blood groups.*

Since 1628 when William Harvey explained the circulatory system in animals, scientists and physicians were aware that blood from one animal species was incompatible with the blood from another animal species. When incompatible blood types are mixed during a transfusion, the blood will clot, blocking blood vessels, which leads to death. At the beginning of the twentieth century, Karl Landsteiner demonstrated that only blood serum from certain types of patients could be mixed with blood from others whose blood had some similar characteristics. He found that the plasma (liquid portion of the blood) from some human donors would form clots in transfusions for a person with A-type but not for a person with B-type blood. Thus, a person with A-type blood could provide blood that was safe for another person with A-type blood. He also found that some other types of blood were incompatible with a person with B-type blood, and that some types of blood would clot for both A- and B-type people, and blood from still other people would not clot either A- or B-type people. This resulted in the classification of blood into the four groups: A, B, O, and AB. Only people with O blood can donate to most people from the other groups, but only in an emergency; a definite match for the other types is required. The understanding of blood grouping has increased and has been used to determine parenthood long before DNA testing. Due to Landsteiner's efforts, blood transfusions are safe.

**LANGEVIN'S CONCEPT FOR USE OF ULTRASOUND:** Physics: *Paul Langevin* (1872–1946), France.

*High frequency electrical currents can cause piezoelectric crystals to produce short ultrasound wavelengths mechanically.*

Paul Langevin built on the work completed by Marie and Pierre Curie as related to piezoelectric crystals. Pierre Curie realized that when a mechanical force was applied to a piezoelectric crystal, an electric current is generated between the two sides of the crystal. He used this process, in reverse, to measure the amount of radiation (strength) of the radioactive element on which he and Marie Curie were working. Langevin theorized that if a variable (alterable) electric current could be sent across one side of a crystal to the other side, the crystal would vibrate rapidly, thus producing sound waves shorter than those that can be heard by the human ear (e.g., ultrasound). He also was aware that sound waves of high frequency travel better under water than do light waves. Therefore, objects under water should be detectable at greater ranges using sound waves rather than light waves. His theory was later applied to the development of a system called echolocation, which during World War II became known by the acronym *sonar* (**SO**und **N**avigation **A**nd **R**anging). Sonar used ultrahigh frequency sound waves generated by the piezoelectric crystals to detect enemy submarines. Since then, sonar has been used as an invaluable tool in the field of oceanography. Not only can it detect objects, such as schools of fish and sunken ships, but it can also be used to measure the contour of the ocean's floor.

**LANGLEY'S THEORIES OF THE NERVOUS SYSTEM:** Physiology: *John Newport Langley* (1852–1925), England.

1. *The autonomic nervous system (ANS) is the part of the nervous system that controls* ***homeostasis*** *of the body.*

2. *There are specific sensors in the nervous system that act as receptors for specific types of drugs.*

John Langley coined the term "autonomic nervous system" (ANS) in 1898. This term includes the "sympathetic nervous system" (SNS) and the "parasympathetic nervous system" (PNS). Sympathetic nerves originate in the vertebral columns near the middle of the spinal cord. Their formation begins at the first thoracic segment (chest area) of the spinal cord, thus the SNS has a "thoracolumbar outflow" of nerves that extends downward to the third lumbar area (lower back area) of the spinal cord. The sympathetic nervous system is often used to describe the term "fight or flight" as a response to a perceived danger. This is the common way to explain the "sympatho-adrenal response" of sympathetic fibers and glands that secrete acetylcholine that, in turn, secrete adrenaline (epinephrine) and noradrenaline (norepinephrine) that prepare smooth muscle response for action. To a lesser extent, these automatic responses also take place prior to simple everyday movements of the body such as walking, waving your arms, eating, and so forth that may or may not require conscious response. Other functions performed by the sympathetic nervous system are the rate of the heartbeat, the level of blood pressure, and other automatic regulatory functions that are performed without intervention of conscious thought. In addition, the SNS is responsible for the following: widening bronchial passages; decreasing peristalsis movement of the large intestine, constricting blood vessels, dilation of the pupils in the eyes; erection of the dermal papillae, commonly known as "goose bumps"; and perspiration.

In 1870 Langley demonstrated that the use of an extract from a plant containing the drug pilocaine could slow the rate of the heartbeat that was exactly the reverse reaction of the drug atropine on the heart rate, and that the effects of both of these drugs were not dependent on the functioning of the vagus nervous system. Likewise, pilocaine stimulates the formation of saliva in the mouth whereas atropine inhibited the production of saliva. In the early 1900s Langley demonstrated that the drug nicotine causes contractions in muscles, while the drug curare causes the contracted muscles to relax. These experiments led Langley to the theory that drugs do not act directly on muscles but rather on an accessory substance that is actually the recipient of the drug stimuli and that, subsequently, transfers the contractual material to the receptive substance of the muscle.

**LANGMUIR'S THEORIES OF CHEMICAL BONDING AND ADSORPTION OF SURFACE CHEMISTRY:** Chemistry (Physical): *Irving Langmuir* (1881–1957), United States. Irving Langmuir was awarded the 1932 Nobel Prize for Chemistry.

***Electrons surround the nuclei of atoms in successive layers:*** *The electrons surrounding a nucleus progress in number from two, located in the closest layers (orbit), followed by additional layers containing eight, eight, eighteen, eighteen, and thirty-two electrons successively.*

It was understood for some years that because the atom is neutral, it must have as many negative electrons as positive protons. It was also determined that the electrons in the outer layer (orbit or shell) are held with the weakest force to the positive nucleus. In other words, their "energy level" is less than the electrons located in the inner orbits closer to the nucleus. Therefore, these outer electrons must be responsible for different atoms combining in specific ratios to form molecules, or for similar atoms to combine to form simple diatomic molecules, such as $O_2$ or $Cl_2$ (*see* Figure S2 under Sidgwick). Two models for the structure of atoms were proposed by Niels Bohr's quantum concept of a "solar system" atom and Gilbert Lewis' idea that electrons are shared as "bonds" to form molecules. Both of these models were based on Langmuir's "layered" structure for an atom's electrons (*see also* Bohr; Lewis).

***Langmuir's adsorption theory:*** *The adsorption of a single layer of atoms on a surface during a catalytic chemical reaction is controlled by the gas pressure if the system is maintained at a constant temperature.*

While working as a research scientist at General Electric's research laboratory in Schenectady, New York, Irving Langmuir developed light bulbs containing inert gases (such as argon) that did not oxidize the bulb's filaments. He also lengthened the life of the bulbs by using tungsten filaments, which further reduced oxidation. As a result, he theorized that electrons from the metal filament interact with monolayers (single layers) of atoms or molecules adsorbed to the surface. His theory is related to the **adsorption** of single layers of an element (usually a gas such as hydrogen) on the surface of another element (note that a*d*sorption is not the same as a*b*sorption). For example, the chemical reactions that take place inside an automobile's catalytic converter occur when the hydrogen compounds formed by the burning of the hydrocarbons in gasoline are a*d*sorbed on the platinum metallic beads inside the converter. This means the exhaust gases resulting from combustion are spread on the surface of the platinum, where they are converted to less toxic gases. These hydrocarbon atoms and molecules are not a*b*sorbed as a sponge a*b*sorbs water but rather obey the laws of surface chemistry (a*d*sorption).

**LAPLACE'S THEORIES AND NEBULAR HYPOTHESIS:** Physics: *Marquis Pierre Simon de Laplace* (1749–1827), France.

***Laplace's theory of determinism:*** *What affects the past causes the future.*

In 1687 Sir Isaac Newton published his laws of motion, which were deterministic and mechanical in that they explained the movements of objects on Earth as well as celestial bodies. Aware of perturbations and irregularities in the motions of planets and other heavenly bodies, Newton also believed the universe would end if these irregularities were not somehow corrected. The Marquis de Laplace believed these irregularities did not indicate the presence of some massive destructive force because they were not cumulative. In other words, forces generated by these perturbations did not combine as one big force, ending in disaster. Rather, they were of a periodic nature and occurred at regular time intervals. He believed the future is determined by past events. This theory became known as *Laplace's demon*. In essence, it states that what has affected things in the past will also cause the future. Laplace thought that if all data were known and analyzed, this information could then be stated in a single formula and there would be no uncertainty. Thus, the future would be caused by the past. This is an old concept that could not be supported once new and more difficult equations were explored. The concept of the past causing the future can be expressed by linear equations or straight-line conceptualizing; new nonlinear equations and methods of reasoning provided more branches or alternative causes resulting in a multitude of possible effects (e.g., chaos theory).

***Laplace's nebular hypothesis:*** *The solar system was formed by the condensing of a rotating mass of gas.*

The concept of swirling bodies in the universe originated with the ancient Greeks, but it was Laplace who tried to establish his nebular hypothesis by using Newtonian principles and mathematics. His concept stated that a ball of gas formed the sun, and from this the planets were "thrown off" their normal circular orbits, which in turn "threw off" their moons from their normal orbits. The nebular hypothesis has been updated as a rotating cloud of gas that cooled and contracted, and as it rotated centrifugal force resulted in matter forming individual rings of matter, which further contracted to form planets and moons. The great mass of leftover condensing gas formed the sun and other bodies in the solar system.

***Laplace's theory of probability:*** *Mathematics can be used to analyze the probability (chance) that a specific set of events will occur within the context of a given set of events.*

Probability is the likelihood of a particular cause resulting in the occurrence of a particular event (effect). Today, most scientists apply probability theory to the study of many fields of science, such as thermodynamics and quantum mechanics. The probability scale ranges from 0.0 that indicates an event is highly uncertain or unlikely to occur (or will not occur) to the probability of 1.0 for high certainty or it is likely that an event will or did occur. The terms "possible" and "impossible" are not measurable and thus have no meaning when considering probability as related to an event.

*See also* Fermat; Gauss; Ulam

**LARMOR'S THEORIES OF MATTER:** Physics: *Sir Joseph Larmor* (1857–1942), Ireland.

***Larmor's theory of electron precession:*** *An electron orbiting within the atom will wobble when subjected to a magnetic field.*

Sir Joseph Larmor's concept of matter was a synthesis referred to as the electron theory of matter. This was a rather radical description for the structure of atoms, the nature of matter, and the electrodynamics of moving bodies that relate to kinetic energy. The *Larmor precession* describes the behavior of an orbiting electron when moving through a magnetic field. The axis of the electron actually changes its angle (precession) while moving in the field and thus appears to wobble. Larmor then calculated the rate of energy that radiated from an accelerating electron, an important concept for future work in particle physics.

***Larmor's concept of the aether:*** *Space is filled with an aether partially composed of charged particles.*

Sir Joseph Larmor was one of the last physicists who attempted to justify the existence of an **aether** (or ether) in space, believing that it was necessary as a medium to provide a means for waves (e.g., light, radio, and other electromagnetic waves) to travel from one point in space to another. (His reasoning was that water waves require the water as a medium to travel from one place to another.) He also believed aether must contain some electrically charged particles in order for matter and light to traverse space. These concepts were accepted in classical physics, but Einstein's theories of relativity made the concepts, such as the aether, invalid. Even so classical physics made some valuable contributions to the field.

**LAURENT'S THEORIES FOR CHEMICAL "EQUIVALENTS" AND "TYPES":** Chemistry: *Auguste Laurent* (1807–1853), France.

***Laurent's chemical equivalents:*** *A definite distinction between atoms and molecules exists based on their equivalent weights.*

Auguste Laurent classified molecules composed of two atoms, such as oxygen, hydrogen, and chlorine, as *homogeneous compounds that become heterogeneous compounds when they are "decomposed."* Laurent's work established the relationships of the elements' atomic weights to their other properties and characteristics. This concept of atomic weight as being related to an element's chemical properties provided a key to Mendeleev's arrangement of the elements in his **Periodic Table of Chemical Elements**.

***Laurent's type theory:*** *Organic molecules with similar structures can be assigned to a classification of "types."*

From the days of alchemy, scientists placed different chemicals and minerals into separate categories for the purpose of classification. These grouping were usually based on the "types" of color, physical consistency, or reactions with each other, but usually not on their basic elementary structures. Laurent's work with chemicals provided the distinctions needed to develop types of compounds, which were arranged according to his concepts of their structures. For example, he considered water to be one type of compound and alcohol another. Although both are composed of hydrogen and oxygen atoms, each is different in structure. His theory of types was helpful in classifying organic compounds but was not adequate for describing their different structures. Nevertheless, it was a step in the right direction toward understanding the structure and nature of organic compounds.

**LAVOISIER'S THEORIES OF COMBUSTION, RESPIRATION, AND CONSERVATION OF MASS:** Chemistry: *Antoine Laurent Lavoisier* (1743–1794), France.

***Lavoisier's theory of combustion:*** *The gas emitted when cinnabar (mercuric oxide ore) is heated is the same as the gas in air that combines with substances during combustion (burning).*

Antoine Lavoisier was a meticulous scientist who at first believed in the "phlogiston" theory (*see* Stahl). However, his experiments soon indicated there was another explanation. When he burned sulfur and phosphorus in open air, they gained weight, while other substances lost weight. He concluded that something in the air, not the "phlogiston" in the substance being burned, was involved in combustion. In 1774 Joseph Priestley determined that cinnabar, when heated, emits a gas that made a candle burn more brightly and a mouse that was placed in this air more lively. Taking this further, Lavoisier believed the gas produced by cinnabar was the same as one of the gases in air that combined with substances as they burned. He later named this gas *oxygine*—Greek for "acid producer." Lavoisier then proceeded with several experiments, one of which was to repeat Priestley's experiment of burning a candle in an upturned jar placed in a pail of water. As the flame in the candle was slowly extinguished, the water level in the jar rose. He also burned a candle in the gas produced by heating cinnabar. The results were the same; thus, his conclusion was that the gas in air was the same as the gas emitted by burning the cinnabar (mercuric oxide, HgO). After the candle burned out, the gas remaining in the jar was inert and made up a large portion of the volume inside the jar. Because mice could not live in this leftover gas, Lavoisier concluded it would not support life as did his "oxygine." He called this inert gas *azote*, meaning "no life" in Greek. This gas is now called nitrogen.

***Lavoisier's theory of respiration:*** *Animals convert pure air to fixed air.*

Antoine Lavoisier was the first to test experimentally Joseph Priestley's concept that normal air lost its phlogiston during combustion and respiration. First, he placed a bird in an enclosed bell jar. When it died, he tried to burn a candle in the air left in the jar. It would not burn, and according to the science of the day, the air was then pure, or as they said, it was "dephlogisticated" air. In other words, the bird had used up the phlogiston. Conducting other experiments with burning candles, Lavoisier hypothesized that as more air was available to the burning candle, more of it would be converted to "fixed air." In 1756 Joseph Black first prepared and named fixed air, or carbon dioxide. Carbon dioxide gas dissolves in water, as Priestley later discovered when he "invented" carbonated soda water. But the proportion of the gas consumed by the candle to the proportion of the gas left in the jar indicated there was some other gas that made up the remaining air in the jar. At the time it was not known that oxygen composes only about one-fifth of a given volume of air, while nitrogen gas makes up most of the remaining four-fifths. Nevertheless, Lavoisier was the first to measure the amount of oxygen consumed and carbon dioxide emitted through animal respiration. He was also the first to measure the heat produced by respiration and determined it could be compared to the amount of oxygen required to burn charcoal. He changed the concept of phlogiston to the concept of caloric, referred to as weightless fire that changed solid and liquid substances into gases. Science was no longer saddled with the misconception of the phlogiston theory.

***Lavoisier's law of conservation of mass:*** *The mass of the products of a chemical reaction are equal to the mass of the individual reactants.*

When some metals are "roasted" at a high temperature in the presence of air, their surface turns into a powder called an **oxide**. In Lavoisier's time this coating of metallic oxide was referred to as *calx*. The old phlogiston theory explained the loss of weight in the air to the loss of phlogiston. Because metals gained weight during smelting, it was believed the calx combined with the charcoal and that the charcoal contained phlogiston. We now know this is not a true concept. To make careful measurements of the burned substances, Lavoiser invented a delicate balancing device that could measure a tiny fraction of a gram. Using his balance, he conceived his law of conservation of mass which is still valid today. Lavoisier is known as the father of modern chemistry due to his use of step-by-step experimental procedures, making careful measurements, and keeping accurate records.

*See also* Black; Priestley; Scheele; Stahl

**LAWRENCE'S THEORY FOR THE ACCELERATION OF CHARGED PARTICLES:** Physics: *Ernest Orlando Lawrence* (1901–1958), United States. Ernest Lawrence was awarded the 1939 Nobel Prize for Physics.

> *When charged particles are accelerated in a vertical magnetic field, they move in accelerated spiral paths.*

Since the discovery of natural radioactivity, it was known that alpha particles (helium nuclei with a positive charge) were ejected from the nuclei of radioactive substances and could induce other nuclear reactions. A method to increase the acceleration of charged particles by using electromagnetic forces was needed to penetrate and "smash" atomic nuclei to separate their component particles. To achieve the greatly increased speeds for charged particle "projectiles," John Douglas Cockcroft and Ernest Thomas Sinton Walton developed a low-energy linear (straight line) accelerator in 1929. Their "atom smasher" used a voltage multiplier to build up a high-voltage capacity capable of accelerating alpha particles beyond the speed from which they are emitted naturally from radioactive elements. However, these early linear accelerators did not provide the energies required to "smash" the nuclei of atoms to the extent that they produced smaller particles.

Ernest Lawrence proposed a unique design to solve the problem of early linear accelerators' not having adequate energies to interact with heavy nuclei to produce smaller nuclear particles. He constructed two D-shaped metal halves of a hollow circular device with a small gap between the two semicircular Ds with the vertical portion of the Ds facing each other. He named his first model the *cyclotron*, which was only four inches in diameter. By shooting charged particles into the semicircular "Dees" and then applying high-frequency electric fields to the particles, they reached tremendous speeds as they continued to circle in increasing spirals inside the Dees, getting an electromagnetic "push" each time they passed the gap (something like repeated pushes making a swing go faster and higher). As the spiraling particles approached the inside rim of the Dees, they achieved maximum acceleration. At that point they were directed toward targeted atoms, which created smaller bits of subatomic particles and energies as a result of the collisions. Lawrence's early device was the precursor of the current circular accelerators that are several miles in diameter and develop very strong electromagnetic fields aided by cryogenic superconductivity (*see* Figure V2 under Van der Meer). For the current giant atom "smasher," the particles (alpha, beta, and other subatomic particles) are accelerated by a powerful linear accelerator, which then feeds the particles into the giant cyclotron, thus combining a linear device to a circular one. These giant particle accelerators generate very high electron voltages (eV), which are a measure of the energy of the particle. Physicists continue to use these powerful devices to produce many different types of subnuclear particles, which may help explain the basic nature of matter, energy, and life. Lawrence's original cyclotron produced particles with only about 10,000 to 15,000 electronvolts (eV). Advanced particle accelerators are being developed that could reach 300 billion electronvolts (BeV) per nucleon (proton or neutron). These will be much larger and many times more powerful than the cyclotrons used in the latter part of the twentieth century. The greater the power of these new accelerators to increase the speeds at which particles slam into each other, the more information will be obtained to answer questions as to the nature of the universe.

**LEAKEYS' ANTHROPOLOGICAL THEORIES:** Anthropology: *Louis Seymour Bazett Leakey* (1903–1972), *Mary Douglas Nicol Leakey* (1913–1996), and *Richard Erskine Frere Leakey* (1944–), England.

> *To give meaning to where humans are today, we need to explore where we have come from; being able to look backward gives the present a root.*

Louis Leakey was the patriarch of a family of three generations who spent their lives contributing to the evolutionary study of humans. Louis Seymour Leakey married Mary Douglas Nicol Leakey in 1936. She has also made paleontological history in her own right. They had three sons. One son, Richard Erskine Frere Leakey, born in 1944, also became an anthropologist and paleontologist. Richard's wife Meave Leakey and their daughter Louise Leakey have continued the three-generational Leakey dynasty's contributions in paleontology, archaeology, and paleontology.

Louis Leakey was born in British East Africa (now Kenya) in 1903 where his parents were missionaries at the Kabete Mission near Nairobi. Louis found his first fossil at the age of twelve, which was instrumental in forming his life-long interest in archaeology. After graduating from Cambridge University in England in 1926, he returned to Africa to prove or disprove Darwin's theory that this continent was the site of the origin of human evolution. This led him to explore in more detail the large 30-mile-long chasm called the Olduvai Gorge made famous when the German entomologist Wilhelm Kattwinkel (dates unknown) discovered it in 1911. In the meantime Louis' first marriage to Frieda Leakey ended in divorce in 1933 after he fell in love with Mary Douglas Nicol, a twenty-year-old who had no formal education. They made a good team. Together they studied and determined that the stone tools that they had discovered were formed by early humans. Leakey's first major find was a well-preserved jaw of a prehuman called *Proconsul africanus*. During World War II he became a spy for the British government and later acted as an interpreter. He also became interested in writing in the fields of conservation, African natural history, and the psychological behavior of ancient humans. He became famous for his writings about Mary's (along with a collaborator of Leakey's) discovery of what they called *Zinjanthropus boisei*, a large skull with teeth that he claimed was six hundred thousand years old. Later this date was disputed when carbon-14 dating of artifacts proved it to be otherwise. The new date for this find at this particular site was 1.75 million years old.

Mary Douglas Nicol Leakey was born in London, England, in 1913 to a landscape painter who often moved to various international sites. After her father's death, her mother enrolled her in a Catholic convent school from which she was often expelled. About her experiences in the school she said that it was "wholly unconnected with realities of life." Soon after, she began her self-education by attending lectures on geology and archaeology at the University of London. She was a good artist and became an illustrator for several books on archaeology. In the early 1930s she worked at several stone-age "digs" in England, including Windmill Hill near Stonehenge and at Hembury near Devon, where she drew skillful illustrations of stone-age tools. In 1934 she was responsible for her own "dig" at Jaywick Sands in Essex at which time she published her first scientific paper. Mary met Louis Leakey at a dinner party, and less than a year later he asked her to marry him after he had left his wife. In October 1934 he left for Tanzania, and Mary followed him the following April. They were married December 24, 1936. As Louis went his own way, Mary did field work on a Neolithic site near Lake Nakuru, Kenya, where she found iron and stone tools in old home and burial sites. These discoveries led to her recognition as a professional archaeologist. Louis spent more time in London raising money for additional archeologicalal work in Africa while Mary spent the next years at the Olduvai Gorge site. Consequently, around the late 1960s they spent more and more time living separate lives, both professionally and personally. In 1978 Mary made her most famous discovery of well-formed footprints of a child and two adults. These prints proved to be 3.6 million years old and were impressions made in volcanic ash at a site in Tanzania called Laetoli. They belonged to a new species of hominids related to the 3.2 million-year-old skeleton of Lucy that was found in Ethiopia by Donald Johanson (*see* Johanson). This led to a dispute with Johanson as to the ages of these prehumans.

Richard Leakey is the second son of Louis and Mary Leakey born in 1944, who soon followed in the footsteps of his parents when he found his first fossil, parts of an extinct giant pig, at age six. (He has an older brother Jonathan and younger brother Philip. A sister Deborah died in infancy in 1943.) During his school years he was mostly interested in learning how to track animals, and at age seventeen he left school to start a successful photographic safari enterprise. He joined a fossil-hunting group in 1967 that searched the Omo Valley in Ethiopia for fossils. On an airplane trip to Nairobi he observed out the plane's window what looked like a sedimentary rock formation that might contain some fossils. Later he formed a team to excavate the area on the shores of Lake Turkana. This site proved to be very productive over the next thirty years for Richard and his team of paleontologists as they collected over two hundred ancient hominid fossils. Their most famous "find" was the cranium of the almost complete 1.6 million-year-old *Homo erectus* skull discovered in 1984 and which they named "Turkana Boy." After his first marriage failed, he married a fellow paleontologist, Meave Epps, in 1970. After serving as director of the National Museums of Kenya from 1968 to 1989, he joined the Kenya Wildlife Service as its director. In this position he spearheaded the effort to end elephant poaching that helped stabilize the elephant population in that area of Africa. Later in life he felt the lack of a formal higher education had hindered his reputation as a paleontologist and archaeologist. Richard Leakey survived a near-fatal plane crash in Kenya in 1992 but lost both legs as a consequence. Always a polarizing and controversial figure, his career as a politician and activist has been contentious.

Meave E. Leakey's early education was in boarding and convent schools. Later she received several college degrees including a PhD in zoology from the University of North Wales in 1965. She met Richard Leakey when she joined his expedition at a new site on the shores of Kenya's Lake Turkana. They have two children, one of whom, Louise, followed in her parents' and grandparents' fossil-hunting footsteps. The National Museums of Kenya where Richard was the director until 1989 also employed Maeve and Louise. Maeve Leakey made an impressive find in 1999 when she discovered a 3.5-million-year-old lower jaw and skull of what turned out to be an unknown branch of early hominids. She named this new genus *Kenyanthropus platyops*, which stands for "flat-faced man of Kenya." Since joining the National Museums of Kenya in 1989, she has focused on finding evidence of the earliest human on Earth. She is working on sites that are yielding fossils that are between eight and four million years old. One of the oldest she found represents a new species *Australopithecus anamenis*. This finding resulted in the revision of the timeline for the evolution of humans by several millions of years. In some ways Maeve Leakey's accomplishments have surpassed those of her husband and mother- and father-in law. With Louise following in the paths of her parents and grandparents, the Leakey fossil-hunting dynasty continues.

**LEAVITT'S THEORY FOR THE PERIODICITY/LUMINOSITY CYCLE OF CEPHEID VARIABLE STARS:** Astronomy: *Henrietta Swan Leavitt* (1868–1921), United States.

> *The periodicity of the brightness of Cepheid variable stars can be used as a standard to determine the distance to the group of stars (galaxy) in which the variable star is found.*

Henrietta Leavitt is one of science's ignored women who made important contributions to astronomy. The daughter of a Congregationalist minister, Leavitt became progressively more deaf throughout her lifetime, which did not handicap her contributions to science, although her accomplishments were not recognized until late in life. She graduated from Radcliffe College in 1892 and soon after accepted a position as a research assistant at Harvard College Observatory. Her job was that of a human "computer." In those days, educated women performed the work that today is accomplished by modern digital computers to analyze mathematical data. In her case, she was a human "number cruncher." She viewed thousands of photographic plates of the stars made by the astronomers at the Harvard Observatory and recorded and analyzed the data—mainly to measure and catalog the brightness of stars recorded on plates. She viewed thousands of images of stars found in the polar region of the Magellan Cloud. In 1908 she published her conclusion that some of the variable stars showed patterns of brightness. She observed that the brightest variable stars known as **Cepheids** had longer periods of brightness than did the less bright stars. She also proved that this relationship between luminosity and periodicity was predictable. This relationship between a star's brightness and its period of variable brightness was soon recognized, and the value of this discovery was soon used to measure the distance of stars from Earth. In 1913 the astronomer Ejnar Hertzprung used this relationship as a yardstick to measure the distance from Earth of several Cepheids located in the Milky Way. Later, this relationship was again used to measure the distances of the variable stars in the Andromeda galaxy. The analysis of this data proved that stars located at great distances indicated they were located in other galaxies and were not in the Milky Way.

The relationship between a Cepheid variable star's luminosity and its period of variability has been used as a standard measure of a star's distance for about the past hundred years. Sometime later, it was learned that most Cepheids belong to the classification of *population I stars,* and therefore are called Type I Cepheids. There are slightly different types of variable stars known as Type II Cepheids. Cepheids are large bright yellow stars that have an oscillation of their luminosity caused by regular and precise expansion and contraction. Cepheids have periods ranging from one day up to about fifty days and their luminosity doubles from their dimmest to their brightest. This cycle of expansion and contraction is caused by stars' using up their supply of hydrogen fuel, causing instability, and resulting in pulsations between their dimmest to brightest periods. Also, ionization of helium gas in the Cepheid's atmosphere varies with this cycle of the star's atmosphere resulting from the state of its helium gas. The star's ionized helium that is closest to the sun has a greater density and thus is more opaque to the star's light than is the gas further from the sun where it is deionized and thus less dense. This factor sets up a cycle between the two states of helium gas, which can be correlated to the Cepheid's mean density as well as its luminosity and, more important, as a measure of the star's distance from Earth. The precision with which a star's distance can be determined by this relationship between a star's luminosity and periodicity led to its use as a standard "candle" to measure, with some degree of accuracy, the distance of the brighter stars.

Leavitt measured and established the relationship between luminosity and periodicity for many hundreds of Cepheid variables. She also determined that that there is a relationship between the lengths of the periods of brightness. A three-day period for one type of Cepheid exhibits a luminosity eight hundred times that of our sun. Another type, that is, a thirty-day period Cepheid, has a luminosity about ten thousand times that of our sun.

In 1921 the American astronomer Harlow Shapley, as director of the Harvard College Observatory, appointed her as head of stellar photometry at the Observatory. However, she died later that year. Henrietta Swan Leavitt never received much recognition for her theory and discoveries. An asteroid was named the 5383 Leavitt Asteroid and a crater on the moon is named after her. She was considered for nomination for a Nobel Prize by Swedish mathematician Gösta Mittag-Leffler (1846–1927), but because she had already died the nomination was rejected by the Committee.

*See also* Hertzsprung; Shapley

**LE BEL'S THEORY OF ISOMERS:** Chemistry: *Joseph Achille Le Bel* (1847–1930), France.

> *The asymmetric quadrivalent carbon atom can form molecules composed of the same atoms but with different structures.*

Earlier chemists worked on various theories to explain the structure of atoms and how they bonded (joined) with other atoms to form molecules of different compounds. This dilemma was solved when it was determined that carbon had a tetrahedron structure. Joseph Le Bel devised his concept of the asymmetric carbon atom at about the same time as did another chemist, Jacobus Van't Hoff. Both of their concepts were based on the tetrahedron structure of the carbon atom with its four valence electrons arranged something like a three-legged tripod, with the fourth bond pointing up (*see* Figure V3 under Van't Hoff). Le Bel's concept for the structure of the carbon atom was published in 1874, just two months before Van't Hoff published his almost exactly similar discovery. The tetrahedron structure of the carbon atom indicated how other carbon atoms or atoms of other elements, in pairs or individually, combined with carbon to form inorganic and organic molecular compounds. This would produce **isomers** of compound molecules that had the same chemical formula but different physical characteristics, such as boiling points, color, and reactivity. This concept ultimately resulted in the development of organic (carbon) chemistry and explained the myriad existing organic molecules. Although there are some inorganic compounds that contain carbon, for example, carbon dioxide ($CO_2$) and cyanide (CN), all organic compounds contain carbon (e.g., all living matter and products of living matter, including all foods whether sold as organic or not). This is why organic chemistry is referred to as "carbon chemistry."

*See also* Kekule; Van't Hoff

**LE CHATELIER'S PRINCIPLE:** Chemistry: *Henri-Louis Le Chatelier* (1850–1936), France.

> *Any change made in a system in equilibrium results in a shift of the equilibrium in the direction that minimizes the change.*

In essence, Henri-Louis Le Chatelier's principle describes what happens within a system that is in equilibrium (symmetry, parity, stability, or balance) when the factors of temperature, pressure, or volume change. If there is an increase in the pressure, the system decreases its volume to bring itself back into equilibrium. This principle includes the law of mass action and the theory of chemical thermodynamics. Le Chatelier's concept provides scientists with a mathematical interpretation of the system's dynamics and a practical physical means to control what occurs within a system where the changes in pressure and temperature cause the system to readjust its equilibrium. Le Chatelier's principle is invaluable for understanding how to control the mass production of industrial chemicals (e.g., ammonia and hydrocarbon products, such as gasoline).

*See also* Boyle; Haber

Henri-Louis Le Chatelier made another contribution to science in the field of thermometry that was based on German physicist Thomas Seebeck's idea for a thermocouple. In 1826 Seebeck (1770–1831) demonstrated that when two different metals that are placed together and heated, a current will flow between them, and the current will be proportional to the differences in the temperature of the metals forming the junction where the metals meet. Le Chatelier conceived the idea of using an alloy metal for one side of the junction. He successfully placed platinum metal on one side of the junction and an alloy of platinum/rhodium on the other side where the temperature was to be measured. Keeping the platinum metal of the junction at a constant temperature allowed the temperature on the alloy metal side of the junction to be calculated by measuring the amount of current flowing through the junction between the two metals.

Working in his grandfather's mines with the structure of alloy metals and their temperature differences when exposed to heat, Le Chatelier arrived at his principle of how the temperature, pressure, and volume were related to the concept of equilibrium in 1887. His principle was challenged and later replaced by two laws proposed by Jacobus Van't Hoff. The first law states *an increase in pressure will favor the system that has the smaller volume.* The second law states *a rise in temperature favors the system with absorption of heat.* This law explains the equilibrium existing for reversible chemical reactions that are expressed by using a double arrow ( $\Longleftrightarrow$ ) as in the equation expressing the exothermic reaction that take place during the formation of ammonia: $N_2 + 3H_2 \Longleftrightarrow 2NH_3$.

## LEDERBERG'S HYPOTHESIS FOR GENETIC ENGINEERING:

Biology (Genetics): *Joshua Lederberg* (1925–2008), United States. Joshua Lederberg shared the 1958 Nobel Prize for Physiology or Medicine with Edward Lawrie Tatum and George Wells Beadle.

> *If viruses can inject themselves into the genes of bacteria cells to cause infections, then it should be possible to inject genes into animal cells.*

Joshua Lederberg's first experiments demonstrated that bacteria contain genes in their nuclei and at times reproduce by sexual mating, as well as by conjugation. Previously it was believed bacteria reproduced only by "fission," where the mother cells split into two new daughter cells without any interchange of genetic material. This is known as asexual reproduction. Lederberg demonstrated that when crossing different strains of bacteria, a mutant strain would develop randomly, which caused a mixing of genetic material between the two strains. Because the crossed bacteria could develop their own colony of bacteria, sexual mating must be occurring. Occasionally he found that some enzymes were destroyed by what are called *bacteriophages*, which are viruses that enter and infect bacteria, thus causing genetic changes (*see* Figure D7 under Delbruck). Lederberg and Max Delbruck proved that new strains of viruses result when two different strains are combined in a form of sexual reproduction just as occurs for bacteria. Their work led to the new science of genetic engineering, where genes can be recombined by inserting them into bacteria and other cells.

*See also* Delbruck

**LEDERMAN'S TWO-NEUTRINO HYPOTHESIS:** Physics: *Leon Max Lederman* (1922–), United States. Leon Lederman shared the 1988 Nobel Prize for Physics with Melvin Schwartz and Jack Steinberger.

*The two different types of neutrinos are generated by different physical decay processes.*

When beta particles (electrons) were ejected during radioactivity, the end particles exhibited less energy than expected. To explain this seeming negation of the law of conservation of energy, the neutrino was postulated to account for the missing energy. (The Italian physicist Enrico Fermi named the neutrino, which means "little neutral one," in the 1930s.) Even though neutrinos may be considered "nonparticles," they do exist, as do "antineutrons." Both are important to maintain the symmetry and mathematics related to particle physics. Leon Lederman recognized that there are two different decay processes controlled by the weak interaction between subatomic particles that produce neutrinos. One decay process occurs when pions decay into muons ($\mu$) plus neutrinos ($V\mu$). The result is the formation of one type of neutrino that Lederman hypothesized is different from the other type. The second decay process is a form of beta decay, where a neutron ($n^\circ$) is converted into a proton ($p^+$) by ejecting an electron ($e^-$) and a neutrino (V) (*see* Figure F2 under Fermi). In other words, Lederman attempted to find out if the muon-related neutrino was the same particle as the electron-related neutrino. His experiment with his two colleagues, Melvin Schwartz (1932–2006) and John Steinberger, resulted in the identification of the existence of the muon neutrino ($V\mu$), and the ability to distinguish it as a different subatomic particle from the electron plus neutrino combination emitted when a neutral neutron is converted to a positive proton.

*See also* Fermi; Pauli; Steinberger

**LEE'S THEORIES OF WEAK NUCLEAR INTERACTION:** Physics: *Tsung-Dao Lee* (1926–), China and United States. Tsung-Dao Lee and Chen Ning Yang shared the 1957 Nobel Prize in Physics.

*Parity is not conserved in interactions between elementary particles.*

The conservation of parity for the classical laws of physics is the concept that these laws are symmetrically the same for all special axes or coordinates. This means that the results of experiments viewed as a mirror image themselves will produce the same results. Parity was assumed to be a natural universal law of conservation for the classical concepts of gravitation and electromagnetism, for instance, when negatively charged particles are balanced by positively charged particles ($e^-$ and $p^+$). Another example is when the strong force that holds together the nuclei of atoms is in parity with the weak force exhibited by radiation. Or the conservation of parity can be related to the symmetry or the right-handed image to the mirror left-handed image. The laws of nature were long thought to be the same under mirror reflections of right/left and thus were the same under the same conditions in the universe. In 1956 Tsung-Dao Lee and his collaborator Chen Ning Yang discovered that the weak force involving the weak nuclear interaction between elementary particles (gravity and electromagnetism) maximally violated parity. These two physicists and other scientists provided experimental evidence that the right–left symmetry involved in the weak force was not consistent and in fact maximally violates parity. Their work led to the formation of the "Standard Model" for particle physics in 1968 that describes the theory for the two electromagnetic weak interactions developed by Sheldon Glashow, Abdus Salam, and Steven Weinberg.

*See also* Fermi; Glashow; Salam; Weinberg; Yang

**LEEUWENHOEK'S THEORY OF MICROSCOPIC LIFE:** Biology: *Anton van Leeuwenhoek* (1632–1723), Holland.

*Multitudes of living "animalcules" exist in water and other fluids.*

Anton van Leeuwenhoek is sometimes, and incorrectly, credited with inventing the microscope. Leeuwenhoek is best known for developing improved microscopes in the seventeenth and early eighteenth centuries, but he did not invent the instrument. His design consisted of one small lens fixed between two metal plates. The object to be observed was placed on a "pin" that could be focused by moving the object up and down by turning a screw device. During his lifetime, he constructed and sold over five hundred models of his microscopes. When he died, he left 247 completed instruments plus over 170 mounted lenses. Most have disappeared, and only nine of his original microscopes have survived.

He based his theory of "little animalcules" on a lifetime of observing the microscopic world around him. He was the first to observe and describe protozoa in water, bacteria in his own feces, red blood cells, nematodes in soil, rotifers, and ciliates such as *vorticella*, bacteria with different shapes, *spirogyra* (alga), and human sperm. Some of his descriptions were very accurate and led to further investigations. He examined the plaque and sputum from his mouth and the mouths of others and then described the strong actions of these multitudes of "animalcules" found in spittle as "fish swimming in water." This was the first viewing and written description of bacteria. As early as 1684, Leeuwenhoek calculated that red blood cells were twenty-five thousand times smaller than specks of sand. He also made extensive observations of microscopic fossils, crystals, minerals, as well as tissues from a variety of animals and plants. His microscopes were used to view the microscopic world rather than for scientific purposes, whereas his drawings and descriptions proved valuable for future biologists.

*See also* Galileo; Janssen

**LEIBNIZ'S THEORY FOR "THE CALCULUS":** Mathematics: *Gottfried Wilhelm Leibniz* (1646–1716), Germany.

*Finite areas and volumes for curves can be calculated by use of differential and integral mathematical calculations.*

The dilemma of how to determine the area on or within a curved surface had been explored by dozens of mathematicians, philosophers, and scientists since ancient Greece. Leibniz realized a workable notation method (the use of symbols to represent quantities) was required to solve problems related to the areas and volumes of curves. His solution to the notation problem was: $\int y dy = y^2/2$, which is still used today. Leibniz published the results of his calculus in "Mathematical Calculations for the Investigations and Resolutions of Multiple Variables" in 1684, which turned out to be a significant event in mathematics. A major dispute as to the discoverer of calculus resulted when Sir Isaac Newton, who had developed his calculus much earlier in 1665, delayed publication of his calculus until 1687. Therefore, Leibniz is credited with the discovery and development of calculus. It seems that Leibniz learned about Newton's letter to a mutual friend that described his mathematics related to calculus. Thus, he was aware of Newton's procedures for solving the problem. The basis and origin of calculus became a dispute among mathematicians as well as the two principles in question. Their differences were greater than just who invented calculus. Leibniz also disagreed with Newton's theory of gravity. Leibniz, using Aristotle's metaphysical concept of motion, claimed that a body is never moved in nature unless and until another body moves it which results in the first body's motion and continues until another body acts on it. Later it became clear that Leibniz used different notations of symbols to represent quantities in his calculations that were based on his unique invention of differential and integral calculus.

*See also* Newton

**LEISHMAN'S HYPOTHESIS FOR PARASITIC DISEASES:** Biology (Bacteriology): *Sir William Boog Leishman* (1865–1926), England.

***Leishman's hypothesis:*** *Oval bodies imbedded in spleen tissue are responsible for protozoan infections related to parasitic diseases.*

William Boog Leishman was born and educated in Glasgow, Scotland, followed by a tour in the British Army Medical Services in India until 1897 when Leishman returned to England. He became an assistant professor of pathology in the Army Medical School in 1900. While at this post, he developed a new method of staining blood that could be used to identify malaria and other types of parasites. This new stain was a combination of Methylene Blue and eosin and became known as *Leishman's stain*. A few drops of the stain are placed on a slide with a specimen of the blood from a patient where it sets for 20 seconds. Then more drops of a buffer solution at pH6.8 is mixed with the stain. After a short period, it is then washed off before viewing the specimen with a microscope. In 1901 while examining tissue from the spleen of a patient who had died of *Kala-Azar,* he used his stain to identify tiny oval bodies imbedded in the spleen tissue. He hypothesized that these bodies were responsible for the protozoan infection that caused the disease called *Kala-Azar*. Two years later another physician with the Indian Medical Service, Charles Donovan (1863–1951), independently made the same discovery of the protozoan that causes *Kala-Azar*. It was originally named *Leishmania donovani*, but later this category of protozoan became known as *Leishmaniasis*. It was determined that *Leishmaniasis* is transmitted by the bite of the female sand fly that injects its saliva into the victim as it sucks out some blood from its bite while depositing some protozoa. This disease has many names and is found in over eighty-eight countries. It was known as far back as two thousand years ago and is identified by many names, including Kala-azar, black fever, Aleppo or Oriental boils, white leprosy, Andean sickness, sand fly disease, valley sickness, espundia, or Dum Dum fever. It is estimated that over twelve million people are now infected worldwide and several hundred millions more are at risk.

Leishman was also instrumental in the development of a number of vaccines, particularly typhoid which afflicted large numbers of native populations, as well as members of the military serving in indigenous regions. Due to the success of a typhoid vaccine, by 1909 Leishman reported that only five out of nearly eleven thousand vaccinated soldiers died in India, compared to the deaths of forty-six out of almost nine thousand soldiers who were not vaccinated died. Fewer than two thousand British soldiers died of *Leishmaniasis* by the end of World War I.

During World War II in 1943 about one thousand U.S soldiers stationed in the Middle East came down with the cutaneous version of *Leishmaniasis.* There are three distinct versions of the sand fly disease: the *cutaneous* (skin), the *visceral* (internal organs), and *mucocutaneous* (mucus membranes) that exhibit different symptoms. The most common is the cutaneous *Leishmaniasis,* which is caused when the parasite in the saliva of the female sand fly burrows in the wound. The protozoa in the saliva multiply rapidly until there is a visible ulceration, which festers and takes months to heal. The small circular lesions are painless but often leave scars. Although it is recommended that treatment be sought, the wound will heal itself in time. It is not certain, but it is possible that the infection will be with the patient for life.

Even though a significant number of U.S. troops were deployed in Iraq during Desert Storm in 1990, there were only thirty-two confirmed cases of *Leishmaniasis.* One reason is that the major sand fly season is from the end of March to September. Because most of the service personnel involved in Desert Storm left Iraq before the main season of sand fly infestation, they were not infected. The same did not occur during the more recent Operation Iraqi Freedom campaign where hundreds of troop serving in Afghanistan and Iraq were infected. The main treatment center for U.S. military personnel is the Walter Reed Army Medical Center in Washington, D.C. A second center was opened at the Brooke Army Medical Center in San Antonio, Texas. Treatment consists of ten to twenty days of intravenous infusion of the drug called Pentostam. Freezing the protozoan at the infected site with liquid nitrogen can treat mild cases. Even unsightly lesions will take more than a year to heal. At last report about two hundred service men and women decided against treatment and let the lesions called "the Baghdad boils" heal on their own. Although military experts expect an increase in cases of *Leishmaniasis,* with better housing facilities and rotation of personnel experts believe the number of cases in the Middle East will decline. Currently, there is no vaccine available.

## LEMAÎTRE'S THEORY FOR THE ORIGIN OF THE UNIVERSE:

Astronomy: *Abbé Georges Edouard Lemaître* (1894–1966), Belgium.

> *Contrary to Einstein's belief in a static universe, Lemaître believed that the theory of relativity requires an expanding, not static, universe.*

Georges Lemaître was one of the first astronomers to relate relativity to cosmology. He based his nonstatic universe on the supposition that if matter is expanding everywhere within the universe, then there must have been a moment in the past when this expansion began. Although he disputed Einstein's belief in a static universe, Lemaître based his own thesis on Einstein's theory of special relativity of space-time. Lemaître assumed that if we could revert far enough in time, we would see the entire universe as a very compact, compressed point of matter and energy. He also considered radioactivity as the force that caused the original explosion, an idea no longer considered a valid theory for the big bang. Unfortunately, Lemaître did not completely calculate the mathematics for his theory of an expanding universe. From the later 1920s to the early 1940s, his expansion theory was unpopular with other astronomers, who still considered a static universe the preferred model. The most important aspect of Lemaître's theory was not just the expansion concept (which was well known), but the idea that something started the whole process—that is, there was a physical origin to the universe. In the late 1940s, Lemaître's theory of an expanding universe was revived and revised by George Gamow, who named it the "big bang." Today, it is considered one of the most likely explanations for the origin of the universe.

*See also* Einstein; Gamow

**LENARD'S THEORY FOR ELECTRON EMISSION:** Physics: *Philipp Eduard Anton Lenard* (1862–1947), Germany.

*During the photoelectric effect, the speed of electrons emitted is a function of the wavelength of the light (electromagnetic energy) involved.*

Philipp Lenard based his research on the photoelectric effect first detected by Heinrich Rudolph Hertz. Lenard observed that when ultraviolet light "struck" the surfaces of certain kinds of metals, electrons were "kicked" out and could be detected. He designed experiments to determine the cause and found that the speed at which electrons are ejected from certain types of metal during exposure to the light was a function of the wavelength of light used. Further, he found the shorter the wavelength of light used, the greater the speed of the emitted electrons. At the same time, the intensity of the light had no effect on the electrons' speed, but the brighter the light, the greater quantity (number) of electrons emitted. Some years later, Einstein explained the photoelectric phenomenon by relating it to Planck's quantum theory.

*See also* Einstein; Hertz; Planck

**LENZ'S LAW OF ELECTROMAGNETICS:** Physics: *Heinrich Friedrich Emil Lenz* (1804–1865), Russia.

Lenz's law can be written in several forms. They all are special examples of the law of conservation and are extensions of Michael Faraday's Law of induction of a magnetic field by the flow of an electrical current in a conductor. There are various ways of stating Lenz's law. Several examples follow:

1. *The EMF induced in an electric circuit always acts in such a direction, that the current it drives around the circuit opposes the change in magnetic flux, which produces the EMF.* (Note EMF stands for "electromotive force" which is the difference in potential that exists between two dissimilar electrodes immersed in an electrolyte.)
2. *The induced current produced in the conductor always flows in such a direction that the magnetic field it produces will oppose the change that produces it.* (In essence, this version of the law states that in a given circuit with an induced EMF caused by a change in a *magnetic flux*, the induced EMF causes a current to flow in the direction that opposes the change in flux. Note: *magnetic flux* is related to the magnetic induction that is perpendicular to the surface. It is better known as the density of a magnetic field.)
3. *The current induced by a change flows so as to oppose the effect producing the change.* This law is related to the more general law of conservation of energy and is a

special case because if the induced current were to flow in the opposite direction in the conductor, it would be an example producing electrical energy without any work being done (perpetual motion), which is impossible according to the law of conservation of energy.

Heinrich Lenz was born in Tartu (present-day Estonia). After finishing secondary school he attended the University of Tartu. Upon graduation, he was appointed as the geophysicist on two expeditions around the world from 1823 to 1826, where he made important measurements of the climate and physical conditions at various geographic locations. Following these tours he became dean of mathematics and physics at the University of St. Petersburg in Russia where he remained until 1863. He began his studies of electromagnetism as early as 1831. In addition to discovering Lenz's law he also independently discovered a version of Joule's law in 1842, which in Russia, is referred to as the Joule–Lenz law.

*See also* Faraday; Joule; Ohm

**LEVENE'S TETRA-NUCLEOTIDE HYPOTHESIS:** Biology: *Phoebus Aaron Theodor Levene* (1869–1940), Russia and United States.

*The DNA molecule is composed of a string of four nucleotide units consisting of equal amounts of adenine, guanine, cytosine, and thymine.*

In 1909 Levene found that pentose sugar *ribose* is found in the nucleic acid of yeast. However, it was not until 1929 that Levene identified the carbohydrate in the nucleic acid of the thymus. Because its molecule lacked one oxygen atom of ribose, it was called *deoxyribose*. These discoveries prompted Levene to hypothesize that a simple tetranucleotide (a combination of four nucleotides) was responsible for the structures later named *ribonucleic acid* (RNA) and *deoxyribonucleic acid* (DNA). This led to his "tetranucleotide hypothesis" that stated that DNA was composed of equal amounts of *adenine, guanine, cytosine,* and *thymine.* It was Levene who demonstrated that the components of DNA were linked in the order of phosphate-sugar-based units he called nucleotides that formed the backbone of molecules of living organisms. Regrettably, his concept (and thus hypothesis) for the structure of DNA was wrong because he believed that only four nucleotides exist in each molecule. To his credit, Levene recognized that his four-nucleotide concept was not able to account for a system to store the genetic code for living organisms because it was much too simple. Nonetheless, his hypothesis provided the basis for additional study, and the later discoveries that finally led many scientists to work on the puzzle of the structure of the famous double helix of the DNA molecule, which provides the copying mechanism for the genetic material. Crick, Watson, and Maurice Wilkins received the 1962 Nobel Prize for Physiology or Medicine. (Rosalind Franklin, who was Wilkins' assistant and coauthor provided the crystal photos of the DNA molecular structure that provided the clues to the unique helix structure of the molecule.) One of the amazing revelations about the structure of DNA is that one DNA molecule of one human cell, if laid end to end would be about 1 millimeter in length, which is many, many times longer than the cell itself.

*See also* Crick; R. Franklin; J. Watson

**LEVI-MONTALCINI CELL GROWTH THEORY:** Biology: *Rita Levi-Montalcini* (1909–), Italy and United States. Rita Levi-Montalcini shared the 1986 Nobel Prize for Physiology or Medicine with Stanley Cohen.

*A naturally occurring protein molecule in the body of living organisms, known as the nerve growth factor (NGF), stimulates the growth and differentiation of the sympathetic and sensory nerves.*

The nerve growth factor (NGF) is a protein secreted by the target cells of the neurons and is critical for the well-being and maintenance of the sympathetic and sensory nervous systems. This procedure is responsible for the movement of NGF in long-distance signaling of neurons. NGF binds two receptors on the surface of cells that respond to the two types of growth factors, that is, Track A (TrkA) and low infinity nerve growth factor receptors (LNGFR). These factors activate the signaling of nerve impulses between the stimulated cells and the receptor cells. Levi-Montalcini's research has led to advances in the regeneration of various types of nerves. It has been discovered that sensory nerves have greater potential for rehabilitation than do motor neurons. Although some nerve cells can and do regenerate if damaged, it is hoped that research using fetal stem cells will lead to improved treatments to restore or replace damaged nerve cells.

Rita Levi-Montalcini was born into a Sephardic Jewish family of four children raised by a mathematician father and talented painter mother. Her father did not want her to go to college, but she insisted and entered the Turin Medical School. She graduated with an MD degree in 1936. Her career was cut short due to Mussolini's law that barred Jews from academic careers. During World War II she set up a home laboratory in her bedroom where she studied the growth of nerve fibers in chicken embryos. Her family fled to Florence where she set up another home laboratory. In 1946 she received and accepted an invitation to Washington University in St. Louis, Missouri, to spend a semester as a research assistant. She stayed for thirty years. While at Washington University, she did her best and most important work and became a full professor in 1958. In 1961 she became director of the Research Center of Neurobiology in Rome, Italy, where she split her time between Italy and United States. Levi-Montalcini was only the tenth woman ever elected to the U.S. National Academy of Sciences. In addition to receiving the 1986 Nobel Prize (shared with Stanley Cohen), she also received the Albert Lasker Award for Basic Medical Medical Research. In 1987 she received the National Medal of Science, considered by some the highest honor in science in the United States. Still active in Italian politics, she was appointed a Senator-for-Life in 2001 in the Italian senate.

**LEWIS' THEORY OF COVALENT BONDS:** Chemistry: *Gilbert Newton Lewis* (1875–1946), United States.

*When atoms combine to form molecules, they share a pair of electrons, thus forming covalent bonds.*

Ionic bonds, also called polar bonds, were first introduced in the late 1800s and were thought to be one-to-one sharing of electrons from one atom to another. **Ionic bonding**

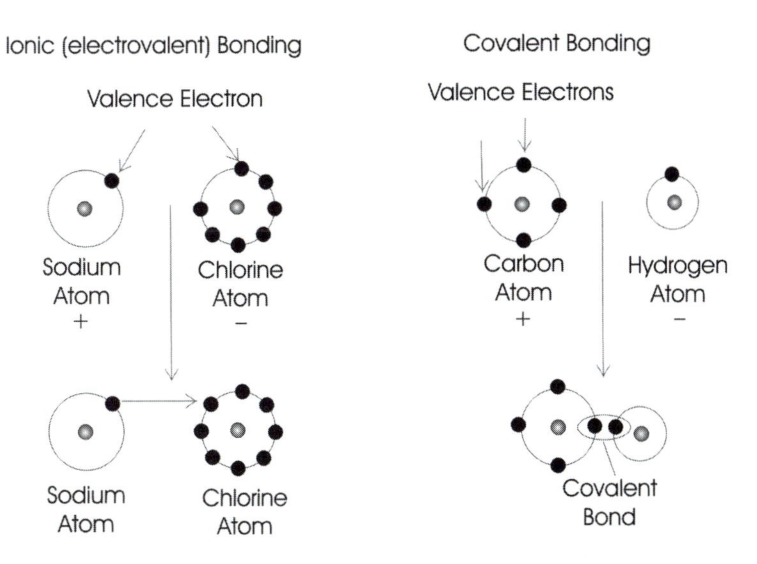


Figure L1. Examples of two types of bonding—*ionic* where one or more electrons are exchanged between atoms and *covalent* where one or more electrons are shared.

occurs when atoms with a dearth of negative charges (electrons) in their outer orbits (valence) naturally attract electrons from other atoms to complete their outer orbit. These atoms now form a molecule of a new and different more stable compound and are now neutral as they have lost their individual electrical valence charge (*see* Figure L1).

However, this concept was not valid for the formation of all molecular compounds. Lewis first conceived the structure of atoms as cubes with the possibility of one electron located at each of the eight vertices (corners) of the cube. More complex atoms were structured with smaller cubes located inside larger cubes. Lewis soon rejected this cube structure, but he still believed that at least eight electrons were required for each neutral atom. This knowledge led to the Lewis–Langmuir octet theory (*see* Langmuir), which provided information about the atomic structure of rare gases in which all eight vertices of their cube are occupied. Therefore, because the atoms of these gases do not gain, lose, or share electrons, they are inert because electrons occupy all the eight vertices of their cube image. This also explained why atoms that did not have a complete octet of outer electrons were available to combine with other atoms, by either ionic bonding or sharing electrons as covalent bonds. All atoms have a natural tendency to attain the same octet formation and thus become more stable (inert). For example, sodium has only one vertex (in its imaginary cube structure) occupied by an electron, while chlorine has seven vertices (in its cube) occupied. Sodium and chlorine act to establish the stable octet structure; thus sodium gives up its one electron, and chlorine accepts it, satisfying both. (For instance, $^{+}Na + {}^{-}Cl = NaCl$, where the salt NaCl is neutral, is an example of ionic bonding, while $^{--}O + {}^{--}O = O_2$ depicts a pair of atoms sharing electrons, that is, covalent bonding to form a diatomic molecule of oxygen gas.) The most common type of covalent bonding is single bonding where just two electrons are shared—one from each partner. There may be double bonds as well, involving four shared electrons, or even triple bonds, with six shared electrons. Lewis and Langmuir's octet theory resulted in a better understanding of the laws of thermodynamics and the periodic arrangement of elements.

*See also* Langmuir; Sidgwick; Thomson

**LIEBIG'S THEORY OF ISOMERS AND ORGANIC COMPOUND RADICALS:** Chemistry: *Justus von Liebig* (1803–1973), Germany.

*Inorganic or organic compounds (molecules) with the same formula can have different structures and thus exhibit different characteristics.*

Justus Liebig was working with Joseph Louis Gay-Lussac when he discovered *silver fulminate*. At the same time, his friends Friedrich Wohler and Jöns Jacob Berzelius had prepared *silver cyanate*. To the surprise of all, both of these compounds had the same formula, but they behaved very differently. Berzelius named this phenomenon ***isomerism.*** Today, this concept is used to develop different chemicals that have the same basic formulas but exhibit many different and useful properties. Isomerism is one of the reasons so many different types of organic (carbon-based) synthetic drugs, manmade fibers (nylon), polymers, plastics, dyes, cosmetics, bleaching agents, and numerous other organic chemical products can be manufactured. Exploring this phenomenon further, Leibig and Wohler used different forms of the benzoyl radical (a form of benzene $C_6H_6$) combined with other elements to formulate their theory of compound radicals. This is an example of a family of similar chemicals that can have additional atoms added to the main radical. Chemists can use the basic benzoyl radical to form different compounds by adding, for example, chlorine (Cl), bromine (Br), hydrogen (H), nitrogen (N), or other atoms to the basic structure.

*See also* Kekule; Lewis

**LINDEMANN'S THEORY OF PI:** Mathematics: *Carl Louis Ferdinand von Lindemann* (1875–1939), Germany.

*It is impossible to "square the circle" to arrive at a rational number for pi (π) by using a straight edge and compass and thus accurately determine the area of a given circle.*

Carl Lindemann was aware of Archimedes' method of using geometry (multiple polygons) to determine the value of pi by "squaring the circle" to determine the ratio of the circumference of a circle to its diameter (or radius). Historically, mathematicians wished to use this ratio (pi) to determine the area of a circle ($A = \pi r^2$). Archimedes arrived at a ratio of 3.142, which is very close to the current accepted value of 3.14159. Lindemann used algebraic methods to prove that an accurate ratio for pi could not be determined by geometric methods using a straight edge and compass because pi is a transcendental number. All transcendental numbers are irrational and a real number is said to be transcendental if it is not an algebraic number. This means that pi is not a root for any polynomial equation with rational coefficients, such as $2/3x^3 - 5/7x^2 - 21x + 17 = 0$.

*See also* Archimedes

**LINNAEUS' THEORIES FOR THE CLASSIFICATION OF PLANTS AND ANIMALS:** Biology: *Carolus Linnaeus* (1707–1778), Sweden.

*Plants and animals of different species can be classified according to similarities within a species as well as differences between species.*

All classification systems use some form of similarities and differences between and among what is being classified. Aristotle's taxonomy or classification of living things was based on the assumption that nature proceeds from tiny lifeless forms to larger animal life with no distinct line of demarcation between classes. He designed what is called the "ladder of life" that assigned each species to a lower or higher step on his ladder (*see* Figure L2).

Until Linnaeus' time, plants and animals were classified from the top down, beginning with large classes and working down to smaller groups. Carolus Linnaeus devised the system of taxonomy based on the concept of species that is still in use today. A major factor in determining what animal or plant belongs in a particular species is whether reproduction is limited within that species. We know today that the vast majority of DNA is the same for all mammals. However, just a small difference of DNA between species prevents cross-fertilization. For instance, chimpanzees and humans share over 98% of

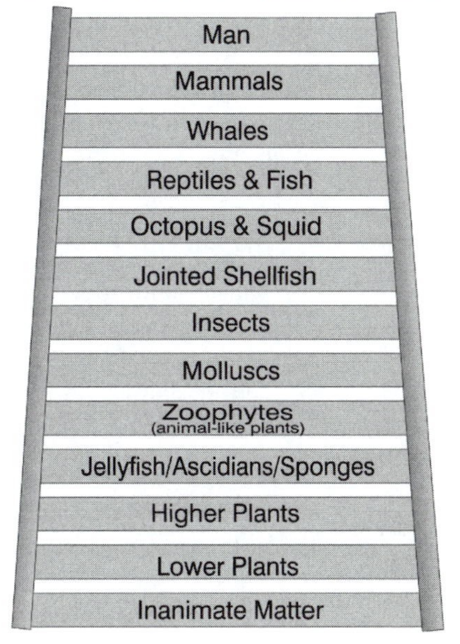


Figure L2. Linnaeus partly based his classification of plants and animals on Aristotle's hierarchy of life that started with man on top and extended to lower plants and finally inanimate matter.

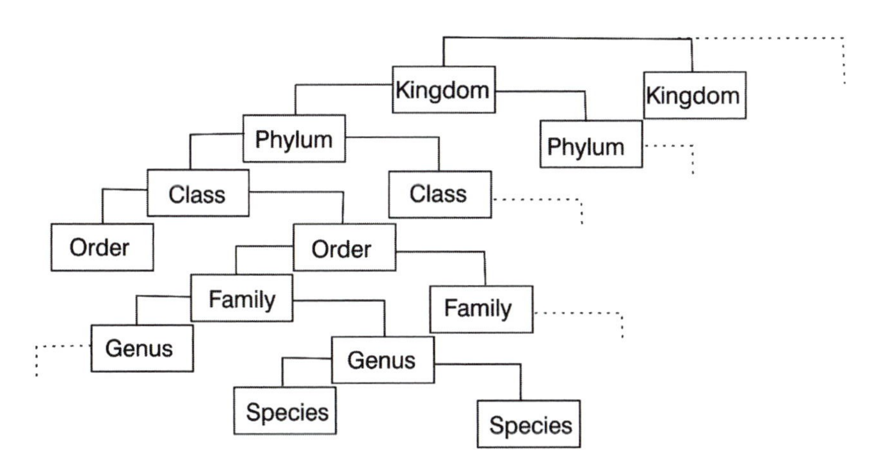


Figure L3. Linnaeus' classification of plants and animals was an advancement from Aristotle's and is similar to what is the current taxonomy.

the same type of DNA. Linnaeus further classified according to similarities within a species as well as differences between species. His concept required a new terminology, for which he used Latin names. Starting with humans, he used *Homo* as the genera for "man" and *sapiens* for the species of "wise man." From here Linnaeus combined similar *genera* into *family*, from family up to larger groups called *orders*, then divided further into *phyla*, and phyla into the two *kingdoms* of plants and animals (*see* Figure L3).

The term "taxon" is meant to encompass all the special traits shared at any of the seven major categories of his taxonomy system. Since then, subdivisions have been added, and with the knowledge of evolution and cellular and molecular distinctions, even finer similarities and differences are used. More recently, biologists have divided the two kingdoms of plants and animals into five major groups. This new classification system consists of three major plant kingdoms and two major animal kingdoms. These distinctions are based on differences in molecular DNA.

*See also* Cuvier; Darwin; Lamarck

**LISTER'S HYPOTHESIS OF ANTISEPSIS:** Biology: *Baron Joseph Lister* (1827–1912), England.

> *Carbolic acid, used during surgical procedures, can prevent and control subsequent infections.*

Joseph Lister based his hypothesis on and credited the work of Francesco Redi and Louis Pasteur. Redi determined that rotting meat developed maggots only when exposed to flies, and Pasteur proved that microorganisms caused such putrefaction. Using this knowledge of airborne microorganisms, Lister sprayed the air of operating rooms with carbolic acid (a derivative of benzene called phenol). To his dismay, deaths from infections following surgery were still over 50%. He then soaked a cloth in carbolic acid and used it to bind an open wound. The wound healed without infection. Following this lead, Lister then soaked all his surgical instruments in carbolic acid, had surgeons rinse their hands in dilute acid, and maintained a clean operating room. These procedures reduced surgical mortality in his operating rooms in 1877 from 50% to about 5%. At first, Lister's hypothesis for controlling infection was not well received; subsequently by the late 1800s antiseptic procedures were standard in all hospitals and doctors' offices. Lister made another contribution when he replaced silk thread, which was used for sutures and could not be easily sterilized, with catgut disinfected by carbolic acid. Today steam is used in autoclaves to sterilize medical instruments, as well as a variety of disinfectants and anti-infectives to prevent infections.

*See also* Pasteur; Redi

**LOCKYER'S SOLAR ATMOSPHERE THEORIES:** Astronomy: *Sir Joseph Norman Lockyer* (1836–1920), Britain.

> *A unique spectral line from the sun's light is produced by a new and unknown element.*

Sir Joseph Lockyer spent much of his life attempting to determine the composition of the sun's atmosphere and its effects on Earth. He and Pierre-Jules-Cesar Janssen devised a method of observing the sun during daylight hours. Up to this time, the only way to view the sun was during a solar eclipse or through a smoky-colored glass produced by holding it over a burning candle that deposits a thin film of carbon on the glass. In 1868 Janssen, using a solar spectroscope, was the first to view a peculiar spectral line in sunlight that did not match other known spectral lines. Although he was the first to see this spectrum, it was later identified and named by Lockyer, who hypothesized that because it existed only in the sun, it should be called helium, from the Greek word *helios* meaning "sun." Helium was considered a hypothetical element until detected on Earth twenty-five years later by Sir William Ramsay. Lockyer made other contributions. Using the Doppler effect (*see* Doppler), he determined the "wind speed" of solar flares. In addition, he determined the temperature of the surface of the sun and that sunspots have a lower temperature than the sun's surface. Lockyer also was convinced the solar atmosphere affected Earth's weather because its orbit is just at the edge of the sun's outer corona. Additionally, he believed the size and number of sunspots affect the amount of rainfall on Earth. At the time, he was unable to examine these phenomena, which today are partially accepted, but not as Lockyer hypothesized.

*See also* Doppler; Janssen; Ramsay

**LORENTZ'S PHYSICAL THEORIES OF MATTER:** Physics: *Hendrik Antoon Lorentz* (1853–1928), Netherlands. Hendrik Lorentz shared the 1902 Nobel Prize for Physics with Pieter Zeeman.

***Lorentz's electron theory:*** *Atoms and molecules are very small, hard bodies that carry either a negative or positive charge.*

James Clerk Maxwell determined that light waves were the result of the vibrations of charged particles (atoms); as these particles oscillated, electromagnetic waves were produced. Hendrik Lorentz's electron theory expanded this theory and was based on the assumption that 1) there is a wave-carrying medium in space known as **aether** and 2) matter (solid, liquids, and gases) was a separate entity from the wave/aether, therefore 3) only electrons could interact between them. He found that atoms with a positive charge "oscillate" in one direction within a magnetic field, and those with a negative charge "oscillate" in the opposite direction. His mathematical theory was developed before there was any proof that electrons existed, but it indicated that light waves were the result of oscillating electrically charged atoms.

***Lorentz force:*** *There is a force applied to moving electrically charged particles when they are in the presence of an electromagnetic field.*

Hendrik Lorentz identified charged particles produced in a cathode ray tube as negative electrons. His theory also explained the Zeeman effect (*see also* Zeeman), which asserted that the spectral lines for sodium atoms split into several closely spaced lines when exposed to an electromagnetic field. This phenomenon was later explained by quantum theory.

***Lorentz's theory for the contraction of moving bodies:*** *Light from moving bodies traveling through the aether caused these bodies to appear to contract in size in the direction of their motion.*

At about the same time Lorentz proposed his theory for the contraction of moving bodies, another physicist, George Francis Fitzgerald, independently arrived at the same concept. Therefore, the mathematics for this phenomenon is known as the *Lorentz–Fitzgerald contraction.* In essence, the theory states that bodies moving through an electromagnetic field contract somewhat in the directions of their motion in proportion to their velocity. This explains why light appears to move at the same speed in all directions at the same time from its source. Einstein used this concept in developing his theory of special relativity (*see also* Einstein; Fitzgerald; Zeeman).

***Lorentz invariant:*** *Natural laws must be invariant to a change in the coordinates (space and time) of any system.*

The Lorentz invariant is sometimes referred to as the Lorentz transformations theory. The theory is based on the mathematics Hendrik Lorentz developed to explain how moving bodies seem to contract. The consequence of these theories is that both space (three dimensions) and time must be equally considered when developing any type of equation that explains the relative motion of matter. The theories of contraction and transformations describe the coordinates that need to be considered for the contraction in the length and the increase in mass of moving bodies at relativistic speeds. They provided the foundation for Einstein's theory of special relativity. Einstein relied on the mathematics of Hendrik Lorentz and also recognized the contributions made by other scientists that aided him in developing his theories of relativity.

*See also* Einstein; Fitzgerald; Maxwell; Michelson; Zeeman

**LORENZ'S THEORY FOR COMPLEX/CHAOTIC SYSTEMS:** Mathematics: *Edward Norton Lorenz* (1917–2008), United States.

*The sensitivity of a dynamic system depends on small initial conditions.*

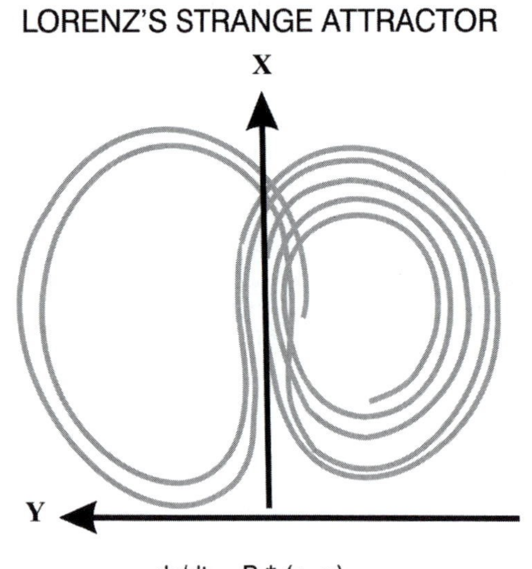


Figure L4. Lorenz's strange attractor depicted as a curve on a plane surface, which is actually a curve in three-dimensional space. The line forming the curve is a single unbroken line that never follows the same path and, in three dimensions, never intersects itself.

Edward Lorenz, a meteorologist, applied mathematics to weather forecasting and climate changes. Using a computer, he analyzed the initial conditions of a weather system with temperature as the only single variable. His original data were carried out to six decimal places but rounded off to three decimal places, a practice used by most scientists and which Lorenz assumed was such a small difference that it would not affect the outcome. This assumption proved to be false because he obtained a different result each time he ran the computer data. At this point, he realized that small initial differences do have a cumulative effect over long periods of time and thus affect events differently. This phenomenon that describes the sensitivity of a system as dependent on small differences in initial conditions became known as the "butterfly effect." The butterfly effect hinges on a number of weather factors, including the temperature and humidity of the air and how the air is flowing. It aptly describes the chaotic systems in which small perturbations result in very different outcomes. The name *butterfly effect* came from a fable about a butterfly whose flapping of its wings created an air flow in China that added to the cumulative air flows around the world, thus causing hurricanes in Florida and snowstorms in Wisconsin. The principles of complex systems, chaos theory, and nonlinear mathematics are used to interpret dynamic systems as diverse as economic cycles, the stock market, population changes, the dynamics of three-dimensional flow of fluid in pipelines, the dynamics of prehistory archaeology, and weather predictions. Edward Lorenz originated the *Lorenz attractor*, a mathematical expression using differential equations to describe how a system settles down, based on the three variables of space orientation (x, y, and z). Diagrams of this concept are looping curves on a two-dimensional surface, but the curves are in three dimensions where the single line never crosses itself at any given moment of time (*see* Figure L4). A point on the line represents a variable of the system expressed as a point in three-dimensional space.

*See also* Penrose; Wolfram

**LOVELL'S THEORY OF RADIO ASTRONOMY:** Astronomy: *Sir Alfred Charles Bernard Lovell* (1913–), England.

*Radio signals collected from outer space can be used to verify the existence of many phenomena, such as quasars with angular diameters of less than one second of arc. They also have the ability to track meteors and comets.*

Bernard Lovell received his PhD from the University of Bristol in 1936 and served on the cosmic ray research team of the University of Manchester until he entered military service during World War II. He used his talents working on wartime aircraft radar for which he received an award. After the war, he continued his work with radar to detect cosmic rays using surplus army radar equipment. Before the war in the United States, Nikola Tesla had recorded extraterrestrial cosmic rays as radio-type signals on his experimental radio equipment. Most of the scientific community did not believe this electronics genius because they rejected the notion that cosmic rays existed. In the early 1930s the Bell Laboratory engineer Karl Guthe Jansky detected radio waves from the center of the Milky Way galaxy. The American radio astronomer pioneer Grote Reber (1911–2002) confirmed Jansky's discovery a few years later. It was not until after World War II that the field of radio astronomy became a major research effort for England and the United States.

Lovell was convinced that extraterrestrial radio waves were a possible tool for advancing theories in the field of astronomy. Lovell installed surplus wartime radar equipment at Jodrell Bank in the Cheshire region in England. Thus began his life-long research investigating cosmic rays, tracking meteor velocities, and comets. He soon realized that he needed a much larger and steerable radio telescope dish that could receive radio waves of at least 30 centimeters. Because this was before the days of "big" research budgets, Lovell had difficulty raising funds for the project, but he persevered. The 250-foot diameter dish could be steered to pick up signals from locations in space from horizon to horizon. In 1957 the Lovell Telescope at Jodrell Bank was used to track *Sputnik*. It has proven to be an excellent investment ever since as it is used for teaching and research by students and engineers in cooperation with other types of telescopes. Although the Lovell telescope has been upgraded and is still in use, a number of more technologically sophisticated and larger radio telescopes have since been constructed.

*See also* Jansky; Tesla

**LOWELL'S THEORY OF LIFE ON MARS:** Astronomy: *Percival Lowell* (1855–1916), United States.

*Canals and oases seen on Mars indicate that it was once inhabited.*

Percival Lowell became interested in the 1877 report by Giovanni Schiaparelli in which he stated that he observed *canali* (Italian for "channels" or canals) on the surface of Mars. Lowell constructed a 24-inch reflector telescope atop a 7,200-foot mountain in Arizona to make use of the area's clear sky. He reported he also saw Schiaparelli's Martian canals and claimed they were built by intelligent beings. He theorized these canals were dug to transport water from melting ice at its poles to the dry central regions of the planet. It is now known that these lines and patches were aberrations in Lowell's lens/mirror system, the "shimmering" of Earth's atmosphere, or the results of a vivid imagination.

*See also* Schiaparelli

In the year 2004 there was a scientific breakthrough that confirmed Percival Lowell's conjecture that at one time there was water on the planet Mars. However, there was no evidence to support this belief that canals were dug to transport the water from the polar regions to the dryer equatorial areas of the planet. The evidence that water once flowed over the surface of Mars was confirmed by two six-wheeled robots—the Spirit and Opportunity—that landed on the red planet's surface in January 2004. Their lifetimes for sending signals back to Earth was estimated to be just three months, but they roamed over many kilometers, even in Mar's subfreezing winters, sending information to Earth for far longer. According to a major U.S. science journal, *Science,* the success of these two robots and the value of the geological data gathered and transmitted to Earth was a major triumph. The robots were even able to determine that the water on Mars was salty, and acidic, and possibly capable of supporting life. NASA hopes to send a robot to the Martian polar ice sheet in 2008 and a mobile laboratory in 2009 to answer the following questions: Where and when water may have flowed in the past on Mars? Where it might be found today? And, what forms and amounts might be available for use? Although more robots will be sent to Mars, the moon, and other planets, the ultimate goal is to, in the not too distant future, send humans to Mars to conduct extensive geological research.

**LYELL'S THEORY OF UNIFORMITARIANISM:** Geology: *Sir George Lyell* (1797–1875), Britain.

*Currently observed geological changes and processes are adequate to explain geological history.*

This basic concept was first expressed by the Scottish geologist and naturalist James Hutton (1726–1797) and John Playfair (1748–1819), the Scottish mathematician and philosopher, some years before Lyell clarified it. However, both neglected to explain fully or examine their concept in detail. Sir Charles Lyell explicitly stated that the same scientific laws and geologic processes that applied in the past, and the present, and that will also apply in the future are, therefore, responsible for the physical and chemical processes that result in geologic changes. This led to his famous saying, "The present is the key to the past." He explicitly rejected the theory of German geologist

Abraham Werner, who believed that a huge deluge of water (the "flood") was responsible for Earth's current topography. Lyell believed the action of the wind, rain, the sea, earthquakes, and volcanoes, rather than some great catastrophe, explained geological history. He rejected the concept of catastrophism, which was first believed to conform to biblical history. Lyell based much of his "uniformitarianism" concept on his study and classification of the strata of ancient marine beds. He observed that the layers of sediment closest to the surface contained shells as well as the remains of animal species still living in modern times. Conversely, the deeper, older strata contained more fossils of extinct species. He divided the rocks containing fossils into three groups, or epochs and named them after ancient geological periods—Eocene, Miocene, and Pliocene—terminology still in use today. Charles Darwin, who developed the theory of organic evolution, relied on parts of Lamarck's and Lyell's earlier works.

*See also* Darwin; Eldredge–Gould

**LYSENKO'S THEORY OF THE INHERITANCE OF ACQUIRED CHARACTERISTICS:** Biology: *Trofim Denisovich Lysenko* (1898–1976), Russia.

Trofim Lysenko followed in the footsteps of I.V. Michurin (1855–1935), a Russian who advocated the acceptance of Lamarckism. Michurianism and later Lysenkoism were the biological and genetic party line (ideology) of the Soviet Union under Stalin. Lysenko, a minor agriculturalist, promoted a new theory based on an old farmers' concept, called "vernalization," as a means to improve the germination of grain. He claimed that by treating grain with cold water, the flowering of the grain would improve, and it would sprout sooner in the spring. Thus, it would take less time to raise a crop and would increase the production of grain to feed the masses. It did not work. Neither this concept nor any of his other ideas used standard controlled experiments, peer review, or other accepted processes and procedures of scientific research. His mistake, as well as those of some others in Russia, was a rejection of the science related to Mendelian genetics. His insisted that this "cold treatment" would not need to be repeated each year because, once used, it became an "acquired" characteristic that would be passed on from one generation of grain to the seeds of the next. Lysenko's ideas seemed to fit the Marxist philosophy, and he soon discredited the president of the Lenin All-Union Academy of Agricultural Sciences, Nikolai Vavilov (1887–1943), who was exiled to Siberia. Lysenko then became head of this all-powerful institution and had complete support from Joseph Stalin and the Communist party. Some, but not all, biologists in Russia believed that the theory of acquired characteristics could be applied in many instances. Lysenko believed in his new but unproven "science" that proclaimed that any desirable acquired characteristics could be inherited under the right conditions. For example, it was supposedly said that if a woman wanted a red-haired child, all she need do is dye her hair red and that characteristic would be inherited by her offspring. Russian biologists who did not support Lysenkoism but rather supported the science of modern genetics, as well as Darwin's theory of natural selection, were designated as reactionary and decadent enemies of the Soviet people. They were, over time, either excommunicated, sent to death camps, or disposed of.

*See also* Darwin; Lamarck; Mendel

# M

**MACH'S NUMBER:** Physics: *Ernst Mach* (1838–1916), Austria.

*There is a ratio that expresses the velocity of an object in a fluid to the velocity of sound in that fluid.*

Ernst Mach believed if information about nature cannot be sensed, it was useless. In addition, he thought discoveries could be made by intuition and accident (serendipity) as well as by using mathematics and scientific methods. These ideas and his concept of motion, which states that the inertia of a body arises from interactions with all of the mass within the universe, influenced the field of quantum mechanics and Einstein's formulation of his theory of relativity. His experimental work with vision and hearing led him to use high-speed photography to detect the shock wave produced in air by a high-speed bullet. This so-called barrier is also created when an airplane approaches the identical speed of sound traveling in cold air, which is about 750 (more or less) miles per hour. This "sound barrier" was first believed to be similar to a wall that must be overcome. However, there is not now, nor has there ever been, a wall to overcome. For example, it is well known that artillery shells, bullets, and thunder all travel faster than the speed of sound and produce shock waves. Air molecules are compressed and produce a shock wave. "Sonic booms" are heard when two shock waves are so close together that they are heard as a single "boom" by a bystander on the ground. The wave front for an airplane is a V-shaped area of compressed air analogous to the V-shaped bow wave produced by a speeding boat in water. The exact speed of the object traveling through a fluid (e.g., air) required to break this "barrier" will depend on the temperature of the air as well as the air's density and moisture content. The denser the medium through which sound travels, the faster it travels; at room temperature, sound travels 1,126 feet per second in air, 4,820 feet per second in water, and 16,800 feet per second in iron. The greater the density of the medium, the faster the sound proceeds

The concept of inertia (the resistance of a body to any change in momentum) is an old and often-confusing concept. Aristotle believed that once a body was in motion, some type of force was required to continue its movement—therefore there could be no action at a distance that could affect the body. Many years later Sir Isaac Newton stated his classic "three laws of motion." The first law deals with inertia that states that an object with mass will remain at rest while objects in motion with a constant velocity will remain in motion at that velocity until an external force acts on the object. His concept of inertia of a body was intrinsic to a body having mass and not dependent on the existence of any other matter. Thus, once a body with mass at rest is acted on by a force and then put into motion, it would continue moving at that speed in the same direction until some other force acted on it to change its speed and direction. Newton formulated his three laws of motion against the concept of absolute space and time. Mach disagreed with this concept of absolute space and time and believed only *relative* motion, not *absolute* motion, exists in the universe. He concluded from this that it is immaterial to think that Earth revolves on its axis or that the stars in the sky do the revolving and that Earth is motionless. This became known as "Mach's principle" by cosmologists. Mach believed that our world was nothing but sensations, thus resulting in "economy of thought," which was the simplest way to explain science phenomena. This led to the concept of "logical positivism" which had an influence on Einstein in the development of his theory of relativity.

through that medium. Whenever the speed of an object exceeds the speed of the sound in a particular medium through which the object is traveling, a shock wave is produced. At 0°C the speed of sound traveling through dry air is 331.4 meters per second at sea level. The Mach number varies for airplanes flying at different altitudes. At higher altitudes, the air is colder, thinner, and dryer than at sea level; thus the sound barrier is reached at different speeds at different altitudes.

The Mach number that Ernst Mach devised is the ratio of the velocity of an object, such as an airplane, to the speed of sound in air though which it travels. Mach numbers below 1 are referred to as subsonic flows of fluid; numbers greater than 1 are supersonic flows of fluid. An airplane flying at a speed lower than a Mach number of 1 will be traveling in subsonic flight. Once the airplane exceeds Mach 1, it has reached supersonic velocity, and the so-called sound barrier will be broken. As an example, if an airplane travels 1,500 miles per hour and the speed of sound in air through which the airplane is flying is 750 miles per hour, the ratio is 1,500/750 = Mach 2. The airplane overtakes the wave fronts in the front as well as in the rear of the airplane, producing overlapping wave fronts.

*See also* Aristotle; Einstein; Newton

## MAIMAN'S THEORY FOR CONVERTING THE MASER TO THE LASER:

Physics: *Theodore Harold Maiman* (1927–2007), United States.

*The variable wavelengths produced by the maser can be altered to produce shorter visible coherent wavelengths.*

The word "maser" is an acronym derived from the term "microwave amplification by stimulation emission of radiation" and was developed about the same time by three physicists, Charles Townes at Columbia University, and Drs. Nikolai Basov (1922–2001) and Aleksandr Prokhorov (1916–2002) of the Lebedev Institute in Moscow, Russia. The maser is a device that produces coherent electromagnetic radiation as the result of stimulated emission within a limited portion of the electromagnetic spectrum. Originally, the microwaves produced by the maser were very weak, but this was soon overcome which made it more useful for a number of applications, such as atomic clocks and radio telescopes. Therefore, some physicists have replaced the term "microwave" with "molecular" to represent their use when working with molecules that are the basis for kinetic energy.

This background provided Harold Maiman with the idea that the range of wavelengths produced by the maser could be extended to provide much shorter and visible wavelengths. The shorter white light wavelengths can travel much longer distances than those produced by the maser. While at the Hughes Research Laboratories in Malibu, California, Maiman designed an instrument that used a ruby-red cylinder-type crystal with a mirror-like finish at both ends of the cylinder. These mirrors enabled the light that entered the cylinder to "bounce" back and forth in the cylinder. The cylinder then became a resonant cavity. This provided a means for the flashes of incoherent white light of the maser to change into a pulsating beam of coherent monochromic (one color) light that could travel great distances without spreading into a wider beam. In 1961 Iranian-born physicist Ali Javan (1926–) and colleagues at the Bell Laboratories in Murray Hill, New Jersey, produced the first continuous coherent beam of light. These improvements led to the naming of this type of optical source as a LASER, which is the acronym for "light amplification by stimulated emission of radiation." Of some interest is that more recently the word "laser" (no longer capitalized) has become a standard word in the English language and thus included in most dictionaries. The verb of the word is now "lase" as in "he has been lased."

In the 1960s and 1970s scientists and engineers sought out problems for which they believed these new lasers would help to find solutions. There are many forms of lasers developed for use in industry, science laboratories, the medical field, the military, and other areas in society. Some examples: supermarket barcode scanners, laserdisc players, laser printers, surgical instruments, and in various forms of communications. There are two different types of basic lasers. One uses a high-peak output that produces high energy, short pulses, such as the Yag laser, which is a crystal composed of yttrium aluminum garnet ($Y_3\ Al_5\ O_{12}$) used for eye surgery, and the continuous wave type that produces a constant output, used mainly for communication and the cutting of hard materials. Another advantage of lasers is that it is one of the few instruments that provides more output power than the input power required to operate it. Charles Townes, Basov, and Prokhorov shared the 1964 Nobel Prize in Physics for their work on the maser. Maiman was nominated twice for a Nobel Prize but never won. He did receive the 1983/84 Physics Prize and became a member of the National Inventors Hall of Fame and the National Academies of Science and Engineering.

## MALPIGHI'S THEORY FOR THE DETAILED STRUCTURE OF ANIMALS AND PLANTS: Biology (Anatomy): *Marcello Malpighi* (1628–1694), Italy.

*Living materials are glandular in organization, and the largest internal organs consist of tiny glands whose function is both the separation and mixing of juices.*

Malpighi was one of the first physicians to use the microscope for research and to study the internal structure of plants and animals. As a physician using his microscope,

Born in Italy in 1628, Marcello Malpighi entered the University of Bologna in 1646. Despite the deaths of both his parents when he was twenty-one years of age, he managed to complete his education. His career in anatomy and medicine was continued as a professor of theoretical medicine at the University of Pisa in 1656. Later in 1659, he returned to the University of Bologna where his colleagues were critical of his microscopic research primarily because of a lack of understanding. Leaving this hostile situation, he accepted a professorship in medicine at the University of Messina in Sicily in 1662 where he and his research were welcomed. While there, he did some of his best microscopic research and was able to identify and describe taste buds in the mouth and how they were the termination of nerves that sent signals to the brain. He also described the optic nerves, some structures of the brain, and in 1666 was the first to not only see blood cells, but also determine that the red color was intrinsic to these cells. Malpighi had a long career. However, despite his many accomplishments and honors, including an honorary membership in the Royal Society in London, his last years were fraught with challenges. He was in ill health, and his many enemies, who opposed his somewhat radical ideas, burned down his house destroying his microscopes, papers, and manuscripts in 1684. In recognition of his status as a great physician/researcher, Pope Innocent XII appointed him as the papal physician in 1691. He was also named count and elected to the Italian College of Doctors of Medicine. He died in 1694 after suffering a stroke.

he made important observations in his study of the tissues and organs of plants and animals, including the human body. He is considered a pioneer in using this instrument for physiological and anatomical studies. His microscopic studies of living organs, such as the brain, spleen, liver, and urinary system, as well as larger organs, were historic. These studies led him to conclude that minute glands exist at the tissue level as well as the organ level. He theorized that these glands were designed for the separation as well as mixing of vital juices. He used his microscope for some of the first examination and study of interspecies similarity and differences of tissues and organs. He contributed several insights to the field of comparative anatomy, particularly for the skin, kidneys, and liver of various species. For instance, he discovered that silkworms and other insects do not have lungs to breathe. Rather, there are small holes in their skin called tracheae. He also discovered that food, produced by sunlight action on the green chlorophyll in leaves of plants and required for plant growth, comes downward from the leaves of trees. Most important, he was the first to observe capillaries and discern their relationship to the movement of blood between arteries and veins. For many years physicians, including William Harvey, did not understand how blood flowing through arteries entered the veins for the return trip to the heart. Malpighi's observation provided the basis for a more accurate understanding of human physiology.

Marcello Malpighi is best known as an Italian physician and biologist who pioneered the science of microscopic anatomy of living things. His research with the microscope was enormously useful in future studies in other fields, such as physiology, embryology, and general medicine.

**MALTHUSIAN POPULATION CATASTROPHE THEORY:** Biology (Economics): *Thomas Robert Malthus* (1766–1834), England.

*The power of population is indefinitely greater than the power in Earth to produce subsistence for man. Populations, when unchecked, increase in a geometrical ratio.*

Malthus was an economist interested in the biology related to the growth of population outstripping the potential supply of food. The theory has received many names, such as "Malthusian dilemma," "Malthusian limit," "Malthusian disaster," and "Malthusian check on population." The theory is based on the mathematical concepts of geometric and linear progression of biological growth. The population (of any animal species) increases geometrically (i.e., 2, 4, 8, 16, 32, etc.) if no restrictions are placed on the reproduction process, whereas the increases in natural resources supporting the animal species is a linear arithmetical progression (i.e., 1, 2, 3, 4, 5, etc.). During the early period of the industrial age that began in the early nineteenth century Malthus was concerned with the decreasing death rate and increasing birth rate in Europe as the population moved from an agrarian to an industrial society. He stated two postulates. First, that an adequate supply of food must be available to sustain human existence, and second, that the natural passions between the sexes will lead to offsprings (geometric progression). After all, reproduction is the main purpose of all life; otherwise most species, including humans, will become extinct. In essence, the Malthusian population theory states that any population growing faster than the resources (food) necessary to provide for the increasing population will experience catastrophe. This is why Malthus considered the geometric (or exponential) progression of growth in the human population an impending catastrophe. One upside of biological population theory states that if the reproductive potential of virtually any organism or species greatly exceeds Earth's capacity (land, water, or food) to support all the potential offspring, many will die and, consequently, species diversity will be preserved. Thus, in the long run, the evolution of new species may be introduced. Many biologists (and economists) believe that Malthus' theory was too pessimistic, at least for humans, because it did not take into account technological advances in agricultural and food technologies such as the "green" revolution of the 1960s and, more recently, the genetic engineering of plants and animals that have led to increases in Earth's agriculture and husbandry capacity to support growing populations. Some Central and Eastern European countries, Central and Eastern as well as Southern Africa (due to HIV), England, and the United States have a negative growth in their *native* populations. However, through the emigration of people from countries that are overpopulated, these regions, with the exception of Southern Africa, now have not only a positive growth in population, but also the consequential social and economic problems resulting from seemingly unchallenged immigration. Even so, unchecked breeding by humans will in time outstrip our technological ability to support the numbers predicted to be about nine billion people within the next hundred years. Even today, there are regions of the world where the increase in population restricts the quality of life of the inhabitants of those regions. Although there are individuals and organizations that recognize the problems related to excessive population growth, unfortunately there are many more humans who ignore warnings of future consequences. Because Malthus' was an economist, his theory is considered by some to be more mathematical than biological and, therefore, is not completely accepted by all biologists.

**MALUS' LAW FOR THE POLARIZATION OF LIGHT:** Physics: *Étienne Louis Malus* (1775–1812), France.

*When a perfect polarizer is placed in a polarized light beam, the intensity of the light that passes through the polarizer is expressed by the following equation:*

$$I = I_0 \cos^2 \theta i :$$

where $I$ is the intensity of the beam of polarized light; $I_0$ is the original intensity of the light beam; $\theta_i$ is the angle between the light's starting plane of polarization.

Malus graduated from a French military school and attended the École Polytechnique graduating in 1796 as a military engineer. He served as an "examiner" in physics where he was able to pursue various areas through his research projects, mainly in optics. In 1808 he made a discovery, partly by accident and partly by using his inquiring mind. By viewing the image of a house through a crystal of Iceland spar (transparent calcite used in optical instruments), he noticed that by rotating the crystal the reflected rays of the sun were extinguished, but only when the crystal was held in certain positions. Because he was a firm believer in the Newtonian corpuscular concept of light that claimed that particles of light had poles, Malus believed that the planes in the crystal's structure oriented the poles of the light particles (corpuscles), thus causing the obstruction of some of the light. He named this phenomenon "polarization." This was the first time the word was used to describe this odd behavior of light passing through a crystal.

Despite conducting research with various types of glass, Malus was unable to produce any polarization of light because of the poor quality of glass at that time. However, by using high-quality natural crystals he developed his law that gives the initial and resulting intensity of light when a polarizing crystal is placed into the path of the incident beam of light. Thus, he was able to arrive at his law related to the intensity of polarized light. Since Malus' initial endeavors, research has been conducted on many types of materials to determine the effects of polarizers.

Some examples are:

1. *Absorptive polarizers*, including the "wire-grid polarizer," that can polarize electromagnetic waves resulting from electric fields.
2. *Beam-splitting* polarizers that split a single incident beam of light into two beams of different polarizations.
3. *Polarization by reflection* occurs when light reflects at an angle from an interface between two transparent materials—depending on the orientation of the plane from which the light is reflected.
4. *Birefringent polarizers* are dependent on the birefringent properties of quartz or calcite crystals where the beam of light is split by refraction into two rays.
5. *Thin film polarizers* are special types of glass on which a substrate of special optic coatings are applied to the film, causing them to act as beam-splitting polarizers. They can be either a thin plate of glass, or a wedge of glass with the film attached in a particular orientation.

The equation related to Malus' law indicates that real polarizers are not perfect transmitters of light components. For instance, if two polarizers are placed one on top of the other, the orientation angle between their polarizing axes is determined by the $\theta_i$ in Malus' law. In theory, if the two polarizers are crossed, no light should be transmitted. However, in practice because no polarizer is perfect, the light is not completely blocked.

**MANSFIELD'S THEORY OF MAGNETIC RESONANCE:** Physics: *Sir Peter Mansfield* (1933–), England. Peter Mansfield shared the 2003 Nobel Prize in Physiology or Medicine with the American chemist Paul C. Lauterbur (1929–2007).

*The nuclei of atoms have a magnetic moment that can be detected and used to form an image of living tissue.*

Peter Mansfield based his work on that of Felix Bloch (1905–1983) and Edward Purcell who discovered that the nuclei of some atoms have a **magnetic moment**. The protons and neutrons of nuclei have spins that are not paired. This creates an overall spin on the particles, generating a magnetic dipole along the spin axis, which is a fundamental constant of physics referred to as the *nuclear magnetic moment* ($\mu$). This can be compared to a wobbling spinning top whose axis circumscribes a precessional path, similar to the precession of Earth's axis. This wobble is created by an unbalanced spin caused by an external torque on the nucleus' axis, which results in a resonance. Absorption of electromagnetic radiation by most atomic nuclei (particularly organic compounds) in response to strong magnetic fields causes the nuclei to radiate detectable signals. The process become known as nuclear magnetic resonance (NMR) because it involved the nuclei of many types of atoms of the substance exposed to the electromagnetic radiation. This process was first used as a spectroscopic method for analyzing the atomic and nuclear structure and properties of matter. Later, the radiation produced by the resonating nuclei was detected and recorded by computers to form a two dimensional spectroscopic image of living tissue. The British electrical engineer and 1979 Nobel Laureate Godfrey Hounsfield (1919–2004) used NMR to develop a computer-aided tomography (CAT) scan that could form an image of human tissue and was less intrusive than X-rays. Mansfield improved the process by altering the manner in which magnets affected the spin of the nuclei so that a three-dimensional image could be produced. This became known as magnetic resonance imaging (MRI), which was designed to produce three-dimensional images of cross-sections of any part of the human body. However, even today confusion still exists between NMR, MRI, and nuclear energy. They are *not* the same. NMR is used as a spectroscope to view characteristics of molecules, and MRI is used to produce three-dimensional pictures of human tissues. Neither produces harmful nuclear radiation. Rather, they use powerful magnetic influences to "excite" the atomic nuclei of various types elements and to make three-dimensional images of tissues in human bodies by causing the nuclei of atoms to resonate (oscillate) differently and thus emit distinct frequencies (signals). Computerized instruments record these signals and analyze the results, thus scientists can study details of molecules and physicians are able to distinguish between healthy and diseased tissue. The word "nuclear" refers to the oscillation of the nuclei of atoms of various elements that make up living cells. The N of NMR has been eliminated from the term because of the public's ignorance and fear of nuclear radiation. It does not mean "radioactive," as with nuclear radiation produced by either nuclear fission or fusion. Thus, NMR has been replaced by the current term MRI of the body for diagnostic and therapeutic procedures.

*See also* Rabi; Ramsey

**MARCONI'S THEORY OF RADIO TELEGRAPHY:** Physics: *Guglielmo Marchese Marconi* (1874–1937), Italy. Marconi shared the 1909 Nobel Prize for Physics with the German inventor and physicist Karl Ferdinand Braun (1850–1918).

*Using Hertzian radio waves, it is possible to create a practical system of "wireless telegraphy."*

Marconi failed the entrance exams for the Italian Naval Academy and the University of Bologna. Nevertheless, he had a keen interest in science, especially electricity, and was permitted to attend lectures and laboratories at the university. He constructed his own laboratory in the attic of his home in Pontecchio, Italy, where he built his own equipment to explore what at that time were known as Hertzian waves or electromagnetic radio waves. He followed the footsteps of several other famous scientists of that period: Heinrich Rudolf Hertz of Germany, Edouard Eugene Desire Branly (1844–1940) of France, Oliver Joseph Lodge (1851–1940) of England, and Aleksandr Popov (1859–1906) of Russia. Heinrich Hertz constructed a "radiator" that consisted of two rods with a spark gap between them. Each had a capacitor at their ends to store the electricity until it reached the level where it was strong enough to create the spark between the gap, thus discharging the electricity stored in the capacitor. Hertz also designed a receiver for the electromagnetic waves produced by the spark that consisted of a loop of wire with a gap. This loop received the spark (electromagnetic wave) produced by current stored in the capacitor of the sender. The radiator sender and receiver were not connected.

This makes Hertz (or possibly Popov) the inventor of a device to send wireless radio signals—not Marconi as assumed by many people. The French consider Branly, a professor of physics, to be the inventor of the wireless telegraph by sending a spark to a receiver that responded by forming a coherent path of loose zinc and silver filings that detected the reception of the sending transmitter. Lodge, the head of physics at the University College of Liverpool, improved Branly's device by shaking the metallic crystals between sparks so the next sparks could be detected. This became the method for sending and receiving early wireless telegraph signals. Although Lodge received a patent for his device, he used it to communicate with the dead, and that was the last that was heard from him, or the dead.

It was Marconi who had the greatest insight into the potential for his device for wireless communications. At a young age Marconi made a number of unique improvements to the systems of others and patented a successful radio telegraph system at the age of twenty-two. His first units could only telegraph a dot-and-dash message a few feet, then thousands of feet, and soon a few miles. In 1901 he successfully transmitted and received wireless messages across the Atlantic Ocean. Although he did not invent wireless transmission of telegraph and radio, he made the advancements that significantly improved the process of radio telegraphy. He soon formed the Marconi Wireless Telegraph Company in England that was taken over by the British General Post Office in 1910.

**MARGULIS' ENDOSYMBIOTIC CELL THEORY:** Biology: *Lynn Margulis* (1938–), United States.

*Primitive single-celled prokaryotes, (cells with no nuclei) engulfed other cells, and, if both survived, they evolved by symbiotic pairing in a cooperative relationship over millions of years into eukaryotes (bacteria-type cells) that contain nuclei and other mitochondria (internal cellular structures) that drive evolution.*

Lynn Margulis was born in Chicago in 1938 and attended the University of Chicago where she received her undergraduate degree. She was married for a time to astro-biologist Carl Sagan before divorcing in the 1960s. She also attended the University of Wisconsin and later received her PhD degree from the University of California at Berkeley in 1965. She was appointed an assistant professor at Boston University where she presented her ideas in a paper titled "The Origin of Eukaryotic Cells." Her views were not well accepted by the biology community, and her paper was rejected by several scientific journals. However, in time, her insights were acknowledged and are now taught in high school biology classes. She was appointed a distinguished professor in the Department of Geosciences at the University of Massachusetts in 1988. Margulis was inducted into the World Academy of Art and Science, as well as the Russian Academy of Natural Sciences, and the American Academy of Arts and Sciences during the late 1990s. In 1999 she received the National Medal of Science Award. Along with several other scientists she has publicly expressed her doubts that HIV causes AIDS. In a July 2006 book review (on *Oncogenes, Aneuploidy, and AIDS* by Harvey Bialy) that she coauthored with her colleague James McAlllister and which appeared on Amazon.com, she claimed this connection is based on moralizing and obfuscation and that though this approach to the viral infection may make for "good marketing," it is not based on good science. She is also interested in *autopoiesis,* which is the physiological outlook as an alternative to what is known as mechanistic neo-Darwinism, and James Lovelock's theory of *Gaia* that claims that "Mother Earth" is regulated by life and its environment.

Lynn Margulis based her theory on the work and concepts of **symbiosis** proposed by other scientists in the late nineteenth and early twentieth centuries. Nevertheless, her theory of **endosymbiosis** was the first to be based on direct microbiological observation, whereas earlier versions were based on biological and zoological observation. Biological observation at the microscopic level led to greater understanding of the new field of evolutionary biology. Margulis' theory posits that about 3.5 billion years ago ancient bacteria-like **prokaryotes** (cells that had no nuclei or membrane surrounding them) were somewhat diversified in functions. Through a symbiotic process of cooperation, these ancient bacteria cells combined with larger cells and became modified with a membrane that surrounded internal structures, known as the genetic material of **mitochondria**, chloroplasts, and other inner cell particles. The nucleus was also surrounded by a membrane, thus forming a new organism, namely, the more advanced **eukaryotic** cells with their own DNA that became more diverse in their functions and became the driving force of evolution. Margulis' endosymbiotic theory is impelling current concepts about the human genome as major portions of this genome (DNA) originated either from bacteria or viral sources—some of these sources were ancient, others more modern. This symbiotic, or possibly parasitic, relationship is being investigated as the driving force for genetic change in all organisms, including humans. In summary, Margulis' theory states that the development of eukaryotic cells are a symbiotic combination of primitive prokaryotic cells.

**MARTIN'S THEORY OF CHROMATOGRAPHY:** Chemistry: *Archer John Porter Martin* (1910–2002), England. Martin and his colleague the British biochemist Richard L.M. Synge shared the 1952 Nobel Prize for Chemistry.

*A mixture (solution) of different chemicals can be separated and analyzed according to their differences in partitioning behavior between either a mobile phase or a stationary phase based on the components of the mixture's electric charge, relative absorption rate, or degree of solubility.*

Archer Martin did not actually invent the original technique for chromatography. Rather, it was the Russian botanist Mikhail Semyonovich Tsvet (1872–1919) who discovered the technique as he was researching the composition of chlorophyll in 1901. Tsvet used a liquid-adsorption column to separate different pigments from green plants by using calcium carbonate, but he had no idea of how the physical process of chromatography worked. That honor goes to Archer Martin and Richard Synge (1914–1994) who were jointly awarded the 1952 Nobel Prize in Chemistry for their theory explaining the partition method of separating different molecules by utilizing the basic concept of a *mobile phase* and *stationary phase* for the original paper form of chromatography. There are two major theories of chromatography, the *plate* and *rate* processes.

The *rate theory* is based on the speed at which a substance is retained as it moves in a chromatographic system. It can be expressed by the equation:

$R_f$ (Rate or Retention Factor) = Distance the sample is moved by the compound used in the solvent divided by the distance the sample is moved by the solvent.

The other theory is called the *plate theory*. Its equation follows:

K (the partition coefficient) = Concentration of the solute in the stationary phase divided by the concentration of the solute in the mobile phase.

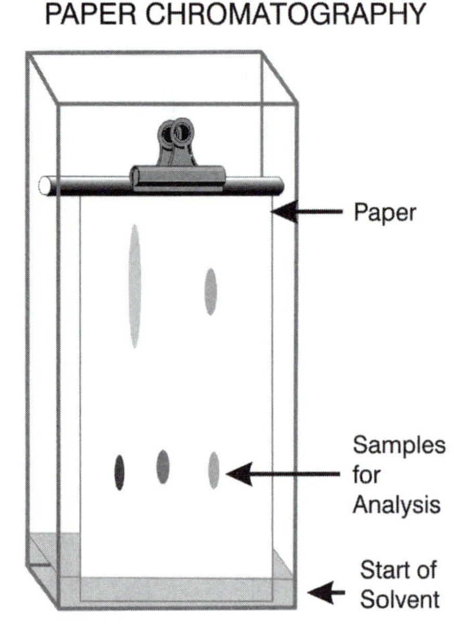


Figure M1. The movement of the samples on the paper determines the rate factor that is compared to a standard used in testing the sample.

Some examples of techniques for separating very different molecules, as well as procedures for separating even very similar molecules, follow.

1. *Paper chromatography* is an original technique where a sample in a solution phase is placed onto a strip of special chromatography paper. The paper is placed on end in a jar with a shallow pool of the solvent. The jar is then sealed to prevent evaporation of the solvent that rises through the paper to meet the sample mixture that travels up the paper along with the solvent. Different molecules of the sample travel different distances up the paper according to their molecular composition. The amount of movement of the components of the sample determine the rate or retention factor as expressed in the above equation, and thus can be compared to a standard that assists in the identification of the unknown sample being tested.
2. *Column chromatography* utilizes a vertical glass column filled with a silica-gel-type of permeable solid as the support to prevent the liquid sample from flowing too freely down the column. The sample

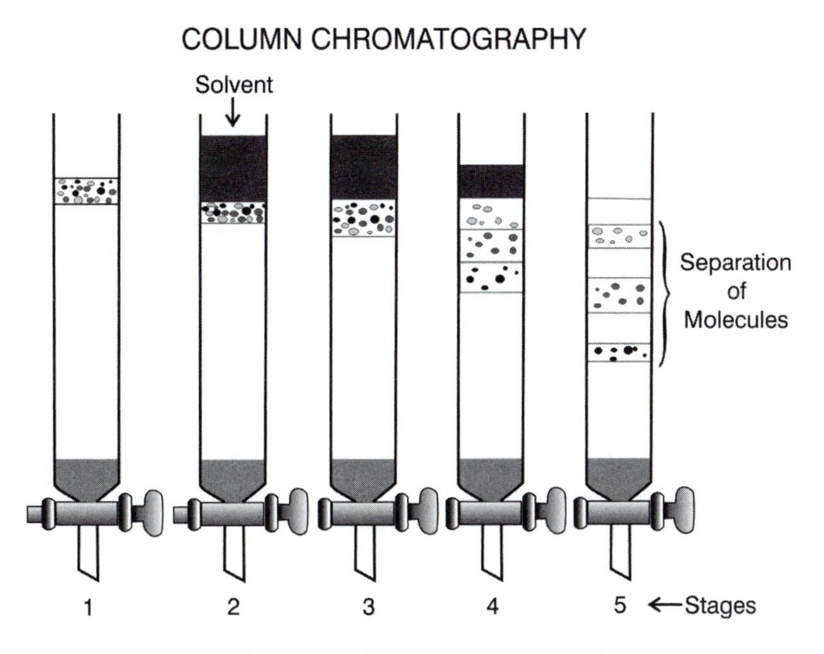


Figure M2. There are several versions of column chromatography that compare the rate of separation of the various molecules as does paper chromatography, but is faster.

to be tested is placed on top of this support substance and a liquid solvent is placed at the top of the column. Gravity moves the solvent vertically along with the sample downward at different rates through the medium and is collected at different exit rates from the column for analysis using the K (partitioning coefficient) equation. There are a number of versions of this technique, including one called "flash column chromatography" where the solvent is driven through the column by applying pressure on it.

3. *Thin layer chromatography* is similar to paper chromatography but involves a different type of stationary phase than paper. A thin layer of absorbent-like silica-gel or cellulose is coated on a flat inert surface. It provides faster "runs" of samples than regular paper chromatography.
4. Other types of chromatography are high performance liquid chromatography; ion exchange chromatography, size exclusion chromatography; affinity chromatography; gas-liquid chromatography; countercurrent (liquid-liquid) chromatography; centrifugal partition chromatography; several variations of the above are also used to analyze specific types of molecules.
5. A similar analytical technique is called *electrophoresis* (*Electro* = "energy" and *Phoresis* = "to carry across"). It is used to separate macromolecules of proteins and DNA based on their size and the inherent small electric charge on molecules. The process uses a small electric charge across a colloidal gel with the negative pole at one end of the gel and the positive pole at the other. The test molecules are placed in the gel and forced across a span of the gel by the electric charge. The molecular particles with a slight positive charge (+ cations) are attracted to the negative end of the gel substrate while the molecules with a slight negative charge (anions) are attracted to the positive end. This method is

used for analyzing amino acids, peptides, proteins, nucleotides, and nucleic acids (DNA).

**MATTHIAS' THEORY OF SUPERCONDUCTIVITY:** Physics: *Bernd Teo Matthias* (1918–1980), United States.

*Superconductivity of a material depends on the number of outer electrons on the atoms of that material.*

Bernd Matthias was born in Frankfort, Germany, at the end of World War I. His father died when he was young, but his mother was a lifelong influence on his career by creating a free intellectual and forgiving atmosphere that influenced his approach to scientific research and his lifestyle. At the age of fourteen his mother, sensing the coming rule of the Nazis, sent him to college in Switzerland. He later received his PhD degree in 1943 from The Federal Institute of Technology in Zurich, Switzerland. He emigrated to the United States in 1947 where he spent some time at Massachusetts Institute of Technology (MIT) and soon after moved on to Bell Labs in Murray Hill, New Jersey. From there he spent a year at the University of Chicago experimenting with the techniques of low-temperature physics. Matthias was an unusual experimental physicist whose discoveries were based on experimentation rather than on theory. He believed that no theorist ever predicted the existence of a new superconductor. His unusual techniques and personality led some theoretical physics to disparage his method as "schmutz" physics, meaning "dirty physics." He was also called an "alchemist," a reference to the ancient philosophers/scientists who believed in the "philosophers' stone" and the transmutation of base metals (*see* Paracelsus). Matthias relished this association as a modern-day alchemist because he loved to prepare and experiment with new materials and to explore their potential as low-temperature superconductors. One reason for his success was that intuitively he used Mendeleev's **Periodic Table of Chemical Elements** to assist in his discovery of different materials, and he attributed his many discoveries to this intuition as well as an understanding of the simplicity of nature. He is said to have commented that if a physics formula was over one-fourth of a page long, to forget it; it is wrong because nature is not that complicated. He enjoyed the process of discovering anything new in physics, particularly if it was not based on theory. During his lifetime he was the leading discoverer of cooperative phenomena in solid materials, particularly crystals.

Bernd Matthias' theory related to attempts to cause the electrons located on the outer shell (orbit) of atoms, or even free electrons, to flow as an electric current without resistance at temperatures much higher than absolute zero. This is known as superconductivity, and it occurs when the electrical resistance of a solid disappears as it is cooled to the "transition temperature," which for most metals and alloys takes place below 20 kelvin (K) or about −253°C. In 1911 Heike Kamerlingh-Onnes, when liquefying helium, also discovered that mercury, the only metal existing in a liquid state at room temperature, became superconductive at about 4 K (−269°C). At normal room temperatures, electrons in metal conductors collide as they flow through wire, causing a resistance to the flow of electrons (current). Thus, the wire heats up (as in the filament of an incandescent light bulb). What caused supercooled metals to lose their resistance to electricity was not known until the early 1950s, when Matthias began experimenting with various metals and alloys. By observing the behavior of several samples, he determined the number of electrons in the outer orbit of atoms was one factor and the crystalline structure of the material was another. Matthias made a compound of niobium and germanium ($Nb_3Ge$), which became superconductive at the unexpected high temperature of 23 K. Since then, physicists have attempted to determine the transition temperatures of numerous alloys and compounds, thereby causing superconductivity at much higher temperatures, because cooling to near absolute zero K is extremely difficult and expensive. Because liquid helium is used to cool metals to near absolute zero, its use is limited to small applications (e.g., cooling MRI supermagnets). Experiments were conducted using the less expensive element nitrogen, which becomes liquid at −196°C, instead of helium. Several ideas have been proposed to develop high-temperature superconductivity. One is to work with films of newly developed materials that can pass on free surface electrons with little or no resistance to the flow of current. A possible application of the film technique is a high temperature superconductor for computer switches and components that can run exceedingly fast without producing much heat. Reports of the discovery of so-called high-temperature superconductivity are very promising, but not yet confirmed. One goal for the future is the development of low-cost electrical transmission lines without the conversion of electricity to heat caused by resistance within the wires. Another is to use supercooled magnets to levitate magnetically driven trains.

*See also* Kamerlingh-Onnes

**MAUNDER'S THEORY FOR SUNSPOTS' EFFECTS ON WEATHER:** Astronomy: *Edward Walter Maunder* (1851–1928), England.

*When there is a minimum of observed sunspots, there is a corresponding long, cold period on Earth.*

While examining ancient reports of dark spots on the sun by astronomers, Edward Maunder realized there were no reports of similar activity on the surface of the sun for the period from the mid-1600s to the early 1700s. It was also determined, from other reports, that this was a period during which Earth experienced lower temperatures than usual. Using this material, Maunder developed a statistical analysis demonstrating that when there is a dearth of sunspots, a prolonged cold spell on Earth occurs. Maunder's "minimum" is still used along with more sophisticated techniques to aid in determining long-term climate changes on Earth. Current theories related to climate change are more concerned with the warming of Earth rather than the cooling correlation with sunspots. Currently, the tendency is to attribute global warming and the one-degree centigrade increase in global temperature over the past century to human behavior, activity, and resource consumption. Not all computer programs designed to determine causes of global warming include all the possible variables that affect global temperatures. It is not always recognized that Earth has gone through many short and long periods (cycles) of cooling and warming in past centuries. The global warming dilemma requires less politics and more science before the problem is completely understood.

**MAUPERTUIS' PRINCIPLE OF LEAST ACTION:** Physics: *Pierre-Louis Moreau de Maupertuis* (1699–1759), France.

*Nature chooses the most economic path for moving bodies.*

Pierre-Louis Maupertuis' principle of least action was a forerunner of all later theories and physical laws dealing with conservation, such as the conservation of energy and the concept of entropy. Maupertuis conceived this principle to explain the path that light rays travel, but it seemed applicable to all types of moving bodies. It basically states that nature will take the easiest, shortest route to move things from one point to another. Maupertuis was also interested in applying this principle as a means of unifying all the laws of the universe and thus arriving at proof for the existence of God. Maupertuis' principle was widely applied in the fields of mechanics and optics. A similar principle was proposed by Leonhard Euler in his form of calculus dealing with mathematical variations. Pierre de Fermat also used Maupertuis' principle to describe how light is refracted according to Snell's law, which states that light takes the least time possible when traveling from a medium of one density to a medium of a different density.

Maupertuis also theorized in areas of evolution with his study of the nature of biparental heredity. He based his theory on his detailed study of the occurrence of polydactyly (extra fingers) on several generations of a family in Berlin. He determined that the polydactyly trait was a mutation and could be transmitted by either the male or female parent to offspring. He was able to determine the mathematical probability that the trait would occur in future members of families afflicted with this anomaly. This was the first scientifically accurate record for the transmission of a hereditary trait in humans. He also wrote a book based on his microscopic examinations of embryos in which he challenged Jan Swammerdam's theory of "preformation." This discredited theory states the germ cell (gamete), either the ovum or sperm, contains a tiny homunculus of a human figure. Those who believed that it was the female gamete egg cell (ovum) that contained a fully formed organism of its kind are known as "ovists," whereas those who believed the homunculus is imbedded in the male gamete (sperm) are known as "spermatists." In either case, the tiny person (homunculus) does not start to grow until fertilization occurs. The development of the embryo involves merely an increase in size. In other words, prefomationists believed a homunculus, which is a tiny version of the adult, was housed in the germ cell, whereas Maupertuis argued that the embryo undergoes distinct stages of development as well as increases in size.

*See also* Euler; Fermat; Snell; Swammerdam

**MAXWELL'S THEORIES:** Physics: *James Clerk Maxwell* (1831–1879), England.

***Maxwell's kinetic theory of gases:*** *All gases are composed of large numbers of particles (atoms or molecules), all of which are in constant random motion.*

James Clerk Maxwell determined that the stability of Saturn's rings could be explained only if the rings consisted of a multitude of very small solid particles, a theory that has been accepted ever since. From this reasoning, he formulated his kinetic theory of gases, which states that heat is the result of molecular movement. His theory is also accepted today with modifications that incorporate relativity and quantum mechanics. Maxwell realized he could not predict the movement of a single tiny molecule. But on a statistical basis, the laws of thermodynamics, first proposed by Nicolas Carnot, could be explained as molecules at high temperatures (rapid movement) having a high probability of movement in the direction of other molecules with less movement (lower temperatures). Maxwell and Ludwig Boltzmann arrived at a theory that relates the "flowing" motion of gas molecules to heat being in equilibrium. In other words a hot cup of coffee, if left sitting, will become cooler—never hotter unless more heat is added, and in time the coffee's temperature will be in equilibrium with

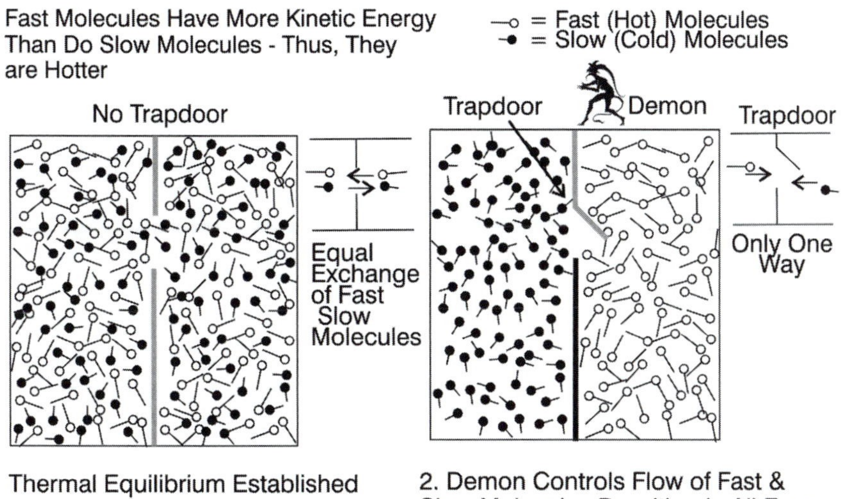


Figure M3. The demon allowed only fast molecules to collect on one side and slow molecules on the other side. Since no equilibrium is established between the two sides, the second law of thermodynamics is violated.

the temperature of its surroundings. Maxwell and Boltzmann also formulated a mathematical expression that indicated what fraction molecules had to a specific velocity. This expression is known as the Maxwell–Boltzmann distribution law (*see also* Boltzmann; Carnot).

***Maxwell's demon paradox:*** *The demon is a tiny hypothetical entity that can overcome the second law of thermodynamics, thus making possible a perpetual motion machine (i.e., its energy will never escape into useless heat).*

James Clerk Maxwell's work with the kinetic theory of gases led him to speculate on what became known as his "demon paradox." He created his mythical demon from the Maxwell–Boltzmann statistical distribution law that describes the properties of large numbers of particles that, under certain statistical conditions, were inconsistent with the second law of thermodynamics. That is, heat does not flow from a colder body to a hotter body without work (energy) being expended to make it do so (e.g., an air-conditioner). Maxwell proposed an intellectual hypothesis where an **adiabatic** wall separates each of two sealed, equal-sized enclosed compartments (*see* Figure M3). The trapdoor separating the left and right sides of container number 1 is open and the molecules can freely go from side to side. Thus equilibrium is established, and the second law of thermodynamics is upheld. The trapdoor for container 2 is controlled by the "demon," which allows only fast (hot) molecules to enter the right side of the container and only slow (cold) molecules to go through the trapdoor to the left side (these are closed systems; no gas can enter or leave the boxes). The "presence" of the tiny "demon" that sits by the trapdoor in container 2 opens the door for molecules of specific high speeds to go in only one direction and the slow-speed molecules to only go in the other. The object of the experiment is for the demon to collect in container 2 all the faster/hotter (on the average) molecules inside the room on the right side and the slower/colder ones (on the average) in the room on the left side. The paradox is there are more fast-moving, hot gas molecules in one room than the other without the input of any outside energy. Thus, the concept of conservation of energy has been satisfied because no kinetic energy was lost, but the concepts of entropy and thermal equilibrium have been violated. However, in the demon's experiment, we now have imbalance of kinetic (heat) energy because the demon allowed only the high-energy (hot) molecules to go into one side of the number 2 container. This appears to be a violation of the second law of thermodynamics.

The paradox is resolved by recognizing that the demon is doing work by opening the doors and that he requires information about the speed of the molecules in order to open the door at the right time. The input of information is responsible for the decrease in entropy of the system. Thus, the second law of thermodynamics is not really violated. Of course, perpetual motion is unobtainable, but this was not understood until the laws of thermodynamics became known. Maxwell's treatment of entropy on a statistical basis was an important step in realizing that to obtain knowledge of the physical world, one must interact with it.

***Maxwell's theory for electromagnetism:*** *Magnetism and electricity produce energy waves, which radiate in fields with differing wavelengths.*

Familiar with Michael Faraday's theories of electricity and magnetic lines of force and expanding on the mathematics developed by Faraday; Maxwell combined several equations that resulted in the establishment of direct relationships in the fields produced by magnetism and electricity and how together they affect nature. Once the equations for magnetic and electric fields were combined, he calculated the speed of their waves. Maxwell concluded that electromagnetic radiation has the same speed as light—about 186,000 miles per second. At first, Maxwell accepted the ancient concept of the existence of the aether in space. (It was thought that light and other electromagnetic waves could not travel in a vacuum; therefore the concept of "ethereal matter" in space was invented but never verified.) Maxwell believed electromagnetic radiation waves were actually carried by this aether and that magnetism caused disruptions to it. Later, in 1887, Albert Michelson demonstrated that any material body such as aether in space was unnecessary for the propagation of light. Maxwell's equations were still valid, even after the aether concept was abandoned. Maxwell concluded there were shorter wavelengths and longer wavelengths of electromagnetic radiation next to visible light wavelengths on the electromagnetic spectrum (later named *ultraviolet* and *infrared*). He further concluded that visible light was only a small portion of a "spectrum" of possible electromagnetic wavelengths. Maxwell then speculated and predicted that electromagnetic radiation is composed of many different (both longer and shorter) wavelengths of different frequencies. This concept developed into the electromagnetic spectrum, which ranges from the very short wavelengths of cosmic and gamma radiation to the very long wavelengths of radio and electrical currents (*see* Figure M4).

The theory of electromagnetic radiation is one of the most profound and important discoveries of our physical world. It has aided our understanding of physical nature and resulted in many technological developments, including radio, television, X-rays, lighting, computers, iPods, cell phones, and electronic equipment. Maxwell combined his four famous differential equations ("Maxwell's equations"). These four rather simple mathematical equations could be used to describe interrelated nature and behavior of electric and magnetic fields. They described the propagation of electromagnetic waves

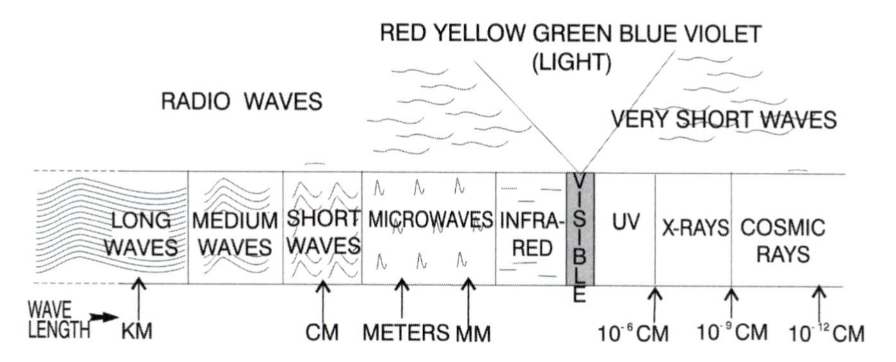


Figure M4. James Clerk Maxwell hypothesized the existence of electromagnetic radiation with longer and shorter wavelength than visible light that is located near the middle of the scale. This idea developed into the electromagnetic radiation spectrum for frequencies from very long radio waves to extremely short X-rays and cosmic radiation.

(radiation) in a form of the wave equation. This was the first time the constant for the velocity of light waves (c) was used; it later became an important constant in Einstein's theories of relativity and his famous equation, $E=mc^2$. Maxwell experimented in many areas, and his accomplishments were substantive, including an explanation of how viscosity of a substance varies directly with its temperature. Maxwell's contribution to the physical sciences was not only significant but his theories were among the few from his day in history that held up following the evolution in knowledge that began with the advent of the new science of relativity by Albert Einstein.

*See also* Einstein; Faraday; Michelson

**MAYNARD SMITH'S THEORY OF EVOLUTION:** Biology: *John Maynard Smith* (1920–2004), England.

*Game theory is related to evolutionary stable strategies (ESS).*

Maynard Smith formulated his theory of biological evolution by using concepts of the theory of games established by John von Neumann in the 1940s. The explanation follows: If there is competition between two animal societies, then an evolutionary stable strategy (ESS) occurs when a majority of that population prevents a mutant (different) strategy from invading the stable strategy. The odds are greater that a stable society of organisms will

> John Maynard Smith borrowed the particular game theory often referred to as "Hawk and Dove" from the Pentagon that is sometimes utilized in the decision-making process to invade—or not to invade—another country. Maynard Smith related this game idea to evolutional survival odds to predict when an animal should fight to defend its territory and food supply, or when to hide or leave the territory. This theory was tested with various animals, including spiders, birds, and even humans. It concerns those that attack and those who defend a territory. When applied to evolution, the payoff in the game is reproductive fitness, whereas a loss is the potential reduction of reproductive fitness.

be a mixed society rather than a single (unmixed) society. Maynard Smith published his theory in a book titled *The Theory of Evolution* in which he presented the reasoning that supports and revealed the mechanisms of evolution. His application of game theory to evolutionary strategies added to a better understanding of the history of evolution. In 1982 he published *Evolution and the Theory of Games* describing the unusual game known as "Hawk and Dove" which is considered a classic theoretical game model. His theory of "ESS" is considered a division point between the old and more modern understanding of evolution.

Maynard Smith wrote in a clear and intelligent manner that even nonbiologists could understand. He was a good teacher who would take time to discuss his ideas with students and colleagues.

*See also* Von Neumann

**McCLINTOCK'S THEORY OF CYTOGENETICS:** Biology: *Barbara McClintock* (1902–1992), United States. Barbara McClintock received the 1983 Nobel Prize for Physiology or Medicine at the age of 81.

> *Genes can move around within cells and modify chromosomes, thus restructuring the genetic qualities of a species.*

Barbara McClintock's moving **genes** were referred to as "jumping genes" because they could transpose (change position or the order of genes) within chromosomes. She traced the evolutionary history of domesticated maize to determine the genetic ancestor of the grass we now know as corn. Much of her work was accomplished by using stains to aid in identifying the ten chromosomes found in maize (corn). These chromosomes were large enough to be viewed by a microscope, enabling her to identify and distinguish the different chromosomes from each other. By planting seeds from corn growing one year to the next, she tracked mutant genes over several generations. She found that in addition to single genes that were responsible for color (pigmentation) in the corn, she found groups of genes linked together that caused other mutations. She referred to these as "controlling elements," which dictated the rate for the on-off switching action of other genes. McClintock discovered that these controlling genes could move within a single chromosome, or they could "jump" to other chromosomes and control their genes. McClintock's work with genetics came just twenty years after Mendel's principles of heredity were accepted. She provided an extensive account of her theory of the process of "transporter" of genes in her 1951 paper "Chromosome Organization and Genic Expression" which was largely ignored. Although her work was ignored primarily because it was far advanced for the 1940s, this was also a period during which Mendel's principles of heredity were generally not understood and were also not generally accepted. McClintock's concept of groups of genes working together and controlling other genes came before the discovery of DNA that advanced our understanding of evolution. Today, we know that her work, although not recognized at the time of its discovery, explained one of the mechanisms of evolution.

*See also* Mendel

**McMILLAN'S CONCEPT OF "PHASE STABILITY":** Physics: *Edwin Mattison McMillan* (1907–1991), United States. Edwin McMillan shared the 1951 Nobel Prize for Chemistry with Glenn T. Seaborg.

*Using variable frequencies of electrical impulses in a cyclotron, it is possible to compensate for the increase in mass of accelerating subatomic particles, thus increasing their speeds.*

Edwin McMillan was aware of the problem that Ernest Lawrence's cyclotron experienced with the ever-increasing "speeds" obtained during acceleration of subatomic particles. According to Einstein's theory of relativity, the greater a particle's acceleration, the greater is its increase in mass. This is the reason why particles with mass can never reach the speed of light because the particle would become more massive than the entire universe—if there existed enough energy in the universe to accelerate it to such a velocity. Lawrence's cyclotron used a fixed frequency of electrical stimulation to accelerate the beta, alpha, or other subatomic particles. As these particles spun around, faster and faster in the cyclotron, they increased in mass and thus became out of phase with the frequency of the electrical impulse, which resulted in a limit to their speed. McMillan's solution was to use a variable frequency that could change as the mass of the particles changed. This led to a new device, the synchrocyclotron (also known as the synchrotron). It was so named because it could synchronize the frequencies required to maintain the increasing speed of the particles within the cyclotron to the regions of hundreds or even thousands of megaelectronvolt (MeV equal to 1,000,000 volts) and continue to accelerate them to even greater energies, thus enabling physicists to explore the "split" particles and radiation resulting from collisions between particles. McMillan and Seaborg shared the 1951 Nobel Prize for the isolation of the elements neptunium and plutonium.

*See also* Einstein; Lawrence; Seaborg

**MEISSNER EFFECT:** Physics: *Fritz Walther Meissner* (1882–1974), Germany.

*The weak magnetic field decays rapidly to zero in the interior of a superconductor metal as the temperature reaches absolute zero.*

Walther Meissner was a mechanical engineer known as a "technical physicist" rather than a "theoretical physicist." In the early 1930s he was working with the German physicist Robert Ochsenfeld (1901–1993) when they established the largest helium-liquefier (to achieve near absolute zero temperatures). They used this device to discover the damping (slowing down or reducing of an effect) of magnetic fields in superconductor materials. They unexpectedly discovered the phenomenon that became known as the "Meissner effect" when they realized that when adjacent large cylindrical crystals of tin had their temperatures reduced to 3.72 K (−452.97 F), their natural magnetic fields disappeared from their interior. This is their critical temperature point (Tc) and indicates the beginning of superconductivity as well as perfect diamagnetism (superconductivity occurs at temperatures that are near absolute zero). In other words, the magnetic flux is expelled from a superconducting metal when it is cooled in a magnetic field to below its critical temperature as the temperature approaches absolute zero and superconductivity is occurring. This discovery led to a deeper understanding of the laws of thermodynamics as well as establishing the theory of superconductivity and the theory of electrodynamics of superconductivity by the brothers Fritz London (1900–1954) and Heinz London (1907–1970), two German-born physicists.

It might be mentioned that later there were some limitations to the Meissner effect. First, the magnetic field is not totally expelled as the metal crystals are cooled, but rather a very thin layer remains on the surface where the current continues to flow. This thin surface layer "screens" the internal portion of the metal from the magnetic field, thus offering less resistance to the flow of current on the surface of the conductor. Second, the Meissner effect is not observed in impure samples, nor certain types of crystals, nor flat round discs of the metal.

**MEITNER'S THEORY OF NUCLEAR FISSION:** Physics: *Lise Meitner* (1878–1968), Austria. Lise Meitner received the 1966 Enrico Fermi Prize from the U.S. Atomic Energy Commission for her work on nuclear physics.

*As the nuclei of uranium absorb neutrons, the nuclei become unstable and "fission" into two smaller lighter elements; in addition, they produce extra neutrons and radiation.*

Lise Meitner, an Austrian physicist who emigrated to Sweden in the 1930s, collaborated with Otto Hahn on nuclear physics research. In 1938 Hahn and the German

Lise Meitner was an unusual woman who struggled all her life to make a scientific name for herself in an exceptionally patriarchal field. She was the first woman to enroll as a student in the University of Vienna in 1901. Early in her academic life she showed an interest in science and mathematics. Her first break came when she pursued these subjects under two great teachers, the famed Austrian physicist Franz Exner (1849–1926) and Ludwig Bolzmann. Meitner had great skills in these areas, and her doctoral dissertation was an unusual experiment related to one of the theories of James C. Maxwell. Because of her gender the only career open to her was teaching. Thus, she decided at the age of twenty-eight to move to Berlin, which was the center of physics explorations. She planned to spend a few years there but stayed over thirty years until 1938 when Germany no longer allowed Jews to enter professional positions. She became an assistant to Otto Hahn while in Berlin, but with no title and no salary. Together they became known as leaders in the transformation of one element into another and, in particular, the heavier elements that were radioactive. She formed friendships with Max Planck, Niels Bohr, Max Born, Wolfgang Pauli, James Chadwick, and Albert Einstein. Her work with the nuclei of uranium was based on James Chadwick's research and discovery of the neutron and Enrico Fermi's theory of shooting neutrons into the nuclei of atoms to explore their structure. Fritz Strassman, Otto Hahn, and Lise Meitner decided to bombard a sample of uranium with neutrons to study the radiation that resulted and to synthesize other elements. During the late 1930s it became too dangerous for her to remain in Germany, but the Nazis refused to issue her a passport. Because she was now a well-known physicist outside her native country, many people helped her to escape to Sweden. Dirk Coster (1889–1950), the Dutch physicist who held a prominent post at the Groningen University in the Netherlands was of particular assistance. She continued to collaborate with Hahn via the mail wherein he informed her that he and Strassman had created the radioactive isotope of barium. She and her nephew Frisch calculated that the process of "fission" should result in enormous amounts of energy. Hahn wrote up the results for publication, but they could not include her name as an author because she was a political refugee. Hahn soon disregarded Meitner's role in this discovery and thus she was not included in the 1944 Nobel Prize that was awarded to Hahn. Although they continued to write to each other, she was disappointed in not receiving credit for her work. Hahn and Meitner had a contentious relationship. She also was critical of him and other German scientists for their failure to confront the racial prejudices of the Nazis. Lise Meitner died at the age of 89 in 1968.

chemist Friedrich (Fritz or Fuzzy) Strassman (1902–1980) became perplexed when they "shot" neutrons into nuclei of uranium. To their amazement, the uranium nuclei became lighter rather than heavier when absorbing neutrons. Hahn contacted Meitner and requested assistance in solving this problem. In 1939 Meitner and her nephew Otto Frisch, who at the beginning of World War II escaped Germany to live in the Netherlands, solved the problem together. As the nuclei of uranium atoms absorb neutrons, their nuclei increased their atomic mass to the point that they become unstable due to the excess of neutrons. Each of the unstable uranium nuclei undergoes fission (splitting) into two smaller fragments of almost the same size. While the uranium nucleus is splitting, at the same time it ejects two or three other free neutrons as well as energy in the form of radiation. Frisch named this process *fission*, after the process that cells undergo when dividing. It was never understood why Meitner and Frisch were never recognized for the importance of their discovery of a chain reaction of fissionable unstable uranium nuclei for the production of energy (Otto Hahn received the 1944 Nobel Prize for Chemistry, Not long after, this concept was used by the United States to develop the atomic bomb.

*See also* Chadwick; Fermi; Frisch; Hahn

## MENDELEEV'S THEORY FOR THE PERIODICITY OF THE ELEMENTS:

Chemistry: *Dmitri Ivanovich Mendeleev (also spelled Mendeleyev)* (1834–1907), Russia.

*There is a definite repeating pattern of the properties of elements based on the elements' atomic weights and valences.*

**MENDELEEV'S PERIODIC TABLE OF THE CHEMICAL ELEMENTS**

| GROUPS / PERIODS | 1 | 2 | 3 | 4 | 5 | 6 | 7 | 8 | 9 | 10 | 11 | 12 | 13 | 14 | 15 | 16 | 17 | 18 |
|---|---|---|---|---|---|---|---|---|---|---|---|---|---|---|---|---|---|---|
| | IA | IIA | IIIB | IVB | VB | VIB | VIIB | VIII | VIII | VIII | IB | IIB | IIIA | IVA | VA | VIA | VIIA | VIIIA |
| 1 | 1 H 1.0079 | | | | | | | | | | | | | | | | | 2 He 4.00260 |
| 2 | 3 Li 6.941 | 4 Be 9.01218 | TRANSITION ELEMENTS | | | | | | | | | | 5 B 10.81 | 6 C 12.011 | 7 N 14.0067 | 8 O 15.9994 | 9 F 18.9984 | 10 Ne 20.179 |
| 3 | 11 Na 22.9898 | 12 Mg 24.305 | | | | | | | | | | | 13 Al 26.9815 | 14 Si 28.0855 | 15 P 30.9738 | 16 S 32.066(6) | 17 Cl 35.453 | 18 Ar 39.948 |
| 4 | 19 K 39.0983 | 20 Ca 40.08 | 21 Sc 44.9559 | 22 Ti 47.88 | 23 V 50.9415 | 24 Cr 51.996 | 25 Mn 54.9380 | 26 Fe 55.847 | 27 Co 58.9332 | 28 Ni 58.69 | 29 Cu 63.546 | 30 Zn 65.39 | 31 Ga 69.72 | 32 Ge 72.59 | 33 As 74.9216 | 34 Se 78.96 | 35 Br 79.904 | 36 Kr 83.80 |
| 5 | 37 Rb 85.4678 | 38 Sr 87.62 | 39 Y 88.9059 | 40 Zr 91.224 | 41 Nb 92.9064 | 42 Mo 95.94 | 43 Tc (98) | 44 Ru 101.07 | 45 Rh 102.906 | 46 Pd 106.42 | 47 Ag 107.868 | 48 Cd 112.41 | 49 In 114.82 | 50 Sn 118.71 | 51 Sb 121.75 | 52 Te 127.60 | 53 I 126.905 | 54 Xe 131.29 |
| 6 | 55 Cs 132.905 | 56 Ba 137.33 | ★ | 72 Hf 178.49 | 73 Ta 180.948 | 74 W 183.85 | 75 Re 186.207 | 76 Os 190.2 | 77 Ir 192.22 | 78 Pt 195.08 | 79 Au 196.967 | 80 Hg 200.59 | 81 Tl 204.383 | 82 Pb 207.2 | 83 Bi 208.980 | 84 Po (209) | 85 At (210) | 86 Rn (222) |
| 7 | 87 Fr (223) | 88 Ra 226.025 | ▲ | 104 Unq (261) | 105 Unp (262) | 106 Unh (263) | 107 Uns (264) | 108 Uno (265) | 109 Une (266) | 110 Uun (267) | 111 Uuu (272) | 112 Uub | 113 Uut | 114 Uuq | 115 Uup | 116 Uuh | 117 Uus | 118 Uuo |

| Series | | | | | | | | | | | | | | | |
|---|---|---|---|---|---|---|---|---|---|---|---|---|---|---|---|
| 6★ Lanthanide Series (RARE EARTH) | 57 La 138.906 | 58 Ce 140.12 | 59 Pr 140.908 | 60 Nd 144.24 | 61 Pm (145) | 62 Sm 150.36 | 63 Eu 151.96 | 64 Gd 157.25 | 65 Tb 158.925 | 66 Dy 162.50 | 67 Ho 164.930 | 68 Er 167.26 | 69 Tm 168.934 | 70 Yb 173.04 | 71 Lu 174.967 |
| 7▲ Actinide Series (RARE EARTH) | 89 Ac 227.028 | 90 Th 232.038 | 91 Pa 231.036 | 92 U 238.029 | 93 Np 237.048 | 94 Pu (244) | 95 Am (243) | 96 Cm (247) | 97 Bk (247) | 98 Cf (251) | 99 Es (252) | 100 Fm (257) | 101 Md (258) | 102 No (259) | 103 Lr (260) |



Figure M5. Mendeleev's Periodic Table of the Chemical Elements was originally organized according to atomic weights and chemical characteristics. It was later revised based on atomic numbers instead of weights.

Dmitri Mendeleev was not the first to recognize some sort of pattern related to similar characteristics of elements based on their atomic weights, atomic numbers, or their ability to combine with each other. Several chemists recognized that elements seemed to be grouped in triads, or in repeating groups of seven, or with some evidence of "octave" periodicity of their properties (*see* Figure N2 under Newlands). In 1870, one year after Mendeleev published his **Periodic Table of the Chemical Elements,** Julius Meyer conceived a table similar to Mendeleev's. Meyer plotted a chart relating physical and chemical properties of elements with their atomic weights. His work, however, was overshadowed by Mendeleev's publication in 1869. Mendeleev classified the elements by their atomic weights as well as valences, even though some of the valence numbers for the elements conflicted with the arrangement of the atomic weights (*see* Figure M5). He realized after a row of seven there must be another column to complete that segment, so that a new row, based on continuing periodicity of atomic weights, could be recognized in the organized chart (the octet or rule of eight). Mendeleev exhibited great insight by "skipping places" in his periodic table for elements not yet discovered. He called the yet-to-be-discovered elements *eka* (meaning "first" in Sanskrit) elements, which he predicted would fit the blank spaces he provided in his table by predicting atomic weights and properties.

It was later discovered that the few inconsistencies in Mendeleev's periodic table were due to the use of atomic weights instead of atomic numbers (the number of protons in the nucleus). Once this was corrected, the current periodic table proved to be not only one of the most useful but also one of the most elegant organization charts ever conceived.

*See also* Cannizzaro; Dobereiner; Frankland; Meyer; Newlands

**MENDEL'S LAW OF INHERITANCE:** Biology: *Gregor Johann Mendel* (1822–1884), Austria.

*Characteristics of offspring are determined by two factors, one from each parent.*

Gregor Mendel entered the Augustinian monastery at Brünn, Austria, in 1843 where he continued his childhood interest in horticulture that led to his study of the role of hybrids as related to evolution. He was a meticulous experimenter, who began by keeping records of as many as seven different characteristics of parent pea plants in succeeding generations. He was interested in the ratio of specific characteristics that passed from parent plants to offspring plants (*see* Figure M6). He calculated ratios for the inheritance of stem length, the position of the flower on the stem, the color of the unripe pads, the smoothness and roughness of the pea pods, color of seeds, forms of seeds, and cotyledon (seed coat) color.

His work, although not recognized at the time he made his observations, was the first to provide a mathematical basis to genetics. From his observations and calculations, he based his law of inheritance on three theories:

1. As the female parent's egg and male parent's sperm sex cells mature, the formerly paired inheritance factors divide, resulting in just one specific factor for each characteristic from each parent. These single factors are then combined into a new pair during fertilization and are responsible for the inherited characteristics of the offspring. This is now known as the principle of *segregation.*

2. Characteristics are inherited individually. This means that one factor or characteristic can be inherited along with another factor and may be either dominant or recessive (e.g., tall stems with wrinkled pea pods). This is known as the principle of *independent assortment*.
3. Each characteristic is the result of the connection of at least two genes, one from each parent. One of these two factors is always dominant over the other. This is now known as the *law of dominance*.

Mendel's law indicated there was not a blend of inherited characteristics, but rather "fractional" inheritance, which strengthened Darwin's concept of natural selection. Later in life, in 1868 he became the abbot of a monastery, thus reducing time he could devote to his research. It would be many decades later before the value of his work would begin to be appreciated.

*See also* Darwin; De Vries

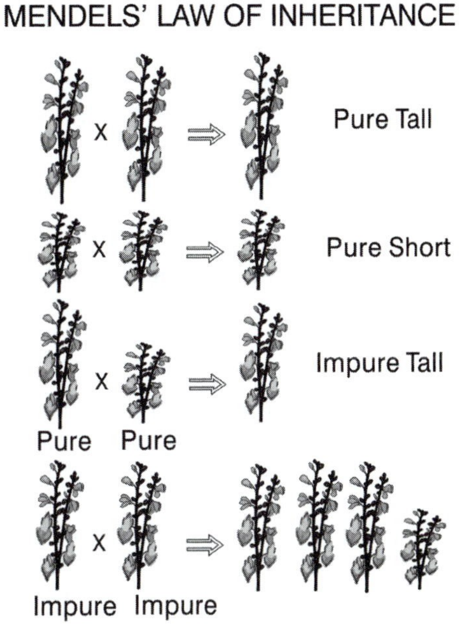


Figure M6. Mendel studied the ratios of plant characteristics to determine the rate of inheritance of these characteristics.

**MERRIFIELD'S THEORY OF SOLID-PHASE PEPTIDE SYNTHESIS:** Biochemistry: *Robert Bruce Merrifield* (1921–2006), United States. Robert Bruce Merrifield was awarded the 1984 Nobel Prize for Chemistry.

*An excess of reagents can wash chains of peptides composed of amino acids to accelerate the removal of side chains and thus protect the peptide molecules.*

Peptides are organic compounds in which multiple amino acids are bonded by peptide bonds, also known as amide bonds. To artificially synthesize peptides it is necessary to join carboxyl groups of one amino acid to other groups of amino acids, using one of two methods: *liquid-phase synthesis* and the *solid-phase peptide synthesis*, known as SPPS.

Robert Merrifield began work on his concept of the SPPS method of peptide synthesis in 1950s. Note: Peptides are similar to proteins but are composed of shorter and less complicated chains of molecules that are mostly proteins. Merrifield realized that if peptides could be synthesized by using less expensive and more rapid methods rather than those that employ complicated laboratory procedures and take months to accomplish, the process could be used to produce many useful commercial and medical products. He experimented with over one hundred different substitute resins and eventually invented a method that applied an ion-exchange process of bonding the amino acids to the insoluble solid support of a polystyrene resin. The beauty of this innovation of using insoluble resins was that it provided a solid support when different solvents were used to wash away impurities. These resins provide a means of introducing an amino acid by the methods of substitution, condensation, or addition-type reactions. In 1964 Merrifield was successful in synthesizing a nine-amino acid peptide that was effective in dilating blood vessels. The steps in this process are many and lengthy—over five thousand steps are involved in forming the final peptide chain. Fortunately, Merrifield and others found a way to automate this process by using a continuous flow method for the reagents that pass through the reaction chamber that hold the resins and peptides. The reagents can "wash" the contents within the chamber and be recycled and used over many times. Using these automated procedures Merrifield was able to synthesize insulin in 1965. The other method of synthesizing peptides, known as the *liquid-phase synthesis* procedure, is the classical approach to accomplishing peptide synthesis. However, today it is only used for large-scale production of industrial-type peptides. Most laboratories prefer the faster, automated Merrifield SPPS method. Merrifield was awarded the 1984 Noble Prize for Chemistry for his work with peptide synthesis.

**MESELSON–STAHL THEORY OF DNA REPLICATION:** Biology: *Matthew Stanley Meselson* (1930–), United States.

*The DNA double-helix molecule replicates, splits, and recombines to repair cells.*

Matthew Meselson and the American molecular biologist Franklin Stahl (1929–) demonstrated that when "semiconservative replication" of DNA (deoxyribonucleic acid) takes place and divides into two new DNA cells, the double helix is also duplicated. The semiconservative aspect of their discovery was accomplished by using the common Escherichia coli cells grown in the presence of the isotopes of nitrogen. A batch of DNA and E. coli was grown in an environment of the isotope nitrogen-15 ($^{15}N$). This batch was then exposed to normal nitrogen-14 ($^{14}N$). Meselson and Stahl then used mass centrifuging to separate the two different weight isotopes by their slightly different densities. When reviewing the results, they discovered three different types of DNA, one type contained the N-15, another type had the N-14, and yet a third type was a hybrid containing equal amounts of the isotopes N-15 and N-14. When the hybrid double strand was heated and the two strands separated, it was found that each single strand could act as a template to form a similar type strand when replicated. The results of their study were published in the *Proceedings of the National Academy of Sciences* in the United States under the title, "The Replication of DNA in Escherichia coli," in 1958.

A few years later in 1961 several researchers used this information to develop more theories related to DNA. It was discovered that RNA and mRNA molecules are stable and serve as a "primer" for DNA that requires a small piece of a new strand so it can complete its synthesis. (RNA is a single strand of nucleic acid that provides instructions in the DNA nucleus and translates this information for the assembly of proteins in the cells.) In other words, for the DNA to replicate, it needs "information" from the RNA molecules to make the correct type of protein cells rather than some foreign cell. It was James Watson and Francis Crick, the discoverers of the double-helix structure of the DNA molecule, who predicted that one strand of the double-helix molecule came from a parent whose DNA was most recently duplicated.

*See also* Crick; Franklin (Rosalind); Miescher

**MESMER'S THEORY OF ANIMAL MAGNETISM:** Medicine: *Franz Anton Mesmer* (1734–1815), Austria.

*The behavior of all things—all living organisms, as well as the heavens, Earth, moon, and sun—are affected by a "universal fluid" that can be received, propagated, and communicated with each other through motion.*

Franz Mesmer based his theory on the three laws of motion as explained by Sir Isaac Newton. He believed that because the moon and sun cause tidal movements, their motions would also cause any earthly object, including humans, to affect each other through some unexplained "universal fluid." As a physician, he applied this theory of fluids and motion to treating patients with magnets, with the idea that the magnets might influence the "fluids" as they do metal. He soon found that the magnets were not needed if the patient was open to his suggestions of how to be "cured." His ideas were not well accepted in France in the late 1700s and resulted in a report on his methods by several scientists who investigated Mesmer's claims. This report stated that magnetism had no medical effect, that Mesmer's "suggestions" seemed to produce nothing but odd behavior in his patients, and a cure with magnetism without imagination did not exist. (This "myth" still exists as some entrepreneur "hucksters" claim that their magnetic devices can cure almost any illness. There is no scientific evidence that "magnetic therapy" is an effective cure for anything except the lack of income of the seller.) Nevertheless, "mesmerism" later became recognized as hypnotism, which was separated from the original discredited concept of animal magnetism. Although Mesmer's theories are no longer considered valid, his legacy remains when someone claims to be "mesmerized" or hypnotized. Hypnotism may affect a person's behavior, but it has never been proven to cure a disease caused by a bacterium or virus. Mesmer's concept of using magnets to cure all types of human ailments has been modernized and is considered by some as a form of alternative medicine, whose efficacy is yet to be proven.

**METCALFE'S LAW:** Computer Science: *Robert Melancton Metcalfe* (1946–), United States.

***Metcalfe's law, which is related to the field of computer and telecommunications networks, states:*** *The value of telecommunications network technologies, such as the Ethernet and Worldwide Web, is proportional to the square of the number of users of the system represented by* $n^2$.

Robert Metcalfe was born in Brooklyn, New York, in 1946 at the beginning of the computer technology age. He graduated with two bachelor's degrees from the Massachusetts Institute of Technology (MIT) in 1969, one in electrical engineering and the other in management from MIT's Sloan School. He then attended Harvard graduate school, earning a master's degree in 1970 followed by a PhD in mathematics in 1973. Although he is involved in many enterprises, Metcalfe is best known for inventing the Ethernet, which is a standard for connecting computers over short distances. His law is responsible for the rapid growth of the Internet, particularly for the Worldwide Web of the Internet.

Metcalfe's law can also be expressed in two ways: 1) The number of possible cross-connections in a network increases as the square of the number of users (computers or other sending-receiving devices) in the network increases. And 2) the value of the network to the total community increases as the number of network users increases. Another way of saying this is: The power of a computer network increases exponentially with the number of computers (or other devices) that are connected to it (*see* Figure M7).

Along with Moore's law that is related to the power of computers (i.e., the number of transistors on an integrated circuit doubles every two years), Metcalfe's law explains the rapid growth of the Internet and the Worldwide Web. Together these two laws

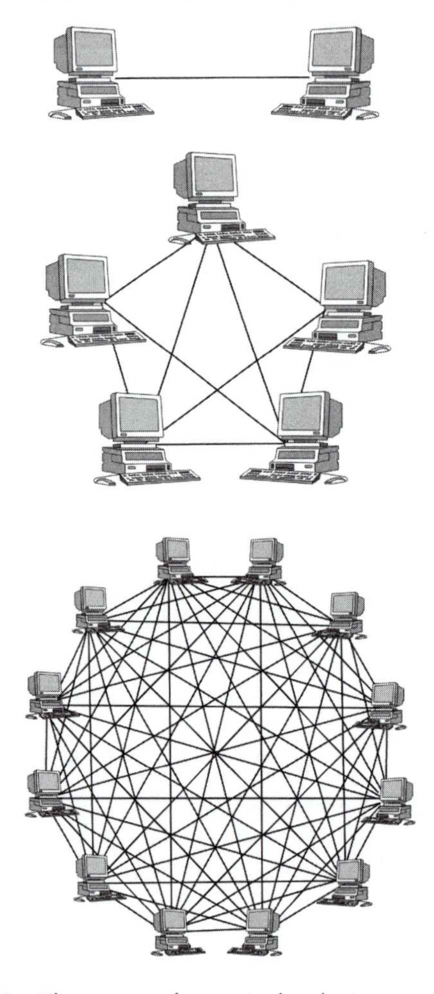


Figure M7. The power of a particular device connected to an interconnected network increases exponentially by the number of similar devices connected to it.

explain the tremendous increase in informational technology in the last quarter of the twentieth century that continues seemingly unabated.

Metcalfe's law has been challenged because it assumes that all connections to a network by all groups are of equal value. If this law was actually applied universally, it would be a great incentive for all networks to use the same technology and merge. However, this is not the way the open market operates. Metcalf's law has been revised to reflect this reality. It now states that the value of a network with $n$ members is not $n$ squared, but rather $n$ times the logarithm of $n$. This revision states that the value of merging networks is not 100% as predicted by $n^2$ and, thus, is more indicative of what happens in real life.

Robert Metcalf is a general partner at Polaris Venture Partners and is a board member of many technology-oriented companies.

*See also* Moore

## MEYER'S THEORY FOR THE PERIODICITY OF THE ELEMENTS:

Chemistry: *Julius Lothar Meyer* (1830–1885), Germany.

> *There are step-wise changes in the valences of elements as related to their atomic volumes and weights.*

Familiar with the work of Stanislao Cannizzaro who related Avogadro's number to the atomic weights of elements, Julius Meyer measured the volume and atomic weights of elements and plotted the results on a graph (*see* Figure M8).

In 1864 Meyer recognized that plotting the values of atomic volume against the atomic weight of elements would produce a graph indicating definite peaks and valleys, which related to the physical characteristics of different elements. Several examples of these peaks of plotted data were exhibited by the alkali metals, such as hydrogen, lithium, sodium, potassium, rubidium, and cesium, and one element not known at that time, francium. One of the most striking examples of periodicity based on properties of the elements was the series of sharp peaks representing the alkali metals. The alkali

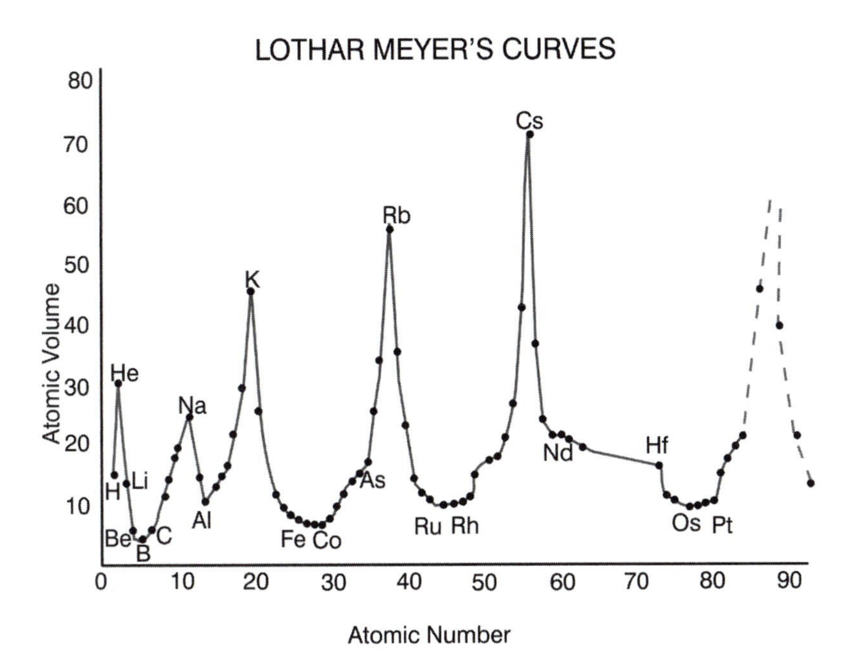


Figure M8. Lothar Meyer plotted the atomic volume against the atomic weights of atoms that results in "peaks" in the graph that relate to the physical characteristics of the different elements.

metal elements not only showed the beginning of a period but also had the greatest volume of the elements in their particular period. He also related the valences of elements that appeared at similar points on the graph to their chemical characteristics. This graph, referred to as *Lothar Meyer's curves*, was, in essence, the basis for the modern **Periodic Table of the Chemical Elements** (*see* Figure M4 under Mendeleev). Unfortunately, Julius Meyer did not publish his work until 1870, one year after Dmitri Mendeleev published his periodic table.

*See also* Avogadro; Cannizzaro; Mendeleev; Newlands

**MICHELSON'S THEORY FOR THE "ETHER":** Physics: *Albert Abraham Michelson* (1852–1931), United States. Albert Michelson was awarded the 1907 Nobel Prize in Physics.

> *If there is an aether, then the speed of light from space traveling directly toward Earth should be less as Earth moves toward the light source. Also, the speed of light traveling at right angles to Earth's motion should be greater than the speed of light coming toward Earth.*

The ancient concept of an **aether** (or ether) was used to explain the existence of some kind of matter in outer space because a pure vacuum was thought to be impossible. Because ancient scientists believed that nature abhors a vacuum, there must be some type of "matter" in space rather than a vacuum. More recently, the aether was considered as something beyond Earth's atmosphere that could carry electromagnetic

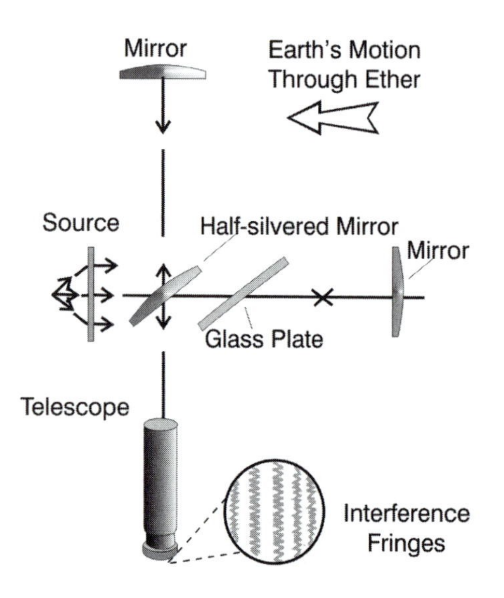


Figure M9. A diagram of Albert Michelson's interferometer designed to determine if an ether (aether) existed in space. If so, it might be detected by comparing a split beam of light as the Earth moved through this hypothetical space medium.

waves (light, radio waves, etc.), similar to how air molecules carry sound waves. Along with his colleague, the American scientist Edward Morley (1838–1923), at Western Reserve College (now Case Western Reserve University) in Ohio, Albert Michelson invented an instrument called the interferometer, designed to split a beam of light by using a half-silvered mirror, which allows half of the light from a source to be transmitted and the other half to be reflected. Each split beam then proceeded to separate mirrors arranged on arms of the apparatus, so that the split beams were again combined at the point at which they would interfere with each other to produce a characteristic pattern of fringes (*see* Figure M9).

The type of patterns formed depended on the time it took for each of the two beams to complete the trip from the source, through the split mirror, and return. The apparatus could be adjusted so that the light could approach at 90°, which then should produce a different fringe pattern. Because it was believed that the aether had no motion, as Earth moved through the aether (as sound moves through air molecules), the light coming directly toward Earth would be slower than the light coming toward Earth at a 90° right angle. After many repeated experiments using this interferometer, Michelson found no difference in the speed of light with his instrument despite the direction from which the light's speed was measured. This was the death knell for the aether concept. Michelson also used his interferometer to measure the diameter of various bodies that could be viewed with telescopes. He did this by comparing the light emitted from both sides of the planets and stars he observed. He continued his research on the speed of light and believed that the figure for the speed that light travels should be used as the standard for measuring length of objects and distance instead of the platinum meter stick that was then kept in Paris, France. This idea was adopted in 1960 when the wavelength of light from the inert gas krypton was accepted as the standard measure instead of the metric meter.

*See also* Einstein

**MIESCHER'S NUCLEIN THEORY:** Biology: *Johann Friedrich Miescher* (1844–1895), Switzerland.

*Independent animal cells found in the pus from open wounds can be separated into proteins and acid molecules.*

Using discarded bandages that contained yellow pus that oozed from healing wounds, Miescher filtered the viscous substance and discovered that it could be separated into independent animal cells (proteins and acid molecules).

He isolated a material that he called "nuclein," which today is known as DNA. Miescher was an eager young researcher hired by the German physiologist and chemist Felix Hoppe-Seyler (1825–1895) to study the chemistry of cell nuclei. He used leucocyte cells that were generally difficult to obtain in adequate numbers for study. However, they were known to make up a significant amount of the pus found in the bandages from healing wounds, thus he obtained his supply from a nearby hospital. Miescher developed a salt solution with sodium sulphate to "wash" the cells off the bandages without damaging the cells in the process. After treating the pus with sodium sulphate, he filtered and centrifuged it, and then allowed the residue to sink to the bottom of a glass beaker. He then attempted to separate the cell nuclei from the cell cytoplasm by further acidifying the precipitate that formed. Miescher called it "nuclein" (which is now known as DNA). Through further research he was the first to identify that this molecule contains phosphorus, nitrogen, and sulfur, as well as other elements, such as carbon, oxygen, and hydrogen. This was an unusual animal cell molecule because it contained the element phosphorus. He and his students contributed extensive research on the chemistry of this unusual molecule that eventually led to the important and later discovery that nucleic acid (DNA) is the carrier of inheritance. One of Miescher's other research discoveries established that the concentration of carbon dioxide in the blood was responsible for regulating breathing.

*See also* Crick, Franklin (Rosalind); Meselson

**MILLER'S THEORY FOR THE ORIGIN OF LIFE:** Chemistry: *Stanley Lloyd Miller* (1930–2007), United States.

*Under conditions of primitive Earth, the correct mixture of chemicals along with the input of energy could spontaneously form amino acids, the building blocks of life.*

As a graduate student of Harold Urey, Stanley Miller conducted experiments at the University of Chicago to demonstrate how life could have started on Earth at an early stage of existence. He theorized that the primitive atmosphere on Earth was the same as now exists on some of the other planets. For example, Jupiter and Saturn are very rich in methane gas ($CH_4$) and ammonia ($NH_3$), as well as possibly water and lightning, and these conditions could be responsible for the formation of life. In his laboratory Miller attempted to recreate a primordial environment where an **autocatalytic** process might facilitate the formation of **prebiotic** life to from organic molecules from inorganic elements. Miller combined ammonia, methane, hydrogen, and water vapor and subjected the enclosed mixture to discharges of high-voltage electricity. After a period of time he analyzed the mixture by using paper chromatography and detected several organic substances including hydrocyanic acid, formic acid, acetic acid, lactic acid, as well as urea. In addition, there were two basic amino acids: glycine and alanine. This experiment was conducted many times with various mixtures and sources of energy. A number of scientists saw this as the basis for an explanation of how life began because Miller produced some complex organic molecules. Unfortunately, additional experiments and the further production of some organic molecules came nowhere near forming a substance that could reproduce and maintain its metabolism. The current thinking is that life was formed by a random process, such as are the processes of mutation and evolution, and that it may require the application of some new, possibly unknown, principles to form complex organic life from inorganic substances, or that life arrived on Earth from some source in outer space.

*See also* Darwin; Margulis; Ponnamperuma; Urey

**MILLIKAN'S THEORY FOR THE CHARGE OF ELECTRONS:** Physics: *Robert Andres Millikan* (1868–1953), United States.

*By indirectly measuring the effects on electrons by an electrical field whose intensity is known, the charge on the electron can be calculated.*

Robert Millikan knew of James J. Thomson's 1896 discovery of the electron, as well as Thomson's use of the Wilson cloud chamber to compare the charge of an electron to its mass (e/m) and arrive at an approximate charge for the electron, which he stated as $4.744 \pm 0.009 \times 10^{-10}$ electrostatic units, which was a somewhat inaccurate measurement. Today the constant for the electron charge is stated as $\pm 1 \times 10^{-19}$ Coulombs. Millikan conceived of a device similar to the Wilson cloud chamber that used an electrostatic charge instead of a magnetic field to measure the charge of the electron (*see also* C. Wilson). This was the classical "oil drop" experiment. He atomized tiny oil droplets, which, due to the "friction" of falling, obtained a charge of static electricity as they fell through a small opening between two charged plates. Millikan created a variable electric potential between the plates (+ and − charges) and exposed this area with a light beam enabling him, with the use of a microscope, to observe what occurred as the charged drops fell through the small opening between the charged plates. When the current was off, there was no charge on the plates, thus permitting the oil drops to fall at a constant rate due to gravity. As he adjusted the charges on the plates, the oil drops were deflected up or down according to several factors (the electrical potential between the charged plates, gravity, the electron's mass, and its electrical charge). He calculated the basic charge on an electron, a constant unit in physics, to be $4.774 \times 10^{-10}$ ($\pm$ 0.009) electrostatic units. With this information and using Thomson's data for the e/m formula, Millikan determined the electron has only about 1/1836th the mass of a hydrogen ion (a proton).

*See also* Compton; Thomson; Wilson (Charles)

**MINKOWSKI'S SPACE-TIME THEORY:** Physics: *Hermann Minkowski* (1864–1909), Germany.

*Any event occurring in both local space and time exists in the fourth dimension of space-time.*

In 1907 Hermann Minkowski proposed his theory known as "Minkowski space." His space-time concept provided the mathematics for local (measurable continuum) events occurring simultaneously that led to Einstein's general theory of relativity published in 1916. In addition to the three dimensions of space (x, y, and z, or width, height, and depth), Minkowski's theory included the fourth dimension of time; thus *space-time* represents the inertial frame of reference for all bodies in motion. This concept indicated that such a phenomenon would account for a curvature of space-time, which accounts for gravity.

*See also* Einstein

**MINSKY'S THEORY OF ARTIFICIAL INTELLIGENCE (AI):** Mathematics and Computer Science: *Marvin Lee Minsky* (1927–), United States.

*Artificial brains could be "grown" and self-replicated to be intelligent and with the ability to learn as do human brains through the interactions of nonintelligent parts.*

Marvin Minsky's theory is based on his work with several other computer theorists related to how artificial brains could be "grown" by a process somewhat similar to the development of human brains. The key seems to be based on understanding how the human brain programs language, behavior, emotions, and so on, and how such human activities are related to the working of the brain's randomly "wired" neural networks. His theory was based on the concept that a computer could be "wired" and programmed to replicate the similar randomly "wired" neural network of the human brain.

Minsky rejects the need for better supercomputers as means to build better artificial intelligence (AI) computers; rather what is needed is a better understanding of what software to use with them. The best computers today are faster than the human brain in processing information, and such computers could act as AI "brains" if we knew what software to develop and use. Today, vast amounts of information are available via computers, but in many ways computers are very limited. Computers cannot answer commonsense human questions based on emotions or even those that are self-evident in the day-to-day interactions among people. Computers have no remorse. Computers associated with robots can perform repetitive mechanical tasks in factories and solve complicated mathematical equations. Although they can read, they can neither understand nor explain a simple story in a book. They can beat the best human chess players, but they cannot run a successful household for the simple reason that running a successful home

Marvin Minsky did most of his work related to artificial intelligence with colleagues at The Artificial Intelligence Laboratory (AI LAB) at the Massachusetts Institute of Technology (MIT) that he and fellow computer scientist John McCarthy (1927–) established in 1959. In 1974 Minsky introduced a new concept for how the human mind can understand a number of different things about a specific topic. He called his theory "frames" for a collection of specific knowledge about a topic stored in the mind. For instance, a "fish" frame would store all kinds of information about fish, thus not requiring the mind to remember and recall each and every aspect about fish already known. He has spent most of his career at MIT working on his theories. In addition to his research in AI, he has contributed to the fields of mathematics, cognitive psychology, linguistics, optics, in particular using computers in robots for mechanical manipulations, vision, and understanding language. His inventions include several robotic devices for hands and limbs, the "Muse" which is a musical synthesizer, and the Confocal Scanning Microscope that is an optical machine that has superb image resolution. He also constructed SNARC (Stochastic Neural-Analog Reinforcement Computer), the first randomly wired neural network learning machine. In 1985 he published an unusual book *The Society of Mind* that was composed of 279 one-page ideas related to his theories. Each page was in some way related to another page and was a solution to a problem related to a specific psychological phenomenon.

requires a wider range of abilities, including the ability to adapt to more types of random chaotic situations than any known chess movements.

**MISNER'S THEORY FOR THE ORIGIN OF THE UNIVERSE:** Astronomy: *Charles William Misner* (1932–), United States.

*The universe began in a nonuniform state, which in time became uniform as it expanded due to natural forces and physical laws.*

Charles Misner based his "mixmaster" model for the origin of the universe on the idea known as the *horizon paradox.* He theorized that in the beginning, all forms of matter were very much mixed up, with no or very little order. As the universe expanded, forces such as friction and gravity affected this diverse mixture and formed a more homogenous, isotropic, and uniform universe. His evidence was the horizon paradox, which states that the universe, when viewed from Earth, is so huge that it appears, even to incoming microwave signals, as a very uniform structure—just as when viewing the horizon of Earth, the distant landscape on earth appears more uniform than when viewing the same scene up close (e.g., viewing Earth from airplanes). Misner's mixmaster concept and the horizon paradox provided support for the big bang inflationary theory for the origin of the universe.

*See also* Gamow; Guth

**MITSCHERLICH'S LAW OF ISOMORPHISM:** Chemistry: *Eilhard Mitscherlich* (1794–1863), Germany.

*Substances with identical crystalline forms (isomorphism) also have similar chemical compositions and formulas.*

Early in his academic career, Eilhard Mitscherlich studied oriental languages in Germany but then became interested in medicine while in Göttingen, Germany, in 1817. For reasons unknown, his interests turned to a then-popular area of study in chemistry known as crystallography. He discovered that similar compounds tend to crystallize together. He, therefore, surmised that if compounds crystallize together, they are most likely to have a similar structure. He came up with this concept after he demonstrated that manganates, chromates, sulphates, and selenates are all isomorphous. In other words, the formula of the just-discovered selenates could be deduced from the well-known formulae of the other three (manganates, chromates, and sulphates). This idea was confirmed when the atomic mass of selenium was discovered in the early 1800s. Mitscherlich stated his "law of isomorphism" as: If compounds crystallize together, they are probably of similar structure (have the same formula). During this time he worked in Stockholm with his mentor the famous chemist Jöns Jakob Berzelius who along with Wilhelm Hisinger is credited with the discovery of selenium. Upon his return to Germany in 1822, he was appointed to the head of the department of chemistry. While continuing his research with crystallography, he also studied organic chemistry, microbiology, and geology. He was the first to produce benzene by heating calcium benzoate and furthered the development of the industrial process of extracting cellulose from wood pulp. Mitscherlich's law of isomorphism was very useful to Berzelius and other chemists in identifying the formulas of other elements and determining their atomic mass. The law also aided Dmitri Mendeleev in determining the atomic weights (masses) of elements as a format for arranging individual elements in his **Periodic Table of the Chemical Elements**. Later the elements were arranged by their atomic numbers (number of protons in the elements' nuclei) rather than atomic masses.

## MOHOROVICIC'S THEORY OF THE EARTH'S INTERIOR STRUCTURE:
Geology: *Andrija Mohorovicic* (1857–1936), Croatia.

*There are definite boundaries between Earth's crust and mantle.*

Based on his observation of an earthquake that occurred in 1909 in his native Croatia, Andrija Mohorovicic believed there was a boundary between the layers of Earth. Using a seismograph, he recorded waves from an earthquake as they penetrated the deep areas of Earth and compared them with the waves that traveled on the surface. He discovered the waves from a deep layer traveled back to Earth faster than did the surface waves, because the deep mantle layer was of greater density than the crust at the surface (the denser the medium, the greater will be the speed of sound and vibrations traveling through it). He concluded that there must be a relatively abrupt separation between these two layers with the mantle starting about 20 to 25 miles below the surface (crust). It was later determined the crust under the oceans is much thinner—only about three miles deep. The continental surface crust ranges in thickness from 22 miles in valleys to about 38 miles under mountains. The mantle is about 1,800 miles thick, and the outer core is about 1,400 or 1,500 miles thick, with an inner core of about 1,500 miles in diameter. This discontinuity between the crust and mantle was named after Mohorovicic and is sometimes referred to as the "Moho."

In 1960 the National Science Foundation (NSF) sponsored a project to drill into the Mohorovicic discontinuity. At the time the deepest well drilled on Earth was about 10 km (6.2 miles). Therefore, to reach the Moho, it would be necessary to drill about three times that depth on land. The project was abandoned due to a lack of resources and the realization that the tremendous heat at that depth would destroy most drilling equipment. Temperatures increase about 15° to 75°C (59° to 167°F) for every kilometer drilled into Earth; therefore, at just a bit deeper than 25 km the temperatures would be almost 2,000°C. Iron melts at 1,536°C.

## MONTAGNIER'S THEORY FOR THE HIV VIRUS: Medicine: *Luc Montagnier* (1932–), France.

*A number of biomolecular mechanisms may be responsible for a depletion of lymphocytes in HIV-infected individuals.*

In 1983 in Africa, Luc Montagnier and his colleagues discovered and isolated the human **retrovirus**, named HIV-1, which is related to AIDS. Later his team discovered a second retrovirus, HIV-2 (animal retroviruses were known, but they were not generally associated with humans). Montagnier's theory states that HIV exhibits characteristics of a retrovirus, which is the main mechanism that reduces the bacterial-viral-fighting lymphocytes in the human immune system, thus allowing **AIDS** to develop. He also investigated several other possible mechanisms that could relate the virus to AIDS. In the meantime, Robert Gallo of the United States, using a sample of Montagnier's retroviruses, discovered two viruses, one similar to HIV-1 and HIV-2, which Gallo named HTLV-1, and later another variety named HTLV-3, both of which were found in T-4 cells (special lymphocytes of the immune system). The viruses Gallo identified were not exactly the same as Montagnier's but may have been mutations of HIV. A dispute arose over who discovered what and when, which eventually was settled by naming Montagnier's virus LAV and Gallo's as HTLV-3. In 1986 it was agreed that all varieties of the retrovirus would henceforth be called HIV.

*See also* Baltimore; Delbruck; Gallo

**MOORE'S LAW:** Mathematics and Computer Science: *Gordon Earl Moore* (1929–), United States.

***Moore's law states:*** *The number of transistors on integrated circuits will double in complexity every eighteen months.*

This empirical observation was made in 1965 and is attributed to Gordon Earl Moore who cofounded the Intel Corporation in 1968. Moore did not name his statement "Moore's law," but rather his statement was given that name by Carver Mead (1934–), a professor and computer scientist at the California Institute of Technology. By 1970 it was common to consider Moore's law as a reference to how rapid advancements in computing power is related to per unit cost of transistors because complex transistors are also a measure of computer processing power. Moore claims that he originally stated that the cycle of technology of chip improvements would occur every two years instead of every eighteen months. He said he was misquoted. However, it is now and forevermore stated as "every eighteen months."

Moore's law as related to computers and transistors is not the first historical statement that pertains to improvements of computer devices and price per component of unit items resulting in today's greatly reduced price, and ever increasing power, of personal computers for home use. Historically, the picture unfolded as follows:

1. Computers were mechanical devices that were utilized in the 1890 U.S. census.
2. Mechanical type computers evolved into Alan Turing's relay-based machine that cracked the Nazi enigma code in World War II.
3. The introduction of the vacuum tube for use in early computers was used in a CBS radio broadcast that accurately predicted the election of General Dwight Eisenhower as president of the United States in 1952.
4. As newly invented transistors improved, they were used in launching satellites into space in the early 1960s.
5. The subsequent revolution of integrated circuits led to their increasing complexity, thus the advent of integrated circuits that led to Moore's law and personal computers.

Some technical experts believe that Moore's law will break down by the year 2020 due to the development of transistors that are only a few atoms or molecules in size.

Even though Moore's law only makes predictions regarding the computer and related components, many people mistakenly believe it applies to all forms of technology, which it does not. The only reason that the costs per chip are kept low is not necessarily related to Moore's law. The processes related to the design, research and development, manufacturing and equipment, testing, and labor, are all involved in the production costs that are in the *reverse* of Moore's law. Therefore, the more complex the chips become, the more the total production cost. Total costs to develop a new chip are in excess of $1 million. Once the first chip is produced at $1 million plus, the manufacturing companies can then begin to sell many more of the powerful chips at a much lower cost, thus making a profit. Moore's law also seems to not only apply to integrated chips but also to the access speeds of hard drives read only memory (ROM) and the capacity of the computer's random access memory (RAM). Some technologists claim that, under Moore's law, everything can become better and better, and thus it is a violation of Murphy's law, which has many different laws and corollaries about things that may not turn out as expected. An example of this is part of Murphy's Law that states: *Anything that can go wrong will go wrong.*

*See also* Babbage; Noyce; Shockley; Turing

**MOSELEY'S LAW:** Physics: *Henry Gwyn Jeffreys Moseley* (1887–1915), England.

*There is a distinct relationship between the X-ray spectrum and the proton number of chemical elements. When exposed to X-rays, each element with its specific atomic number produces a unique spectrum of wavelengths.*

Henry Moseley, using X-ray spectrometry, examined the lengths of electromagnetic waves emitted by different elements when exposed to X-rays. He observed that each element produced its own specific wavelength, which he examined by using crystal diffraction. His data indicated that, each element might be considered a separate integer that is proportional to the square root of the frequency of its specific wavelength. We now refer to this integer as the *atomic number,* which is the number of positive protons in an atom's nucleus that determines a specific element's physical and chemical properties. Moseley's data improved Mendeleev's **Periodic Table of the Chemical Elements** by arranging the elements in the table according to their atomic numbers rather than their atomic weights. The crystallographer Rosalind Franklin used X-rays to create patterns of molecules to determine the arrangement of atoms in their molecular structure in 1952. She made microphotographic pictures of the DNA molecule that gave Watson and Crick the insight to determine that the DNA was shaped as a double-helix molecule.

*See also* Franklin (Rosalind); Mendeleev

**MULLER'S THEORY OF MUTATION:** Biology: *Hermann Joseph Muller* (1890–1967), United States. Hermann Muller was awarded the 1946 Nobel Prize for physiology of medicine.

*X-rays and other ionizing radiation can induce chemical reactions that produce genetic mutations.*

Familiar with Mendelian heredity and the concept of genes as the carriers of inherited characteristics, Hermann Muller experimented with the fruit fly *drosophila*. His early research indicated that raising the temperatures of the eggs and sperm of fruit flies led to an increase in the rate of mutations. In 1926 he discovered that X-rays would also cause mutations. Some mutations were recessive, but mostly the mutated genes were dominant and thus harmful and passed onto offspring. He concluded that mutation was a chemical reaction and could be caused by exposing the eggs and sperm to a variety of chemicals and forms of ionizing radiation. Muller was concerned with the increased exposure of humans to all types of ionizing radiation (medical X-rays, nuclear radiation, cosmic radiation, ultraviolet, etc.). He believed excessive amounts of exposure to radiation caused genetic mutations—some positive, but mostly negative—that would be passed onto future generations. There is an ongoing debate as to the consequences of excessive mutations in the general population.

*See also* Lysenko; Mendel

**MULLIKEN'S THEORY OF CHEMICAL BONDING:** Chemistry: *Robert Sanderson Mulliken* (1896–1986), United States. Robert Mulliken was awarded the 1966 Nobel Prize for Chemistry.

> *Nuclei of atoms produce fields that determine the movement of electrons in their orbits. Thus, the orbits of electrons for atoms combined in molecules may overlap and include two or more molecules.*

Robert Mulliken formulated the concept of molecular orbits in which the valence electrons located in the outer orbits are not bound to any particular atom but may be shared with several different atoms within a molecule. Familiar with Niels Bohr's quantum electron orbital model of the atom, Mulliken and his colleague, the German physicist Friedrich Hund (1896–1997), applied quantum mechanics to explain how the valence electrons are delocalized in the molecular orbit where the bonding (combining) takes place (*see* Figure S2 under Sidgwick). In other words, the orbiting electrons of isolated atoms become molecular orbitals that may represent two or more atoms for each molecule. Thus, the energy of bonds could be determined by the amount of overlap of atomic orbitals within the molecular orbitals. This may be one reason that there are so few "free" atoms of elements in the universe. They are mostly joined in a great variety of molecular compounds. The concept of electronegativity that Mulliken devised is related to chemical bonding. *Electronegativity* is the ability of a specific atom within a molecule to attract electrons to itself and thus enable the molecule to carry an extra negative charge. To explain this phenomenon, he developed the formula $\frac{1}{2}(I + E)$, where $I$ is the ionization potential of the atom and $E$ is the atom's affinity for electrons.

*See also* Bohr

**MULLIS' THEORY FOR ENZYMATIC REPLICATION OF DNA:** Biology/Biochemistry: *Kary Banks Mullis* (1944–), United States. Kary Mullis shared the 1993 Nobel Prize for Chemistry with Canadian microbiologist Michael Smith (1932–2000).

***Mullis' theory states:*** *Using a controlled temperature applied to a segment of the DNA molecule will cause the helix to unravel, resulting in fragmentations that then can be reproduced as unlimited copies in vitro.*

The original process for replicating and identifying DNA fragments was slow and cumbersome and required the use of living organisms. It was a time consuming effort. Kary Mullis developed the "polymerase chain reaction," now referred to as PCR. Mullis' PCR method began in vitro, meaning "outside the living organism" and thus could be accomplished more rapidly in a simplified laboratory environment using only heat, reagents, and test tubes.

After receiving his BS in chemistry from the Georgia Institute of Technology in 1966, he continued his education receiving his PhD in biochemistry from the University of California, Berkeley in 1972. In 1979 he joined the Cetus Corporation, a new biotech firm in Emeryville, California, where he conducted research on synthesis of oligonucleotides (fragments of DNA or RNA containing fewer than fifty nucleotides) and invented the PCR method in 1983. According to Dr. Mullis, as he was driving through the mountains of Mendocino Country, California, he mentally conceived of a new way to analyze changes in DNA by amplifying individual sections or regions of the DNA molecule. He came up with the idea of taking genetic material from a single molecule of DNA and using what he later called his PCR method that could replicated to over one hundred billion related molecules in a few hours. And, the method requires only the application of heat and a few pieces of laboratory equipment. The PCR method is now used for biological and medical research as well as for forensic purposes. It can be used to detect hereditary diseases, to identify DNA genetic fingerprints, the cloning of genes, paternity testing, identifying mutated genes, diagnostic testing, pharmacogenetic tailoring of specific drugs for specific diseases, and so forth. PCR also can be used in a new type of science known as "paleobiology" which is the analysis of ancient DNA from fossil bones. This idea led to the story and movie of *Jurassic Park* where a variety of dinosaurs were recreated from DNA of fossils millions of years old.

Dr. Kary Mullis received some notoriety as a potential forensic DNA analyst in the 1995 O.J. Simpson murder trial due to his extensive work in this area. He was scheduled to be an expert defense witness. However, he was not called as it was learned that the prosecution planned to discredit Mullis because of his known eccentric behavior, his many failed marriages, legal patent disputes, use of controlled substances (LSD), and in general his unconventional lifestyle. In the end, the defense presented information that convinced the jury that DNA evidence was not viable or reliable.

Although Kary Mullis conceived of the idea for amplifying fragments of DNA to simplify analysis of base pairs, it was his colleagues at Cetus who, over a period of seven years, reduced his ideas into practical laboratory procedures. Thus, many consider the success of the process to be a team effort. Cetus Corporation patented the PCR techniques in Mullis' name in 1983. This was followed by several lawsuits. Hoffmann-La Roche purchased the patent rights in 1992, but the rights expired in 2006, leaving ownership unsettled after that date. Kary Mullis received many awards for his work in biotechnology, including an award from the German Society of Clinical Chemistry in 1990, the National Biotechnology Award in 1991, the Research and Development Scientists of the Year Award also in 1991, the Scientist of the Year Award in 1992, and the Thomas A. Edison Award in 1993. He was inducted into the National Inventors Hall of Fame in 1998. Dr. Mullis has received other awards, including the Nobel Prize in Chemistry in 1993. He has many publications to his credit. His best known is his "memoir" titled *Dancing Naked in the Mind Field.*

# N

**NAMBU'S THEORY FOR THE "STANDARD MODEL":** Physics: *Yoichiro Nambu* (1921–), Japan.

> *Quarks can exist with three values, one of which is an extra quantum number referred to as color.*

The "standard model" is one of several postulates of the quantum theory that is applicable to both submicro (quarks) and macro (supernovae) events. The other two deal with the quanta nature (tiny packets) of energy and the particle-wave nature of matter. Yoichiro Nambu investigated "baryons" which are particles with three quarks exhibiting one-half spin as they interact with the strong force. The regular proton in the nucleus has only two "up" quarks and one "down" quark, with the symbol *uud* used to designate this structure. Baryons comprise three identical quarks referred to as "strange" quarks, with the symbol *sss* designating their structure. Nambu's theory stated that quarks really exhibited a value of 3, with an extra quantum number referred to as "color." The terms "up," "down," "strange," and "color" are arbitrary and do not mean the same as the words are commonly used but refer to the particles' orientations in space and spin directions. The three arbitrary colors are red, green, and blue, which enabled quantum rules to be followed by allowing three up quarks (uuu), three down quarks (ddd) or three strange quarks (sss) to exist as long as each "triple" quark has one of the three different "colors." This is now known as the **Standard Model** for elementary particles (quarks) obeying the dictates of quantum theory. Nambu's theory was expanded to explain the possibility for the absence of free quarks, which, if detected, would exist as massless, one-dimensional entities. He suggested that the reason "free" quarks have yet to be detected is that they are located at the ends of "strings." Thus, this became known as the *string theory*, which had some problems when a string was "cut," which produced a quark and an antiquark pair but not a "free" quark. The string theory was later revised and is now known as the *superstring theory*. The string theory itself has generated much debate because so far it is only a mathematical possibility that has not made any viable predictions, which is required to be classified as a "true" theory.

*See also* De Broglie; Gell-Mann; Hawking; Schrödinger

**NASH'S EMBEDDING THEOREMS:** Mathematics: *John F. Nash, Jr.* (1928–), United States. John Nash shared the 1994 Sveriges Riksbank in Economic Science that is also known as the Nobel Prize in Economics with German economist Reinhard Selten (1930–) and John Harsanyi (1920–2000) from the United States.

***The Nash embedding theorems state:*** *Every Riemannian manifold can be isometrically embedded in a Euclidean space of* $R^n$.

"Isometrically" refers to a mathematical means of preserving the lengths of curves that can result in a way to visualize a submanifold of Euclidean space. The complete explanation of Nash's theorems is extremely technical and beyond the scope of this book. The first theorem for the smooth embedding, known as $C^1$, was published in 1954, and the second theorem, known as $C^k$, was published in 1956. John Nash accomplished the analysis related to these two theorems in 1966.

John Nash had a very unusual childhood, and his adulthood has been troubled and challenging. He did not enjoy working with other people and even at a young age preferred to do things alone and in his own way. At twelve years of age he had a laboratory in his room at home in West Virginia, and while in high school attended classes in a nearby college. He received a Westinghouse scholarship to attend Carnegie Mellon University in Pittsburgh, Pennsylvania, where he began taking courses in engineering, but soon changed to chemistry and later mathematics. After graduation he worked on a Navy research project in Maryland. He then enrolled in Princeton University, Princeton, New Jersey, where he earned a PhD in 1950. Nash was interested in equilibrium theory as related to N-person Games (N = any number), and later in two-person games as competitive economics theory. He married a physics student from El Salvador named Alicia de Lardé in 1957. Shortly thereafter, he began to exhibit the symptoms of schizophrenia. In 1959 Alice Nash had him committed to a mental hospital where the diagnosis of schizophrenia was confirmed. Their son, John Charles Martin, was born while he was a patient, but Alicia did not give the child a name for a year because she thought John Nash should participate in naming their son. Their son, who is also a mathematician, was diagnosed with schizophrenia just as his father. John Nash has another son from a previous relationship, but contact between the two has been sporadic and difficult. John and Alicia were divorced in 1963 but moved into the same house in 1970 and lived independent lives. Alicia claimed he was just an unrelated boarder. At Princeton he became known as "The Phantom of Fine Hall" (the mathematics center), and during the middle of the night he would often sneak into classrooms where he wrote odd equations on blackboards.

John Nash is widely known, primarily because of Sylvia Nasar's 1998 biography of his life titled *A Beautiful Mind*, which was later made into an Academy Award–winning movie. However, the movie bore little resemblance to Nash's real life. John Nash's more recent work at Princeton involves advanced game theories. He still prefers to work on problems of his own selections. In the fields of pure mathematics and economics, he is best known for his *embedding theorem* that describes the abstract Reimannian manifold that can be isometrically shown as a submanifold of Euclidean space.

**NATHANS' THEORY FOR RESTRICTION ENZYMES:** Microbiology: *Daniel Nathans* (1928–1999), United States. Daniel Nathans shared the 1978 Nobel Prize for Physiology or Medicine with two other microbiologists, Hamilton Smith and Werner Arber from Switzerland for the discovery of restriction enzymes.

*Restriction enzymes are mechanisms evolved by bacteria to resist viral attack and restrict infections caused by particular bacteriophages by removal of viral sequences.*

A restriction enzyme cuts the double-stranded helix of DNA by making two slices through ends of the phosphate backbones of the double helix, but without damaging the base pairs. This provides an opportunity for appropriate procedures of molecular biology and genetic engineering to be used.

Hamilton Smith (1931–), the American microbiologist, was the first to identify the restriction enzyme in 1970. This opened the door for microbiologists as a procedure for mapping genes. This development inspired Daniel Nathans in 1969 to work on the simian virus (SV40) that causes tumors. Nathans was able to demonstrate that the virus could be split into eleven unique fragments. He then determined the order of the fragments that indicated the method of fully mapping genes. In addition, the technique provided information that proved helpful to others pursuing research in DNA recombination.

In addition to sharing the Nobel Prize, Daniel Nathans was awarded the National Medal of Science in 1993 and served as president of Johns Hopkins University in Baltimore, Maryland from 1995 to 1996.

*See also* Arber

**NATTA'S THEORY FOR HIGH POLYMERS:** Chemistry: *Giulio Natta* (1903–1979), Italy. Giulio Natta shared the 1963 Nobel Prize in Chemistry with the German chemist Karl Ziegler (1898–1973) for their work on high polymers.

*Adding the proper catalysts during the process of polymerizing straight-chain polymers will produce superior and stronger forms of high polymers.*

A polymer is a chemical compound with a molecular weight comprising individual units that are linked to one another by covalent bonds. A polymer's molecules are bonded together to form three different types of polymer structures, as follows: 1) *monomers* that are structures formed by single molecules, 2) *dimers* are two monomers that blend with each other, and 3) *trimers* formed by a combination of three monomers. The latter types of polymers are generally classified as *high polymers* of which there are two types: 1) *linear* high polymers having two bonding sites comprised of units and formed in a chain arrangement and 2) *nonlinear* high polymers having units, each arranged with three bonding sites. High polymers have a wide application due to their unique properties that include high melting points, resistance to moisture, and other chemicals. They provide a high degree of stability, making them ideal for application when heat is used, such as in sterilization equipment for the medical and dental professions. Some specific examples of the use of high polymers in everyday life are well-known plastics, such as ethylene and styrene, and useful products such as ethylene glycol (antifreeze), glycerin, and divinyl benzene.

The *addition* types of polymers have a repeating molecular unit (similar to the monomers), which are used to produce polyethylene and polystyrene, while *condensation* type polymers have units consisting of fewer atoms that repeat what is contained in a monomer. Examples of the products resulting from condensation polymerization are polyesters and polycarbonates. Natta applied various catalysts to propene (a hydrocarbon chain molecule) to form polypropene, a plastic with superior properties. He also worked on what he called "stereospecific" polymers that had properties of heat resistance and improved strength. Although he was a well-respected chemist, he was criticized for accepting a position at the Milan University in 1938 that opened up as a result of the fascist Italian government's anti-Jewish laws. His colleague, who was Jewish, was forced to resign, and Natta agreed to replace him as head of the Chemical Engineering Department.

## NÉEL'S THEORIES OF FERRIMAGNETISM AND ANTIFERROMAGNETISM:

Physics: *Louis Eugène Félix Néel* (1904–2000), France.

> *Ferrimagnetism is an example of a molecular-field theory for the magnetic ordering in a system containing nonequivalent structures of magnetic ions that act as tiny magnets within the system.*

In 1936 Néel did research on magnetic solids that involved a particular magnetic ordering of the solids' particles that he called "antiferromagnetism." This was just the opposite of "ferrimagnetism" where unpaired electrons spins are arranged differently. In a ferrimagnetic material the magnetic moment of the ions (and atoms) on different areas are unequal as well as opposed, resulting in the material being magnetized (also referred to as spontaneous magnetism). For instance, iron contains the $Fe^{2+}$ (ferrous) ion and the $Fe^{3+}$ (ferric) ion and is an example of common magnetic substances. In the 1930s Néel suggested that a new form of magnetism may exist and may act differently than a regular magnet. He called this "antiferromagnetism," an example of which is the compound manganese oxide (MnO) in which the magnetic moments of $Mn^{2+}$ and $O^{2-}$ ions are equal and parallel but oriented in opposite directions (this is the opposite of ferrimagnetism where the magnetic moments of the ions are unequal).

Heating above certain temperatures can disrupt spontaneous coupling of atomic magnets. This temperature, that is different for each type of antiferromagnetic material, is now referred to as the "Néel temperature." Some antiferromagnetic solids have a Néel temperature near room temperature. For others, the temperature may be much higher or even lower. When antiferromagnetic materials are subject to very low temperatures, they exhibit no response when placed in an external magnetic field because the antiparallel structure of the atomic magnets is not altered. However, at high temperatures far above the Néel temperature, some of the atoms break free of their ordered arrangement and become aligned with the magnetic filed. As the temperature increases, the heat agitation increases (kinetic molecular motion), thus preventing atoms from aligning with the magnetic field. Consequently, the magnetism decreases as the temperature increases.

Néel's contributions to solid-state physics and, in particular, his research with the magnetic properties of various solids have been invaluable in the development and improvement of memory components of modern computers. He also contributed to geology by explaining the weak magnetism of certain rocks, thus aiding in the study of Earth's structure and Earth's natural magnetic field.

**NEHER'S "PATCH CLAMP" TO RECORD SMALL IONIC CURRENTS:** Biology and Physics: *Erwin Neher* (1944–), Germany. Erwin Neher shared the 1991 Nobel Prize for in Physiology or Medicine with Bert Sakmann (1942–), the German cell physiologist.

> *By altering the "voltage-clamp" method that invades the cell, the "patch clamp" method can sit on the surface of the cell, and thus it is possible to measure the small ionic currents generated by ions passing though the cell membrane.*

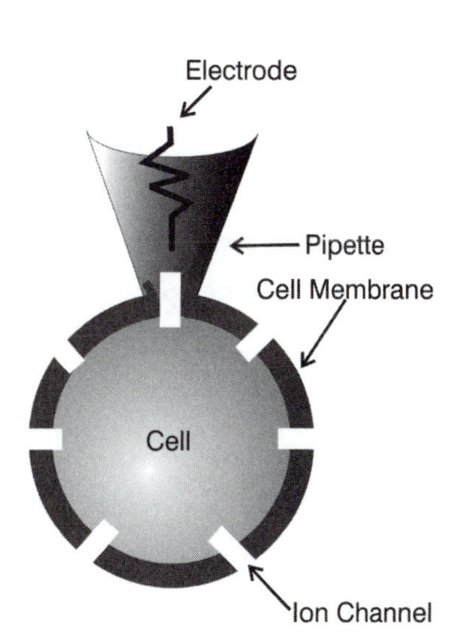


Figure N1. A typical external type of tiny electrode within a very small pipette attached to the outside of a cell to deliver medication through the cell's ion channel.

Earlier studies identified minute channels (holes) in the surface of cell membranes through which ions (charged particles) pass through the cell's wall to the inside of the cell. Early attempts to detect the small electric current created by the flow of ions through the cell membranes were somewhat invasive to the cells. In 1976 Neher and Sakmann detected the minute current created by ions as they pass through just one of many ion channels in the membrane of a receptor muscle cell without the associated thermal "noise" that overshadowed the currents (*see* Figure N1).

There are a number of variations of the basic patch clamp technique depending on what aspect of the cell is under study. Some examples are 1) the *cell-attached*

The amazing process of the "patch clamp" technique is accomplished by using a tiny pipette that has an opening in its tip of just one micrometer. This pipette has a smooth round electrode tip rather than the sharp microelectrodes formerly used to enter the inside of the cell. This new type pipette-electrode is known as a "patch clamp electrode" because it is "patched" to the outer surface of the cell. It does not invade the cell itself. The pipette is usually filled with an electrolyte-like solution such as saline solution. Drugs can also be used. A metal electrode at the other end of the pipette is in contact with the fluid in the pipette as well as the ion channel on the surface of the cell. The patch clamp is gently placed onto the cell membrane. Suction is then applied to bring a small amount of the cell's membrane containing an ion channel inside the tip of pipette. This gentle suction creates a seal with the tip of the electrode. As the ions pass though the ion channel in the cell's membrane, the instruments attached to the other end of the electrode record the tiny electric current the ion generates as it passes through the channel in the surface to the inside the cell's membrane. The detection of this small amount of current proved difficult because each channel is the size of the diameter of an ion. Ions are atoms that have gained or lost electrons, and thus, have one or more + or − charges on their surface. Therefore, ions carry a tiny electrical charge and can be affected by tiny electrical currents. As they pass from the end of the micropipette-electrode, they generate a current of only about $10^{-12}$ amperes. This is an extremely small amount of electricity that proved to be a challenge in the construction of a device that could read these small electrical currents. Thus, the patch clamp system is unique and practical.

patch where the seal is maintained so that drugs can be administered through the ion channel; 2) the *inside-out patch,* where the electrode is withdrawn leaving the patch of membrane attached to the electrode end of the pipette thus exposing the ion channel for study; 3) the *whole-cell patch* increases the suction to break through the membrane and provides access to the inside of the cell as well as improved electrical access to the inside of the cell; 4) the *outside-out patch* is used when the electrode is removed after procedure 3 (above) so that the ion channel can be further investigated; 5) the *perforated patch* is similar to the whole-cell patch (3 above) where a new seal and a new electrolyte solution and an antibiotic fluid in the pipette-electrode are used to significantly shorten the time frame of the experiment. Another variation substitutes the micropipette with a flat patch and is called a *planer patch clamp* which is a small, flat surface material with tiny holes that attaches to the channel cells in the cell's membrane.

**NERNST'S HEAT THEOREM:** Chemistry: *Walther Hermann Nernst* (1864–1941), Germany. Walther Nernst received the 1920 Nobel Prize for Chemistry.

*When the temperature approaches absolute zero, so does entropy approach zero (the kinetic energy of atomic motion cease).*

Walther Nernst's heat theorem was based on experimentation with ions in solution and his attempt to determine the specific heat related to chemical reactions. He measured the heat absorbed in a chemical reaction, which fell along with the chemical's temperature as they both (heat and temperature) approached zero kelvin in value (absolute zero −273.16°C). This theorem is called the *third law* and together with the first two laws of thermodynamics is the formulation of the science of thermodynamics. His theorem deals with the calculation of the *absolute* **entropy**, whereas the second law of thermodynamics only measures the *differences* in entropy.

*See also* Carnot; Clausius; Fourier; Kelvin

**NEWCOMB'S THEORY FOR THE SPEED OF LIGHT:** Physics: *Simon Newcomb* (1835–1909), United States.

*The use of statistical methods can improve the accuracy of the constant for the speed of light.*

In 1880, Newcomb, an employee of the U.S. Naval Observatory in Washington, D.C., was ordered by the U.S. secretary of the navy to measure accurately the speed of light. Newcomb applied statistical techniques to data gathered through repeated measurements of the speed of light. He placed a mirror at the base of the Washington Monument in Washington, D.C. He then proceeded to shine a light from his laboratory onto the mirror, measuring the time it took to make a round trip. From these repeated events he recorded his data as a histogram (a graph representing the statistical means of his data in block form), from which he then considered the distribution and confidence interval for the data to arrive at an average figure for the speed of light. He noticed some so-called outliers, which were measurements way off from the average, which he eliminated. Experimental scientists frequently use this technique to exclude spurious measurements that are often artifacts of the measuring instruments and observer errors.

*See also* Michelson

**NEWLANDS' LAW OF OCTAVES:** Chemistry: *John Alexander Reina Newlands* (1837–1898), England.

> *The chemical elements, when listed by their atomic weights, show a pattern of certain properties repeating themselves after each group of seven.*

John Newlands first stated his law of octaves in 1864, but it was not accepted as anything more than an odd arrangement of the elements. In essence, his law *states that any given element will have similar characteristics to and behave like another related element when organized in rows of seven according to their increasing atomics weights* (*see* Figure N2). Many scientists used various methods to arrange and classify the fifty-five to sixty then-known elements. The "noble" or inert gases found in Group 8 of the modern **Periodic Table of the Chemical Elements** were not yet discovered. Newlands tried something different. He organized these elements into groups by their atomic weights and noticed that similar elements repeated similar properties when listed by rows of seven. For instance, pairs in the second group of seven (by atomic weight) were similar to the pairs in the first row of seven elements. His elements did not include the unknown Group 8, so this repeating of properties reminded him of the seven intervals of the musical scale, where the same seven notes are repeated several times and the eighth note in each octave (row) resembles the first note in the next higher octave. Therefore, he

NEWLANDS' PERIODIC TABLE

SIMILAR

| I | II | III | IV | V | VI | VII | VIII |
|---|---|---|---|---|---|---|---|
| H | | | | | | | |
| Li | Be | B | C | N | O | F | |
| Na | Mg | Ai | Si | P | S | Cl | |
| K<br>Cu | Ca<br>Zn | □<br>□ | □<br>Ti | As<br>V | S<br>Cr | Br<br>Mn | Fe Co Ni |
| Rb<br>Ag | Rb<br>Ag | In<br>Y | Sn<br>Zr | Sb<br>Nb | Te<br>Mo | □ | Rh Rh Pd |

PERIODS

Figure N2. Newlands organized his Periodic Table according to the concept of octaves in the musical scale where each eighth note in an octave was similar to the first note in the next octave (row).

called his organization of the elements the *law of octaves*, which was later corrected to include the eighth group of the inert noble gases. In addition, Newlands' "table of the elements" did not allow blank spaces for yet-to-be discovered elements, and thus it represented a more realistic organization of the elements than did some other arrangements of the known elements. (Later, the periodic table was improved by organizing the elements by their **atomic numbers** rather than their atomic weights; *see* Figure M5 under Mendeleev for a version of the modern Periodic Table of Chemical Elements.) Following the success of Mendeleev's table, Newlands finally published his law of octaves, which he had withheld due to criticisms of his theory. Since that time, Newlands has been given credit for the original concept of the periodic arrangement of the elements by their atomic weights.

*See also*: Mendeleev

**NEWTON'S LAWS AND PRINCIPLES:** Physics: *Sir Isaac Newton* (1642–1727), England.

***Newton's first law:*** *An object at rest will remain at rest, and an object in motion with constant velocity will remain in motion at that velocity unless and until an external force acts on the object.*

This first law refers to the concept of *inertia*, which is the tendency of a body to resist changing its position and/or velocity. Therefore, a body at rest remains at rest, and a moving body will continue to move in its direction with a constant velocity unless acted upon by an external force. This was a major step in revising the ancient concept of motion, which presumed that for a force to move an object, something must be in contact with the object that was "pushing" it. In other words, before Newton's first law, people did not believe in "force at a distance." The accepted belief was that heavenly angels caused the movements of planets or an **aether** in space was pushing them.

***Newton's second law:*** *The sum of all the forces (F) that act on an object is equal to the mass (m) of the object multiplied by the acceleration (a) of the object ($F = ma$).*

Newton's second law explains the relationship between acceleration and force. Acceleration is the rate of change in the velocity of an object with respect to the time involved in the change of velocity. Velocity involves speed of an object and its direction. The second law is expressed as $F = ma$, where $F$ is the force exerted on the mass ($m$) of the object, and ($a$) is the acceleration of the object. To determine the acceleration of an object, its mass would be inversely proportionate to the force acting on it: $a = F/m$. Acceleration and force are considered *vectors* (directional arrows) because both have a direction and a magnitude. Vectors can be added and subtracted, so it is possible to arrive at a sum of several forces acting on an object by adding the magnitude of one vector arrow to the next.

Newton's second law also explains the concept of *momentum*, which is the product of an object's mass times its velocity: momentum $= mv$. The rate of change in momentum equals the strength of the force applied to the object. Momentum explains why a heavy automobile going at the same speed as an automobile of lighter weight has more momentum. For example, a twenty-five hundred pound car going at a velocity (speed) of sixty miles per hour has a momentum equal to $2{,}500 \times 60$; while the lighter car weighing just fifteen hundred pounds going at the same speed has a momentum equal to $1{,}500 \times 60$, which is less. However, if they should meet in a head-on accident at sixty mph, the heavier car will sustain less damage than the lighter car because of its greater mass and thus momentum.

Newton's first and second laws of motion led to the concept of *inertial frames of reference.* An inertial frame occurs when an object that has no external forces acting on it continues to move with a constant velocity. This is why, once you start sliding on ice, it is difficult to stop unless there is a force to impede your progress.

***Newton's third law of motion:*** *When two bodies interact, the force exerted on body 1 by body 2 will be equal to (but opposite) the force exerted on body 2 by body 1.*

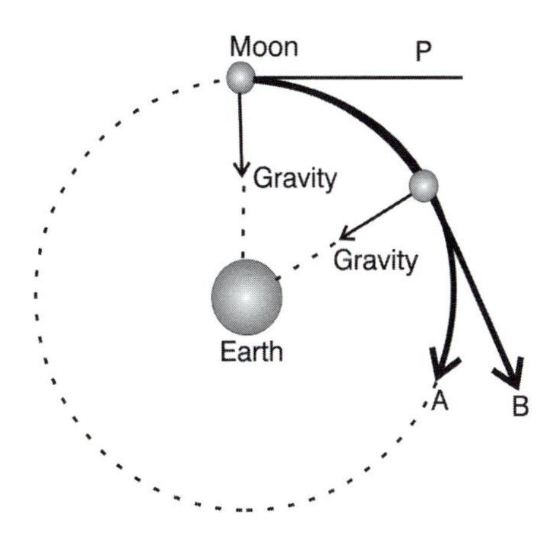


Figure N3. As the moon "falls" toward Earth, due to gravity, its trajectory follows a curved path that misses Earth as the moon follows its trajectory.

Another way to say this is, when one object exerts a force on a second object, the second object will exert an equal but opposite force on the first object. It may be expressed as $F_{1*2} = -F_{2*1}$ and is commonly worded as, *for every action, there is an equal and opposite reaction.* This explains why a pot of soup will remain on a table. As the force of weight of the pot of soup "pushes" down, the table is also pushing up on the pot with an equal and opposite force. Sometimes this relationship can be exaggerated. For instance, when a person walks on the ground, he or she is exerting a backward force on Earth while Earth is exerting an equal and opposite force on the person propelling the person forward. Thus, while the person moves forward, Earth does not seem to move backward because it has a much greater mass than a human. However, if the person and the other object (e.g., a ball the same mass as the person) had identical masses, not much progress would be made by walking. Another example is the equal-and-opposite reaction in a rocket motor. Many people believe the exhaust fire and fumes "push" against the air behind the rocket and thus propel the rocket forward. But there is no air in space, and rockets surely operate in space. The greater the mass of the exhaust exiting the rear of the rocket with tremendous velocities, the greater will be the opposite (and equal) reaction inside the front end of the rocket pushing the rocket forward. Therefore, air (or an aether) is not required for a rocket's exhaust to "push" against. Rather, the faster the gases (and the greater the mass of the exhaust) are expelled, the greater is the opposite reaction to the direction of the gases.

It might be mentioned that Newton's first two laws of motion were related to the concept of inertia first generalized by Galileo. The first law is actually a special case of the second law, and both were based on Galileo's observations. Newton's third law was original. It was never conceived by anyone before him. When considered with the second law, the third law describes the concept of mass in terms of a particular mass's inertial properties. In other words, mass cannot be defined except in terms of its inertial and gravitational properties, along with the concept of force.

***Newton's law of gravity:*** *Two bodies of mass (mass$_1$ and mass$_2$), separated by a distance r, will exert an attractive force on each other proportional to the square of the distance separating them.*

Galileo demonstrated that all bodies in free fall do so with an equal acceleration. In other words, he determined that, disregarding air and other sources of friction, two bodies of unequal weights would accelerate at the same rate when in free fall. He came very close to explaining gravity, but it was Newton who applied the mathematics of his three laws of motion (in particular the third law) to the concept of constant gravitational acceleration as applied to all bodies. The force of gravity is expressed as $F = Gm_1m_2/d^2$ where $F$ is the force, G is the proportional constant for gravity, $m_1$ and $m_2$ are the masses of the two bodies, and $d^2$ is the square of the distance between them. Newton determined that the force of gravity is the cause of the acceleration of a body, and the motion of that body does not change the gradational force. In other words, the rate of acceleration is independent of the force and thus is a property of the body, that is, its mass. Weight is distinguished from mass in the sense that the weight of an object (on Earth) is dependent on the attraction between the object and Earth. In real life, Earth is so much more massive than the object that Earth does most of the attracting. Thus, the more massive the object on Earth, the more it weighs. The force of gravity on an object determines weight. Because the moon has less mass than Earth, its gravity is about one-sixth that of Earth. Therefore, a 180-pound person would weigh only thirty pounds on the moon. The mass of an object might be thought of as the amount of "stuff" in the object; it is the same anywhere the object is located in the universe. In 1665 Newton applied his concepts of motion and gravity to describe why the moon orbits Earth. He figured that the moon was constantly "falling" to Earth as the law of gravity was acting on both bodies. Although the moon is far enough away from Earth, as its inertia tends to move the moon at a right angle in relation to Earth, gravity tends to "pull" the two bodies together. Thus, as the moon "fall" towards Earth, its trajectory follows a curved path that misses Earth as it proceeds in its orbit—eternally falling toward Earth, but missing Earth (*see* Figure N3).

This is the same explanation of why our modern Earth-orbiting satellites remain in orbit. They are continually falling toward Earth, but they are high enough and their velocity (and inertia) is adequate to cause them to miss the surface and just keep falling.

***Newton's theories of light:*** *1) Light is composed of a great multitude of very tiny particles of different sizes, which are reflected by shiny polished surfaces. 2) White light is a heterogeneous composition of these different-sized particles, which form rays of different colors.*

Sir Isaac Newton did not accept the wave theory of light, which many other scientists compared with sound waves. His rationale was that light, unlike sound, could not travel around corners. Therefore, light cannot have wave properties and must be composed of a multitude of very tiny particles.

For the second part of his theory, Newton placed a large "plano-convex" lens on a shiny surface and exposed it to direct sunlight. A distinct pattern of rainbow-like concentric rings of colored light became visible near the edges of the lens. Although this was a good indication that light had wave properties, Newton still maintained his corpuscular (tiny particles) theory of light. He also believed white light was composed of light rays of various colors, each color of which was a different-size corpuscular particle. He was determined to prove sunlight was a heterogeneous blend of a variety of light rays (and particles) and that both reflection and refraction could be used to separate these individual-sized particles and rays. To do this, he covered the window of a room to shut out sunlight, leaving a small opening for a narrow shaft of sunlight, which he directed through a prism. Some of the light rays going through the prism were bent (refracted) more than other rays, and together they produced a color spectrum of sunlight (*see* Figure F8 under Fraunhofer). He then placed another prism in contact with the first and discovered that the second prism caused the individual rays to converge back into the white light of the sun. This proved sunlight was indeed composed of different-colored light rays. However, rather than proving his corpuscular theory, it gave further evidence for the wave theory of light, which was further confirmed when new diffusion gratings were used instead of prisms to demonstrate the wave nature of light. Newton delayed publishing his results for several years fearing adverse criticism of his work (*see also* Fraunhofer; Maxwell).

***Newton's calculus:*** *There is a mathematical relationship between the differentiation and integration of small changes for events.*

Newton referred to this as the "fluxional method," which became known as *calculus*. Gottfried Leibniz independently developed what he called differential calculus. Calculus is the branch of mathematics that deals with infinitesimally small changes. It is used extensively in almost all areas of physics, as well as in many other fields of study.

Today, Newton and Leibniz share the honor for the development of calculus.

*See also* Euler; Galileo; Kepler; Leibniz

**NICHOLAS' THEORY OF AN INCOMPLETE UNIVERSE:** Astronomy: *Nicholas of Cusa* (1401–1464), Italy.

Because he was a bishop and papal advisor, Nicholas of Cusa's theories assumed a religious and metaphysical leaning. Nevertheless, he proposed some rather revolutionary ideas for his time in history.

***His theory states:*** *There is nothing fixed in the universe, yet it is not infinite.*

Nicholas of Cusa did not accept the theories proposed by Aristotle and other astronomers that the universe was composed of a series of crystal domes over a flat Earth, or the alternative—that the universe was one big sphere with Earth at its center. Nicholas proposed a more flexible universe. It was in a state of becoming where nothing was fixed; there was no outer edge or circumference, and the center was not yet established. Another way to look at this is to consider an infinite circle whose infinite circumference is composed of an infinite straight line. He believed that the universe was extremely complex, with everything in a state of flux and in motion. Even so, God's universe was not infinite. He believed that things are finite, but God is infinite. Some of Nicholas' views including that of the solar system were ahead of his time. He also believed Earth, as well as all the other planets, revolved around the sun.

*See also* Aristotle; Copernicus; Eudoxus; Galileo

**NICOLLE'S THEORY FOR THE CAUSE OF TYPHUS:** Biology: *Charles Jules Henri Nicolle* (1866–1936), France. Charles Nicolle received the 1928 Nobel Prize for Physiology or Medicine.

> *The disease typhus is spread by a parasite (body louse) that lives on the patient's body and clothing.*

In the late 1800s Charles Nicolle, director of the Pasteur Institute, was asked to determine the cause of typhus, an infectious disease that had been known to follow wars and plagues as early as 1400 BC. As with measles, people who recovered from typhus were thereafter immune to it. Typhus was also confused with influenza because physicians believed it was spread by "droplet" infection or direct contact with patients'

> The banning of the use of DDT is an example of "good intentions" resulting in "unintended consequences." DDT is an organic insecticide first synthesized by a German chemistry student, Othmar Zeidler (1859–1911) in 1874, although its use as an insecticide was not known until 1939. During World War II DDT was used as an organic insecticide by Allied troops and civilians to control the insects that cause typhus and malaria. Entire cities in Italy were dusted with DDT and the walls in homes were sprayed with it to control mosquitoes. In addition, it was widely used for agricultural purposes. It was considered a rather safe insecticide and was responsible for eradicating typhus in Europe and malaria in Europe and the United States. The progress that improved the health of millions of people came to a halt soon after Rachel Carson published her book, *Silent Spring* in 1962. This book made many controversial claims, among them that DDT causes cancer in humans and destroys wildlife by causing the decline in the population of large birds that is a result of the extremely thin and fragile egg shells preventing reproduction. Her book was a "hit" for many environmentalists because they could now claim that DDT was responsible for many of the health and environmental problems of the world. In years following, many countries banned the use of DDT, even though it is one of the most effective and least harmful of the class of organic insecticides. The results of the ban, mostly in third world countries, are astonishing. It is estimated that over two million people—mostly children—have died each year from malaria since the ban, and that over fifty million people have died because of the ban. It has been proposed that more people have died worldwide from insect borne diseases that could have been controlled by DDT than were killed by Hitler during World War II (Source: Michael Crichton, *State of Fear,* HarperCollins, 2004.) It should be noted that many environmental groups dispute these figures.

clothes or dust. Its death rate was over 50% for adults contracting the disease. Nicolle visited homes where whole families were infected, exposing himself and his coworkers to the infection; several died while gathering information. He observed that though almost all members of a family contracted the disease, once they came to the hospital, it no longer spread. He also noticed a relationship between patients brought to the hospital who were undressed, bathed, and whose clothes were either burned or laundered and the cessation of new infection. As new patients were admitted and went through this procedure, the disease did not spread to others in the hospital. He deduced there must be some type of insect in their clothes or on their bodies.

In 1909 he concluded from this evidence that the culprit was the body louse. Using this knowledge and information, Charles Nicolle experimented with animals, exposing them to the disease carried by the louse. He further established that the typhus germ was not transmitted to a new generation of the louse parasite. Rather, once the germ-carrying adult louse dies, the epidemic ends. He found the guinea pig and some monkeys were susceptible, but not many other animals. Using the blood from infected animals that recovered from the illness, he tried to develop a vaccine to prevent typhus and understand how it affected the immune system. Nicolle's work emphasized the need for better personal and public hygiene to prevent typhus. His discovery that typhus is carried by lice explained why, since ancient times, it is associated with wars. It was not until World War II that the insecticide dichlorodiphenyltrichloroethane (DDT), now banned, was used to control typhus outbreaks in army troops stationed in Italy and other countries, ultimately saving many lives.

**NODDACK'S HYPOTHESIS FOR PRODUCING ARTIFICIAL ELEMENTS:** Chemistry: *Ida Eva Tacke Noddack* (1896–1979), Germany.

*When uranium is bombarded with slow neutrons, artificial elements and their isotopes should be produced.*

All isotopes of elements above atomic number 81 are unstable and radioactive. Ida Eva Tacke Noddack was familiar with Enrico Fermi's work with radioactive elements. Fermi realized that fast neutrons that were slowed could more readily penetrate the nuclei of uranium, thus causing fission to occur. This gave Noddack the idea that by bombarding uranium atoms with slow neutrons new isotopes of elements can be produced. This is exactly what occurred when Otto Frisch confirmed Noddack's hypothesis by bombarding the heavy nuclei of uranium with slow neutrons. This bombardment broke up the nuclei into a few large fragments, which proved to be isotopes of other elements (*see also* Fermi; Frisch).

The story of Ida Eva Tacke Noddack's life in science is similar to that of several other women of science who where more or less ignored for their contributions during their time in history. Eva Noddack attended the Technical University in Berlin where she received her PhD in 1921. She joined her future husband Walter Noddack to work on a chemical research project with the German X-ray specialist Otto Berg in 1925. Their collaboration led to the joint discovery of the isolation of element 75 in 1925. It was named *Rhenium* that is Latin for the Rhine River in Germany. In 1926 she married Walter Noddack, and they worked together until his death in 1960. During the span of their marriage they jointly published over one hundred scientific papers. She soon became interested in the work of Enrico Fermi who was the first to bombard uranium with slow neutrons. Fermi believed that he had added neutrons to uranium atoms to produce atoms of elements heavier than uranium, which he called "transuranic" elements with atomic numbers higher than uranium 92. Ida Noddack challenged Fermi's claim that he artificially produced new heavier elements by adding neutrons to uranium nuclei. She claimed that he did not make new heavy elements but rather, according to her theory, what he really did was artificially split uranium atoms into isotopes of lighter known elements rather than producing atoms heavier than uranium. In 1934 Ida Noddack suggested her hypothesis that by using Fermi's method of shooting slow neutrons into the nuclei of heavy elements, the results would be fragments of those heavy nuclei and therefore would be isotopes of known lighter elements, and not just other heavier closely related elements. She sent the results of her work on the artificial production of isotopes of heavy elements to Fermi, which he pretty much ignored, as he did with most of the science community. Ida Noddack received a number of awards and was nominated several times for the Nobel Prize in Chemistry but never received it. It has been stated that if Germany during the mid- to late thirties would not have ignored her research, Nazi Germany might have easily won World War II by developing the nuclear (atomic) bomb. The same also applies to the work of Lise Meitner in atomic fission that was also ignored by the patriarchal scientific community in Germany during the 1930s.

**NOETHER'S THEOREM:** Mathematics: *Amalie Emmy Noether* (1882–1935), Germany.

*For every differential symmetry generated by local actions, there is a corresponding conserved charge.*

Note: The word "charge" in the above statement is referred to as the "Noether current" that is defined as a nondiverging (nonchanging) vector field.

Symmetry is a basic concept of physical laws and was defined by the German mathematician Hermann Weyl (1885–1955) as: "*That a thing is symmetrical if there is something that you can do to it so that after you have finished doing it, it looks the same as it did before.*" In everyday life there are several types of symmetry. *Bilateral* symmetry is how we are built—each side of our bodies is more or less a mirror image of the other. Also, no matter in what

Amalie Emmy Noether was born in Erlangen, Bavaria, Germany, into a family where her father was a notable mathematician. She showed an early interest in dancing and music at a time when women were not permitted to enter schools in Erlangen. When the local school reversed this restriction in 1904, she enrolled and became known for her publications. In 1915 she moved to the University of Göttingen where she was not allowed to teach. However, she persisted and was finally accepted and even honored until the Nazi racial laws no longer allowed her to teach undergraduate classes in mathematics. In 1935 she fled Germany to Bryn Mawr, Pennsylvania, after she supposedly had an operation in Germany and the doctor reported that she had died during the procedure. She spent the rest of her life at Bryn Mawr and was buried there. Albert Einstein said at her eulogy that she was the most significant creative mathematical genius thus far produced since the higher education of women started.

perspective you look at a cube, it will look the same in all positions. In other words, the symmetry of rotation gives us the law of conservation of linear momentum. *Radial* symmetry is best described by a cylinder that is viewed from the ends, or something going around a central pivot, such as a wheel.

Symmetry is an important concept in physics (*see* the introduction of this book). A major property of symmetry as a physical law is that it is universal. In other words, the laws of physics do not vary with locations in space. When referring to physical laws, no matter what you do to the law, it makes no difference and everything is unchanged. Some examples of the application of Noether's theorem follow:

1. As far as we know, *time* (past, present, and future) does not make a difference in the symmetry of an object in the universe. This invariance gives us the *law of conservation of energy*.
2. When there is no variation with an object's orientation or location in space, the *law of conservation of linear momentum applies*.
3. When the orientation of an object in space does not change with respect to *time*, the result is the *law of conservation of energy*.

**NORRISH'S THEORY OF VERY FAST REACTIONS:** Chemistry: *Ronald George Wreyford Norrish* (1897–1978), England. Ronald Norrish shared the 1967 Nobel Prize for Chemistry with Norrish's student, the British chemist George Porter (1920–2002), and Manfred Eigen.

*An intense burst of light can be used to cause very fast chemical reactions and thus measure and describe intermediate stages of these organic photochemical reactions.*

Norrish's theory describes two types of fast reactions, *type I* and *type II*, that involve a cleavage of organic molecules by a flash of light that results in the rapid production of free radicals.

*Type I Norrish reactions* are photochemical cleavage of aldehydes and ketones into two free radicals (small groups of molecules with a small charge). This occurs when the carbonyl group is excited and accepts a photon that is incorporated into the molecule causing a photochemical reaction to occur. Although of limited utility, Type I Norrish reactions are important for understanding and controlling certain types of organic chemical synthesis.

*Type II Norrish reactions* are photochemical reactions where a flash of light causes intramolecular changes in the carbonyl group by exciting the carbonyl compound to produce a photon radical.

Norrish's development of these types of light-sensitive organic chemical reactions allowed scientists to study the minute intermediate stages of the reactions and thus improve methods of understanding the procedures involved in the reactions.

*See also* Eigen

G. W. Norrish was born in Cambridge, England. He finished the Perse Grammar School there and later received a scholarship to study natural science at Emmanuel College in Cambridge University. World War I interrupted his education as he joined the Royal Field Artillery and subsequently in 1918 became a German prisoner of war. After his release from prison camp, Norrish returned to Cambridge and finished his undergraduate degree in 1921. In the late 1930s he was named professor of physical chemistry and director of the department that he maintained until his retirement in 1965. After retirement many of his former students and colleagues in England and abroad joined together to publish a book in his honor titled *Photochemistry and Reaction Kinetics.*

**NORTHROP'S HYPOTHESIS FOR THE PROTEIN NATURE OF ENZYMES:** Chemistry: *John Howard Northrop* (1891–1987), United States. John Northrop shared the 1946 Nobel Prize for Chemistry with James Sumner and Wendell Stanley (1904–1971).

*If enzymes can be crystallized, their composition must be of a protein nature.*

Several other chemists claimed that enzymes did not have the characteristics of proteins. In the late 1920s the American chemist James Sumner (1887–1955) claimed to have crystallized the common enzyme *urease*, which manifested the characteristics of a protein. In the early 1930s John Northrop and his colleagues were successful in crystallizing several more important enzymes, including trypsin, pepsin, chymotrypsin, and more important, ribonuclease and deoxyribonuclease (related to RNA and DNA). The exact nature of proteins was difficult to determine because of their very long molecular structures. The crystallization provided a means for identifying the structures of the enzyme and confirmed the theory that they were of a protein nature, enabling scientists to study and understand their chemical composition. Later, Northrop proved that bacteria-type viruses (bacteriophages) also consist of proteins and cause diseases by infecting specific species of bacteria. (See Figure D7 under Delbruck for an artist's version of a bacteriophage.)

**NOYCE'S CONCEPT FOR THE INTEGRATED CIRCUIT:** Physics: *Robert Norton Noyce* (1927–1989), United States.

*A series of transistors can be combined on a small single piece of semiconducting material by etching microscopic transistors onto the surface of a chip to form circuits that can integrate the individual transistors.*

Rectifying crystals were in use before the vacuum radio tube was used to control the unidirectional flow of alternating current. They were not very effective, but the concept of using crystals to control the flow of electricity and electromagnetic radiation was not lost on a number of scientists. In the late 1940s, William Shockley, Walter Brattain, and John Bardeen, who were colleagues at Bell Laboratories in New Jersey, used a different substance, a germanium crystal. It was an ineffective conductor of electricity but a good insulator, making it what is now known as a *semiconductor*. Silicon crystals were less expensive and soon replaced germanium as a semiconducting material. It was discovered that if tiny amounts of certain impurities were placed in the semiconductor material, its characteristics could be controlled. These were referred to as solid-state devices, which act as vacuum tubes without having the tubes' large size, generation of heat, or possibility of breakage, and they used very little electricity. Shockley developed this unique device, while the American engineer John Robinson Pierce (1910–2002) gave it the name *transistor*, for its property of transmitting current over a specified resistance (*see* Figure S1 under Shockley). Robert Noyce is credited with the concept of combining a series of transistors onto a small silicon semiconductor chip (about one-fourth square inch or less) to form an integrated circuit. In 1959 Noyce received funding from the Fairchild Corporation to form Fairchild Semiconductor, the first major semiconductor electronics plant in the Silicon Valley of California to exploit the use of the new chip. Later, he formed the INTEL Corporation. In 1959 Noyce filed for a patent for the new chip, even though a few months previously, Jack S. Kilby (1923–2005) of Texas Instruments also filed for a similar patent. After a decade of legal battles, Noyce and Kilby agreed to cross-license their technologies. Kilby later won the Nobel Prize for Physics in 2000 for his part in the invention of the integrated circuit.

*See also* Shockley

# O

**OCHOA'S THEORY FOR THE SYNTHESIS OF RNA:** Biology: *Severo Ochoa de Albornoz* (1905–1993), Spain and United States. Severo Ochoa was awarded the 1959 Nobel Prize for Physiology or Medicine for synthesizing the RNA molecule. He shared the prize with the American biochemist Arthur Kornberg (1918–2007) who synthesized the DNA molecule.

*The enzyme from a type of bacteria found in sewage can act as catalyst to artificially synthesize ribonucleoside diphosphates into the ribonucleic acid (RNA) molecule.*

Ochoa was the first to determine that high-energy phosphates, such as adenosine triphosphate, are responsible for storing and then releasing energy in human cells. While studying the oxidation process that released energy in cells, he also discovered in 1955 the enzyme located in a particular bacteria that could influence the process of synthesizing the RNA molecule. RNA is of importance in the synthesis of proteins in cells. His work also led him to isolate other enzymes that were related to the citric acid oxidation reactions of the Krebs cycle.

Severo was born in a small town in northern Spain and was named after his father who was a lawyer. At age seven the family moved from their mountain home to the town of Málaga where he was educated in a private school in preparation for college. He started pursuing a career in engineering but found the courses too difficult and soon changed to biology. He received his BA from Málaga College in 1921 and two years later entered the University of Madrid's Medical School. He soon realized that his interests and desires were to do biological research. He received a degree in medicine in 1929 and accepted a postdoctoral appointment at the Kaiser-Wihelm Institute in Berlin. In 1931 after he married Carmen Garcia Cobian, they moved to England to continue his studies and research at London's National Institute for Medical Research.

After a few years he returned to Madrid where his research was partially supported by wealthy patrons. During World War II he and his wife, who became his research assistant, moved to Washington University School of Medicine in St. Louis, Missouri, in the United States. He and his wife conducted much of their research at the University's Cori laboratory, known for its outstanding work with enzymes, metabolism, and biochemical reactions that produce energy in cells. Ochoa added to the work of the Krebs cycle as proposed by Hans Krebs that determined which, and how, food is metabolized to provide energy for the body (*see* Krebs). A gel formed when Ochoa added the enzyme from a type of sewage bacteria to nucleotides. He determined that the gel in the petri dish was a synthetic form of RNA.

Many years later it was determined that Ochoa had isolated the enzyme polynucleotide phosphorylase, which actually *catalyzes* the breakdown of RNA rather than, as he originally believed, *synthesizing* RNA.

*See also* Krebs

**ODLING'S VALENCE THEORY:** Chemistry: *William Odling* (1829–1921), England.

*Elements can be grouped according to their analogous properties based on their representative values of replacement.*

The concept of valence, proposed by Edward Frankland in 1854, was unknown to William Odling. Prior to Frankland's valence concept, Odling proposed that during chemical reactions a distinct ratio existed when one element replaced other elements in a chemical reaction. This was a forerunner of the theory of atomic valences, which is the ability of elements to combine with other elements. It can be expressed as the number of univalent atoms with which they are capable of uniting (*see* Figure S2 under Sidgwick). Some elements may be univalent (1); others are divalent (2), trivalent (3), or tetravalent (4) with respect to the number of univalent atoms with which they can combine. Still other elements may possess variable valences (e.g., nitrogen and phosphorus). At first Odling, as well as other scientists of his time, rejected the existence of atoms. At one time it was incorrectly assumed that just one atom of hydrogen combined with one of oxygen to form water. However, after conducting experiments with oxygen, Odling came to believe in the valence theory. He was the first to realize that oxygen had an atomic weight of 16, not 8. This convinced him that oxygen gas had to be diatomic—a molecule composed of two oxygen atoms. He also speculated about a triatomic molecular form of oxygen (ozone, $O_3$).

*See also* Frankland; Sidgwick

**OERSTED'S THEORY OF ELECTROMAGNETISM:** Physics: *Hans Christian Oersted* (1777–1851), Denmark.

*A magnetized compass needle will move at right angles to the direction of an electric current flowing through a wire suspended over the compass.*

Hans Christian Oersted's discovery of the relationship between electricity and magnetism (electromagnetism) was accidental. Aware of the experiments dealing with static electricity and the new form of "flowing" electricity described by Alessandro Volta, Oersted performed various "galvanic" experiments using Volta's cells to produce current electricity. He knew of others who demonstrated that by passing an electric current through water, it could be separated into oxygen and hydrogen gases. He believed that this established the connection between electrical forces and chemical reactions and that water must be a compound, not an element, as had been believed for centuries. Oersted then conducted experiments with this new electricity. He attempted to demonstrate that when a wire was heated by carrying electricity, it would act as a magnet and attract a compass needle. He noticed at once that the needle was not attracted to the wire but rather moved ninety degrees from the direction in which the current was flowing in the wire. Saying nothing about this observation to his students, he continued to turn the current in the wire on and off. Each time he did so, the needle of the compass moved at a right angle to the wire. In 1820 he published his results describing the existence of a circular magnetic field between current electricity and magnetism, now known as *electromagnetism*. The unit for the strength of a magnetic field is named after Oersted and is defined as the intensity of a magnetic field's strength expressed in the centimeter-gram-second (cgs) electromagnetic system of physical units. Oersted's concepts sparked many experiments and theories related to electromagnetism, the end result being our modern "electric" oriented society.

Hans Christian Oersted's PhD degree that he received in 1799 was in the field of philosophy, which was not a field of science as we think of science today. To improve his knowledge of science, he traveled throughout Europe to learn about electricity from various physicists. Upon his return to Denmark he began lecturing and giving demonstrations on electricity to the general public. He was so successful that Copenhagen University gave him a professorship in 1806. Although he believed in the concept of electromagnetism, the idea was unproven. It was during one of these lectures when he noticed a needle was deflected when it was brought close to a wire carrying a current. A compass needle was deflected at a ninety-degree angle when the current was flowing but not when the current was turned off. Oersted was also the first person to determine that a circular magnetic field was formed around the wire when a current was flowing through it. Up to this time most of the scientific world had believed that electricity and magnetism were two completely different phenomena.

*See also* Ampère; Faraday; Henry; Maxwell; Volta

**OHM'S LAW:** Physics: *Georg Simon Ohm* (1787–1854), Germany.

*A unit of electrical resistance is equal to that of a conductor in which a current of 1 ampere is produced by a potential of 1 volt.*

Georg Ohm related electrical resistance to Joseph Fourier's concept of heat resistance, which states the flow of heat between two points of a conductor depends on two factors: 1) the temperature difference between the origin of the point of heat and the end point of the conductor and 2) the physical nature of the conducting material being heated. Ohm speculated how this information related to electricity, and this led to his experimentation with wires of different thicknesses (cross sections). He demonstrated that electrical resistance to current passing through these different wires was directly proportional to the cross section of the wires and inversely proportional to the length of the wires. Almost all effective conductors of heat are also excellent conductors of electricity. The law can be applied to direct and alternating currents. Ohm's law is very versatile and can be used to measure conductance, current density, voltage, **resistors, inductors**, capacitors, and **impedance**. It is stated as R = V/A, where *R* is the natural resistance of the wire (conductor) to the flow of electricity, *V* is voltage, or strength of the electric current divided by A which are the amperes, indicating the amount of the electricity (A also may be expressed as I). There is an expression for conductance (G), which is referred to as the reverse ohm (mho), I = GV, where *I* is the amount of current (same as amps), G is the conductance factor of the wire (how well it conducts electricity), and *V* is the voltage. The symbol Ω represents electrical resistance and is named after Ohm and is the symbol that indicates the amount of resistance to voltage of one ampere of current. Other physicists did not recognize the importance of Ohm's law for some time, but by the early 1840s, the Royal Society in England accepted his law and Ohm as a member.

*See also* Ampère; Faraday; Fourier; Volta

**OKEN'S CELL THEORY:** Biology: *Lorenz Oken* (1779–1851), Germany.

*Living organisms were not created but rather originated from vesicles (cells) that are the basic units of life.*

In 1805 Lorenz Oken theorized that humans and animals not only originated from but were also composed of cells that he called *vesicles*. Until this point, the source of life—its origin and composition—had been the subject of speculation by scientists, philosophers, and theologians over many centuries. Some believed life began with a "primeval soup" or was carried to Earth from outer space, or was derived from self-organizing inorganic molecules that self-replicated to form organic molecular living cells and tissues, or, more acceptable to many people, but less accurate biologically, life was created by a supreme being. The discovery of fossils and the use of a microscope led to further concepts of living tissue. Robert Hooke viewed tiny enclosures in cork bounded by walls that reminded him of the rooms occupied by monks in monasteries; thus, he named them *cellulae*, meaning "small rooms" in Latin. Oken further speculated that these "cells" were the basic units of life, from which all complex organisms were derived and developed, and he theorized that cellular structure was basic for all organic substances. Oken was one of the first of many scientists to contribute to and expand the concept we now know as the cell theory.

*See also* Margulis; Schleiden; Schwann; Virchow

**OLBERS' PARADOX:** Astronomy: *Heinrich Olbers* (1758–1840), Germany.

*If the universe is old, eternal, unchanging, infinite, and uniformly filled with stars, why is the night sky dark instead of bright?*

Heinrich Olbers' paradox has intrigued astronomers, physicists, and mathematicians for decades. It is based on several questions that are still being investigated by scientists: Is the universe finite or infinite? Is it an evolving and expanding universe or a steady-state universe? Are galaxies (groups of stars) evenly distributed in the heavens, or is space nonhomologous? We know that light follows the basic inverse square law. An appreciation of the inverse square law as related to light can be demonstrated at night by shining a flashlight at a one-foot-square white sheet of cardboard held away from the light at several different distances. The illumination on the cardboard is greater at 10 feet than at 25 feet and greater in intensity at 25 feet than at 100 feet. It will be obvious that the light intensity diminishes as the distance between the flashlight and white cardboard increases (the intensity of the light at different distances can be measured with a light meter). Light, over distance, is dispersed and becomes less focused, but it will travel in a straight line forever if not absorbed or distorted (affected) by gravity in space. Therefore, the intensity of light received on Earth from stars is much reduced from its brightness at the source. But at the same time, the average number of stars, at any given distance, increases in number by the square of the distance to Earth. This is the basic distribution of stars in the universe. Therefore, according to one part of the paradox, the night sky should be as bright as the sun. On the other hand, Olbers claimed that the reason the night sky is not as bright as the sun is that interstellar "dust" absorbs the starlight. Today, this is an unacceptable solution to his paradox because the universe is assumed to have come into existence at a finite time, even though it might be infinite in space. It has a beginning, it has history, it seems to continue to expand, and for the most distant and possibly oldest galaxies, light has not had time to get to Earth. Light does disperse over long distances, and galaxies and their stars are not evenly distributed throughout space. Currently, astronomers using the Hubble Space Telescope and the Gemini North telescope installed at Mauna Kea, Hawaii, are examining the question concerning the universe being finite or infinite. The mirror for the Gemini telescope is 8.1 meters in diameter and only 20 centimeters thick. It is difficult to cast a single piece of glass this size without imperfections or cracking. The mirror is the largest single-piece glass mirror ever cast for a reflector telescope (*see* Figure H3 under Hale). It is expected that both of these instruments will locate galaxies at the limit of the speed of light, which means these distant galaxies are receding faster than the speed of light, which may be considered a boundary formed by the limits of just how far we can see into the past using electromagnetic radiation (light, radio, microwaves, X-rays, etc.) originating from the edge of the universe. Or it might mean that we may never be able to "see" the edge of the universe, even if it is finite. Or, if the universe is forever expanding, it may be too young for light to have reached us. Therefore, Olbers' paradox addresses several phenomena of physics and is a puzzle of unknowns.

**OLIPHANT'S CONCEPTS OF ISOTOPES FOR LIGHT ELEMENTS:** Physics: *Marcus Laurence Elwin Oliphant* (1901-2000), Australia.

> *By "shooting" an ion beam at targets of lithium, beryllium, and related elements, new atoms of hydrogen and helium can be created by a process of atomic transformation.*

Marcus Oliphant, known as Sir Mark Oliphant, was influenced by Ernest Rutherford's work with the nuclei of atoms. In 1932, Sir James Chadwick bombarded nuclei

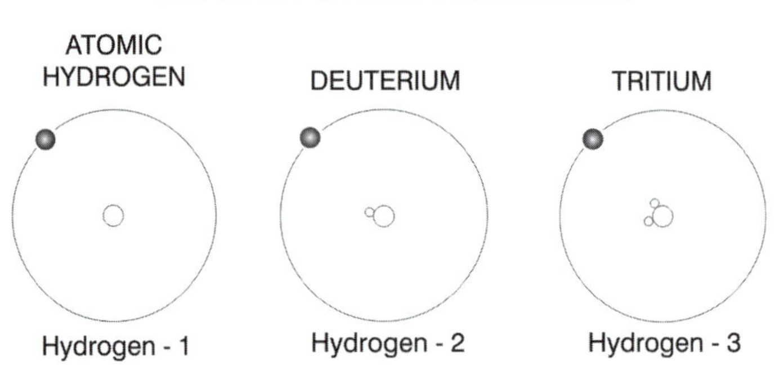


Figure O1. Harold Urey produced heavy hydrogen atoms with one proton and one neutron (deuterons). Marcus Oliphant bombarded deuterons with other deuterons to produce a third isotope of hydrogen—tritium, with two neutrons and one proton. The three isotopes of the hydrogen atom exhibit minor differences in chemical properties due to differences in atomic weights. These differences are minor for chemical reactions involving isotopes of heavier elements. Heavy hydrogen atoms are used to facilitate nuclear reactions.

of atoms and discovered a third basic particle in the atom, the neutron. (The basic particles of atoms are electrons, protons, and neutrons. However, today dozens of subnuclear particles smaller than the proton and neutron have been identified.) Also in 1932, Harold Urey discovered a heavy form of hydrogen called *deuterium*. It was known for some time that the hydrogen atom contained only one particle in its nucleus, the positively charged proton, that made it the lightest of all elements. Urey discovered that the nucleus of hydrogen could also contain a neutron; thus this isotope of hydrogen had an atomic weight of 2 instead of 1, as for ordinary hydrogen. The nuclei of this isotope were referred to as *deuterons*. Oliphant and a colleague bombarded these heavy hydrogen deuterons with other deuterons, producing a new isotope of hydrogen that contained two neutrons plus the proton. Thus, now there were three forms of hydrogen, as follows, H-1, H-2, and H-3 (*see* Figure O1). This third form, composed of one proton and two neutrons, is named *tritium*. It is unstable due to its radioactive nature, with a half-life of about twelve years.

At about the same time, it was discovered that the nuclei of these forms of hydrogen could react with each other, producing other new elements (particularly helium) and releasing tremendous energy. This was the beginning of our understanding of the thermonuclear reactions that take place in the sun, as well as the development of the hydrogen (thermonuclear) bomb. Oliphant moved to the United States and perfected the electromagnetic method for separating rare fissionable uranium (U-235) from the more common form of uranium (U-238). U-235 was used to produce one of the first atomic bombs (the other of the first bombs used plutonium). Oliphant was also the first to realize that nuclear reactors that are used to generate electricity can also produce plutonium. Thus, any country with such a reactor could develop atomic (nuclear) weapons. He became a firm critic of nuclear weapons and an advocate for the peaceful uses of atomic energy.

*See also* Chadwick; Rutherford; Urey

**OORT'S GALAXY AND COMET CLOUD THEORIES:** Astronomy: *Jan Hendrik Oort* (1900–1992), Netherlands.

***Oort's theory for the structure and motion of the Milky Way:*** *Composed of billions of stars, our Milky Way rotates as an entire disk, but the rotation is differential, not uniform, because the outer stars rotate at a slower speed than do the inner stars.*

Jan Oort determined several facts about galaxies, including our Milky Way galaxy. He theorized that the rotating stars in a disk galaxy follow Newton's laws of motion, in particular, the concepts of conservation of energy and angular momentum. Therefore, the outer stars move more slowly than do the inner ones (according to angular momentum) in the gigantic cluster of stars forming a disk galaxy. In addition to this theory of the motions of galaxies, he determined the sun is only thirty-thousand light-years from the center of the Milky Way. This places our sun about one-third of the distance from the outer edge of one of the galaxy's arms. He ascertained, too, that the sun makes one complete revolution around the Milky Way's axis once every 225 million years.

***Oort's comet cloud theory:*** *Comets originate in an area beyond the solar system.*

In 1950 Jan Oort identified about twenty comets with orbits so large that it requires many months, or years, for them to complete each orbit around the sun. He believed they originated from a great cloud or reservoir of over one trillion comets, comet material, and assorted objects that swarm far beyond the edge of our solar system. This cloud was still close enough to be affected by the sun's gravity even though it was about one hundred thousand astronomical units (**AU**) in diameter. In comparison, the minor planet Pluto's orbit is only 40 AU in diameter. An AU is the average distance between Earth and Sun. His theory states that these objects in the Oort cloud are "leftover" matter remaining at the edge of the *solar nebula*, the swirling mass of dust and gas that formed the solar system, planets, meteors, and comets. He suggested there were two kinds of comets, both of which were disturbed from their paths within this cloud by perturbations caused by the gravity of a passing (not too close) star. The path of one type of comet follows a hyperbolic orbit or path through the solar system, meaning it makes a wide sweep but not a closed orbit. Therefore, it appears only once as a comet visible to the naked eye or low-powered telescopes from Earth, never to return again. The paths of other comets become very eccentric, but their orbits are closed. They do not follow a circular path through the solar system but rather eccentric ellipses. This is the most familiar type, such as Halley's comet, and the more recent Shoemaker–Levy comet, which smashed into Jupiter on July 11, 1994. More recently, there is evidence that an inner Oort comet cloud exists beyond the planet Neptune. It is believed it contains about one hundred times the number of comets as does the outer Oort cloud, and these inner cloud comets have a more nearly circular orbit, similar to planets, than do the comets in the outer Oort cloud (*see* Figure K5 under Kirkwood for a similar asteroid belt).

**OPARIN'S THEORY FOR THE ORIGIN OF LIFE:** Biology: *Alexsandr Ivanovich Oparin* (1894–1980), Russia.

*The first living organism subsisted on organic substances, not inorganic matter.*

How life started is one of the oldest philosophical and biological questions. Charles Darwin's theory of evolution did not include an explanation for the origin of life. The concept that life just "arrived" from nonliving substances was known as *spontaneous generation*. Redi, Pasteur, and others disproved this idea that life "just spontaneously started." In 1922 Alexsandr Oparin was the first to theorize there was a slow accumulation in the oceans of simple organic compounds formed from interacting inorganic compounds. He conjectured that the original living organisms were *heterotrophic* because they did not synthesize their food from inorganic materials, as would *autotrophic* organisms, such as some bacteria and green plants. A more recent theory states that energy-producing elements and compounds self-organized into microcomponents of plant and animal cells (mitochondria and chloroplasts) combined through the process of **autopoiesis** to form simple cells.

*See also* Margulis; Miller; Oken; Pasteur; Redi; Virchow

**OPPENHEIMER'S CONTRIBUTIONS TO THEORETICAL PHYSICS:** Physics, *Julius Robert Oppenheimer* (1904–1967), United States.

*As founding father of the school of theoretical physics in the United States, he was called upon to coordinate the multitudinous effort to produce the atomic bomb.*

(NOTE: The complete stories of the development of the first atomic (neutron) bombs and the life of Robert Oppenheimer are fascinating but too long to be incorporated in this volume.)

Julius Robert Oppenheimer was born in New York City on April 22, 1904. His father was a wealthy German textile merchant and his mother an artist. He entered Harvard University in Cambridge, Massachusetts, at the age of seventeen to study chemistry but soon switched to physics. His studies were interrupted after a period of illness, but he graduated summa cum laude in just three years with an AB degree in 1925. Following Harvard he spent a year at the University of Göttingen where he studied with the German physicist Max Born, who was famous for his work in quantum mechanics. Together they published a well-received paper referred to as the "Born–Oppenheimer approximation" that described the separation of nuclear motion from electronic motion in molecules. Oppenheimer graduated form Göttingen with a PhD in 1927 at the age of twenty-two. He continued his studies at several other European universities before he returned to Harvard for a short time. In 1928 he studied at the California Institute of Technology and, at the same time, also accepted an assistant professorship in physics at the University of California, Berkeley. He spent the next thirteen years dividing his time between these two universities. He is credited with important research in areas of astrophysics, spectroscopy, quantum field theory, as well as nuclear physics. He wrote a paper that was the first to suggest the possibility of black holes in the universe.

During the late 1930s many German physicists who were involved in research regarding the process of nuclear fission and its potential for producing great quantities of energy were concerned that the Nazis might use this discovery to develop some form of bomb. Many migrated to the United States and soon became involved in various nuclear research projects around the country. Many people worked on the various aspects related to the successful construction of the first atomic bomb. These physicists believed that Nazi Germany would use the physics of fission to construct new types of extremely powerful bombs. They convinced one of their own colleagues, Leo Szilard, to contact Albert Einstein, who was then considered the outstanding physicist in the world, to write a letter to the U.S. President Franklin Roosevelt. The letter, which Einstein consented to write, implored President Roosevelt to create at the national level some means to study the feasibility of such a project. Roosevelt, to his credit, recognized the gravity of the plea, and soon after the Manhattan Project was established. The rest is history.

Many people worked on the various aspects related to the successful construction of the first atomic bombs. Oppenheimer and his right-hand man, Robert Serber (1909–1997) of the University of Illinois, determined how neutrons moved in a chain reaction and how much U-235 it would take to form a critical mass required for U-235 to fission and explode. After conferring with colleagues, Oppenheimer confirmed that a fission bomb was feasible and could work. He was also convinced that a single centralized laboratory was needed to manage the research and development required for this task. The various procedures for the tasks involved were carried out in three highly secretive sites: 1) Los Alamos National Laboratory in New Mexico was the main research "think tank" and was assigned the task of forming the bomb cores. It was also the final assembly area for the first bombs. 2) The new town of Oak Ridge was built on farmland in Tennessee. Oak Ridge National Laboratories were responsible for the gaseous diffusion and other processes that were used to separate U-235 from U-228, thus producing fissionable bomb material. 3) The Hanford site produced plutonium for a different type of fissionable bomb that used the element plutonium instead of uranium. This site is located in southeastern Washington State on the Columbia River. In addition, there were other research and development (R&D) sites, including the first atomic pile located at the University of Chicago that proved that fissionable material could be controlled, as well as used as an explosive. These joint efforts were called the Manhattan Project, which spread out over the entire United States (as well as some sites in Canada and England). The project employed over 130,000 people and cost what would be over $20 billion in today's dollars. Overall management of this huge effort was assigned to U.S Army General Leslie Groves (1897–1970) who was responsible for the final production of several atomic bombs. Groves had great respect for the knowledge and abilities of J. Robert Oppenheimer and wanted him as the director of all the research efforts involved in the project. However, due to Oppenheimer's youthful sympathizing with the socialist aspects of communism, as well as his marriage to a suspected communist sympathizer, he was placed under investigation in 1943 by both the FBI and the Manhattan Project's internal security branch, primarily because of his past left-wing associations. Despite these suspicions, General Groves considered him too valuable to lose, and Oppenheimer continued to work on the Manhattan Project through 1945, as well as acting as an advisor on other government-related research agencies after the end of World War II.

Because of his questionable political past and his opposition to Edward Teller's project to develop the hydrogen bomb, which was a nuclear fusion bomb many times more powerful than the atomic fission bomb, Oppenheimer's security clearance was revoked in 1953. He never again received government clearance, even though most of his colleagues and the general public considered him to be a great patriot. He returned to academic life but never forgot the sting and consequences of the government's accusations.

**OSTWALD'S THEORIES AND CONCEPT OF CHEMISTRY:** Chemistry: *Friedrich Wilhelm Ostwald* (1853–1932), Germany. Friedrich Ostwald received the 1909 Nobel Prize for Chemistry.

***Ostwald's theory of catalysts:*** *Nonreacting foreign substances can alter the rate of chemical reactions.*

Cognizant of the kinetics (movement of particles) of chemical reactions, their speeds, and equilibrium states, Ostwald theorized that certain substances (catalysts), when added to a chemical reaction, can either speed up or slow down the rate of that reaction; but at the same time, the catalyst will not alter the energy relationship within the reaction nor will the catalyst itself be changed. This concept became extremely important in the development of modern technology for controlling a great variety of chemical reactions, including the platinum beads used in the modern catalytic converter in automobile exhaust systems to reduce harmful exhaust gases to less toxic fumes.

***Ostwald's law of dilution:*** *The extent to which a dilute solution can become ionized can be measured with a high degree of accuracy.*

Ostwald's law of dilution is a means for determining the degree of **ionization** in a dilute solution with some degree of accuracy. A dilute solution is one with a small amount of solute (the substance dissolved) compared to the amount of solvent (the substance dissolving the solute). He patented his process, now known as the Otwald process, which is used worldwide to produce nitric acid by oxidizing ammonia.

# P

**PARACELSUS' CONCEPTS OF MEDICINE:** Chemistry: *Philippus Aureolus Theophrastus Bombastus von Hohenheim (Paracelsus)* (1493–1541), Germany.

*Iatro-chemistry is superior to herbal chemistry for treating diseases and illnesses.*

Early in his career, Philippus Aureolus Theophrastus Bombastus von Hohenheim changed his named to Paracelsus, meaning "equal to or greater than Celsus." Celsus was a first-century Roman physician whom Paracelsus greatly admired. The Roman Galen, the famous herbalist physician, subscribed to the **humoral theory** of disease; his teaching persisted for fifteen hundred years and was considered the authority in medicine during Paraclesus' time. But Paracelsus continually challenged Galen's doctrines as accepted by his contemporary physicians in the early sixteenth century. As a physician, Paraclesus was, in a sense, a compassionate patient advocate. For most of his life and for many reasons, he was an "outcast" in the medical community. One reason was that he usually did not agree with his peer physicians with regard to their accepted methods of practicing medicine. Their disagreements were also based on Paracelsus' concept of viewing the human body from a chemical point of view, not just as a spiritual vessel. Paracelsus' iatro-chemistry (the use of chemicals for medical treatments) was based on his doctrine called *Tria prima*, which was predicated on three basic types of matter: mercury was the spirit, sulfur the soul, and salt the body, with the inflammable sulfur combining the body and spirit into one unit. This, and more, was the basis of medicinal alchemy in the Middle Ages. Alchemy was the study and practice of combining a few basic elements to form the philosophers' stone, considered the key to transmuting base metals (e.g., lead) into gold or to produce the "elixir of life" that was sought as the cure for all illnesses.

Paracelsus was the first link between medieval medical practices and modern scientific medicine and pharmacology. He cured a few powerful men who became his sponsors, which enabled him to continue teaching and to use his unique methods of curing the ill. Paracelsus continued to maintain his belief in astrology, magic, and alchemy but used combinations of chemicals, such as mercury, iron, arsenic, sulfur, antimony, and laudanum (opium), rather than herbs. He was the first physician to try specific remedies for specific diseases and to connect heredity and lifestyle patterns to certain diseases and physical conditions, such as goiter, cretinism, and patterns of syphilis. He criticized his colleagues for their practice of "torturing" the ill by bleeding them, as well as using other inhumane treatments. He denounced local doctors publicly as a "misbegotten crew of approved asses" for their ways of practicing medicine. This attitude, along with the loss of a legal case in which he sued a wealthy citizen for nonpayment of his fee, was used to discredit him. The common people respected him, but after his sponsors died his many enemies in the medical community and universities who threatened his life finally drove him out of the country. For the remainder of his short life, he was a physician to miners in a small town, from whom he learned about metals and minerals, as well as the unique lung diseases endemic to this population. Paracelsus introduced pharmacology, antiseptics, modern surgical techniques, and microchemistry (homeopathy). His work with the miners also qualifies him as the first physician to develop the field of occupational and industrial medicine. He was known for tailoring the dosage of his medications to the amount required to cure his patients. He is famous for a basic biological and medical principle that states: *All things are poisonous and yet there is nothing that is poisonous. It is only the dose that makes a thing poisonous.* Paracelsus' tombstone is a broken marble pyramid located in the cemetery of the Hospital of St. Sebastian in Salzburg in Austria. His epitaph reads, "Here lies buried Philippus Theophrastus, distinguished Doctor of Medicine, who cured dire wounds, leprosy, gout, dropsy, and other contagious disease of the body, and with wonderful knowledge gave his goods to be divided and distributed to the poor."

**PARDEE'S THEORY FOR CELL ENZYME SYNTHESIS:** Biology: *Arthur Beck Pardee* (1921–), United States.

> *A mutant gene can induce dominance on a molecule to suppress production of the enzyme beta-galactosidase.*

Arthur Pardee and his staff crossed a mutant bacteria cell with normal bacteria, which then became capable of synthetically producing a metabolic enzyme crucial to the growth of living cells. Their process produced synthetic beta-galactosidase without requiring outside stimulation. This led to the production of purines and a better understanding of nucleic acids. Purines are double-ring nitrogenous organic molecules such as adenine (A), guanine (G), thymine (T), and cytosine (C), which form the base pairs of the nucleic acids of DNA and RNA. A new pathway was provided for newly formed proteins that can be stimulated

Arthur Pardee participated in a famous experiment, called the PaJaMo experiment that was first published in 1959 in the *Journal of Molecular Biology.* It was named after Pardee and two of his colleagues, François Jacob, and Jacques Monod, at the Pasteur Institute in Paris who performed a series of conjugations—the mating between "male" and "female" E. coli bacteria where the bacteria traded genes. This resulted in restored systems that began production of beta galactosidase almost immediately after the genetic information enabled the bacteria to produce protein.

by growth factors to duplicate their own DNA, and thus continue to divide. The human immune system T-cells, which fight not only infections but also cancer cells, do not normally reproduce, but with this new technique, it became possible to produce numerous T-cells. Pardee's research advanced the understanding of the immune system, enzymes, T-cells, and the HIV virus infection.

*See also* Crick; R. Franklin; J. Watson

**PARKES' THEORY FOR SEPARATING METALS FROM ORES:** Chemistry: *Alexander Parkes* (1813–1890), England.

*Specific metals can be separated from other metals, as well as from impurities found in common ores, by the use of chemical reactions.*

Alexander Parkes was trained as a chemist but became well known as a metallurgist and an inventor of plastics and held over sixty-six patents in his lifetime. He was aware that in addition to the metals that are found in common ores, there are many impurities as well. Thus each type of ore must be treated individually with different chemicals to extract the desired metals. One of the early systems used mercury to extract gold and silver, which are soluble, from their specific ores. These ancient methods were improved as the nature of elements, particularly metals, were applied to metallurgy. Today, the cyanide process is used to extract gold from its ore. The process is efficient but pollutes the environment with toxic cyanide chemical residues.

Parkes developed an extraction method now known as "the Parkes process" to remove silver from lead ores. It was known for years that some lead ores contained a significant amount of silver, but it could not be separated from the ore because it was not soluble by known processes. Parkes' process used molten zinc to "dissolve" the silver from lead ore. This was possible because molten silver is about three thousand times more soluble in molten zinc than lead. Because lead would not dissolve in molten zinc, the silver was "freed." Today, it is known that all metals are combined with other elements and that other chemicals are required to react with them to free the metals from the impurities in ores. The three process used are classified as follows: 1) *pyrometallurgy* where heat is used as in smelting and roasting the ores, 2) *electrometallurgy* where electricity is used in the process of electrolysis to separate aluminum, calcium, barium, magnesium, potassium, and sodium, and 3) *hydrometallurgy*, where various fluid solutions are used to leach and dissolve metals from their ores. Some examples are when copper oxide and copper carbonate compound-type ores are washed with dilute sulfuric acid followed by additional refining processes.

In addition to his patents in electroplating, Parkes received many other patents in the development of plastics. In 1856 he developed a new material he called "Parkesine" that was the first thermoplastic. It was something like celluloid because it was made from nitrocellulose mixed with camphor and ethanol. His company, that was set up to produce his new plastic as a synthetic ivory, failed because Parkesine was very flammable and even explosive. This is understandable because nitrocellulose is used as "gun cotton" and is related to nitroglycerine. His next venture with associate Daniel Spill (1832–1887), a businessman and chemist, was in the development of "Xylonite," an improved form of Parkesine. This patent was involved in a lawsuit with American inventor John Wesley Hyatt (1837–1920) who developed in 1869 what is known today as celluloid. Spill, who brought the lawsuit for patent infringement, was unsuccessful. However, in 1870 a patent judge ruled that Parkes was the first inventor of the process that produced the plastic (Parkesine) later known as celluloid. It was Hyatt's company, The Celluloid Manufacturing Company in Albany, New York, who would achieve commercial success with celluloid, rather than its original inventor, Alexander Parkes. Because the new celluloid was also flammable and had a tendency to crack, it has been replaced over the years by many improved forms of plastic. Today, celluloid is limited in its use, primarily in table tennis balls.

**PASCAL'S CONCEPTS, LAWS, AND THEOREMS:** Physics: *Blaise Pascal* (1623–1662), France.

***Pascal's concept of a barometer:*** *The height of a column of mercury decreases as altitude increases due to a decrease in air pressure.*

Blaise Pascal pursued the concept of atmospheric pressure proposed by Evangelista Torricelli's experiment that demonstrated that a 30-centimeter vertical column of mercury could be suspended in a closed tube. Torricelli was the first to theorize that the mercury was not suspended because a vacuum formed inside the closed tube, but rather by the weight of the air outside the tube (*see* Figure T2 under Torricelli). Pascal set out to prove that the height of this column of mercury was dependent on the weight of the air above the mercury in the dish that contained the column of mercury, and that the weight of air pushing down on Earth varied with altitude. He and his brother first measured the exact height of the suspended mercury in the column at the altitude of Paris, France. They proceeded to move the experiment to the top of a high mountain and demonstrated the column fell (was fewer than 30 cm) as the altitude became greater, proving air above a mountain is less dense and thus exerts less pressure on Earth than at sea level. Thus, he not only confirmed Torricelli's concept of air pressure but also discovered that this phenomenon could be used as a crude altimeter. The concept of the barometer can be used to predict weather conditions based on the fact that warm, stormy air is less dense and thus weighs less per square centimeter than cold, clear, denser air. The altimeter's basis is the fact that air becomes less dense as the altitude increases. Thus, the greater the altitude, the less air weighs per square centimeter. The barometer is vital for weather forecasting, and the altimeter is important for determining the altitude of airplanes (*see also* Galileo; Torricelli).

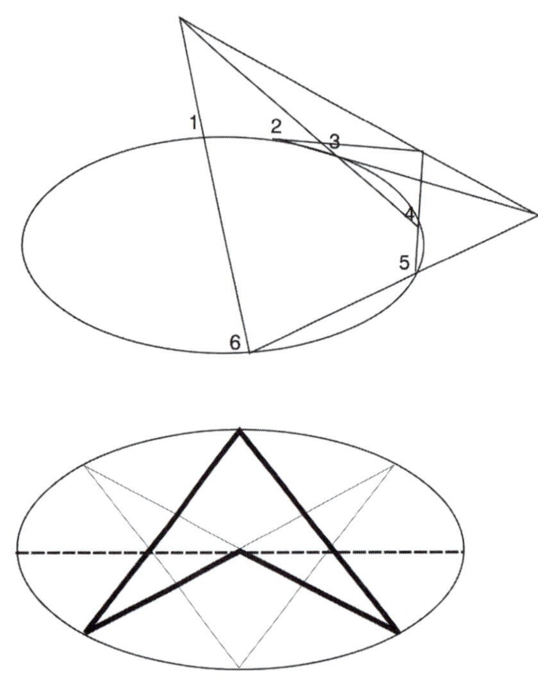


Figure P1. If the lines of a hexagon (six-sided figure) touch the circumference of a circle (conic figure), then the lines of this hexagon can be connected by the lines that meet at a central point inside the conic figure.

**Pascal's law of hydraulics:** *Pressure applied to a contained fluid is transmitted throughout the fluid in all directions regardless of the area to which the pressure is applied or the shape of the container.*

Blaise Pascal based this law on his work with atmospheric pressure and the demonstration, performed by the Flemish mathematician and engineer, Simon Stevin, that the pressure on a fluid of a given surface depends on the height of the fluid above the surface but not on the surface area of the fluid or the shape of the fluid's container (today we know this as the science of hydrostatics). For example, if you fill two 20-foot vertical pipes with water, one of which is just 2 inches in diameter while the other is 12 inches in diameter, and place equal valve taps at the base of each pipe, the pressure of the water coming out of each tap will be equal because the height of the columns of water coming out of each tap is equal, at least at the start of flow from each tap (but their original volumes or surface areas were not originally equal). Pascal's law of hydraulics basically states that fluids transmit pressure equally in all directions. For instance, the pressure exerted on a confined liquid with a plunger having a small 2-cm cross section will exert a greater pressure on a surface area having a much larger cross section. If the second surface's area is twice as great as the first plunger's surface area, so is the pressure exerted on the second area by the plunger. The original pressure is multiplied by the differences in the areas of the cross sections of the two surfaces. He used this law to develop the hydraulic press, which converts a small force into a much larger force. When a person steps on the brake pedal to stop a car, the hydraulic system converts the small force of the foot on the pedal into the much larger force on the brake discs located on each wheel, stopping the car. This force is transmitted through the hydraulic fluid and tubing system to the brake's mechanism on the car's wheels to create adequate friction of the wheel's brake pads or discs, which then stops the car. In this case, a special oil-like fluid is used instead of water because water would freeze and prevent the hydraulic system from working (*see also* Archimedes).

**Pascal's theorem:** *If a hexagon is inscribed in any conic section, the points related to where opposite sides meet will be collinear.*

At the age of sixteen Blaise Pascal had already formulated several theorems of projective geometry that later became known as Pascal's theorem. Viewing a diagram of his theorem is

> Blaise Pacal was the skinny son of a famous French mathematician. As an infant and young boy, Blaise was considered a mathematical prodigy. At age eleven he discovered Euclid's twenty-three theorems with no help from others, and at age seventeen he published an essay on René Descartes. He also developed what some consider was the first digital calculator to help his father who was an administrator in the local town. He later joined the Port Royal Society, a religious community that was controversial and considered by some to be heretical to the teachings of the Roman Catholic Church. In 1656 a Papal Bull was issued, almost excommunicating Blaise and all the others mentioned in the document. Along with his work on air pressure and hydraulics, he was one of the first to use the concept of probability in calculating events. The concept of probability was not proposed as a mathematical theory until the late 1700s when Marquis Pierre Simon de Laplace arrived at the mathematics involved when calculating win–loss odds in games of chance. For example: If a gambler rolls a pair of die twenty-eight times, and 7 comes up each time, the probability that he will roll a 7 the twenty-ninth time is just as great as is the chance that the next person's first roll will result in a 7. Many gamblers do not completely believe this and consider chance more a stroke of luck. In other words, every roll of the dice has the same chance (probability) of arriving at a specific number as every other roll (assuming the die have not been altered in shape or some other way).

easier to understand than a written statement (*see* Figure P1 for two aspects of Pascal's hexagram theorem).

In essence, the theorem states if a hexagon (six-sided figure) is circumscribed onto a conic figure (touches the outside of the circle or oval), then the vertices of this hexagon (lines connecting the opposite points where the sides of the hexagon meet) can be connected by three lines that meet at a central point inside the conic figure.

Pascal in cooperation with the French mathematician Pierre de Fermat formulated the theory of probability that he related to games of chance, but today it is a very important concept in science and mathematics. He later arrived at several statistical procedures for use in mathematics and physics. He also invented the first mechanical adding machine and strongly believed that scientific discoveries result from empirical observations.

*See also* Laplace

**PASTEUR'S GERM AND VACCINATION THEORIES:** Biology: *Louis Pasteur* (1822–1895), France.

***Pasteur's germ theory of fermentation:*** *Fermentation can be inhibited in fermentable (organic) substances by preventing them from exposure to airborne dust particles.*

Louis Pasteur speculated that only living organisms could differentiate between the various shapes of molecules (*molecular dissymmetry*). He concluded that a definite distinction existed between nonliving (inorganic) and living (organic) chemistry. He then studied fermentation and the effects of yeasts on living substances, which he claimed was a chemical reaction involving microorganisms. This prefaced his famous experiment that disproved the ancient concept of the spontaneous generation of life. About two hundred years earlier, Francesco Redi conducted a similar experiment to disprove spontaneous generation, but without the knowledge of the germ theory. Pasteur placed a cooked broth into a sterilized flask structured with a closed curved neck, which prevented air and dust from entering the flask. The broth did not ferment nor develop any bacterial growth. Then he broke off the curved neck, allowing air and dust to enter, which soon caused the broth to ferment. His interest in fermentation was inspired by France's wine and brewery industries, which were unable to control the quality of their wine and beer. Pasteur was asked by these industries to study the problem, which he did, using the results of his famous sterilized flask experiment. He realized "germs" from the air contaminated the wine, causing it to ferment. He also found microorganisms in the lactic acid that result in the fermentation of milk. As a solution, he devised a process whereby the wine was heated but not boiled, and then cooled in closed containers to prevent airborne dust from contaminating the product. He also applied this process of *pasteurization* to milk, which was heated to about 108°F and then quickly cooled. Pasteur also knew that airborne microorganisms also cause some infections; thus he was one of the first to recommend the use of carbolic acid as an antiseptic, the boiling of surgical instruments, and the maintaining of proper hygiene in hospitals.

Louis Pasteur is also known as the father of stereochemistry for his use of the microscope in the discovery of two types of tartrate crystals that are the mirror image of each other. When molecules with the same formula have different arrangements of the same atoms, the molecules are known as **isomers**. Thus, crystals of the same chemical composition can have different geometries. Up to this time only a single type of geometric crystal had been observed. Pasteur then separated the two types of crystals and demonstrated that a special type of plant mold with only one type of molecular structure of the isomer of racemic acid was involved. While investigating why wine and beer eventually soured, he discovered yeasts with two different shapes—one of which formed sour wine and beer; the other formed a good product that did not go sour. This work led to his process known as "pasteurization" that is the process of applying heat to foods (e.g., milk) to kill harmful bacteria, but not hot enough to adversely affect the taste or quality of the food.

***Pasteur's vaccination theory:*** *Attenuated microorganisms from animals with anthrax can be used to inoculate healthy animals, which will prevent the disease.*

This aspect of Louis Pasteur's work is an excellent example of serendipity—an occasion when an unexpected development or insight presents itself to a knowledgeable observer. It seems that after identifying a batch of chicken cholera bacilli under his microscope, he neglected it for several weeks during a hot summer (chicken cholera is similar to but not the same as the waterborne cholera contracted by humans). Even though this was not a fresh batch of his bacteria, he injected it into healthy chickens. These chickens contracted only a mild case of cholera. He then proceeded to infect both the vaccinated group of chickens and a control group (nonvaccinated) of chickens with fresh cholera bacilli. The chickens that received the attenuated (weakened) bacteria contracted only mild cases of the cholera, while the noninoculated chickens died of the disease. Thus, Pasteur discovered how to produce vaccines that could immunize animals against diseases. He then proceeded to heat gently, to about 75°F, the disease-causing bacilli that made sheep, cattle, horses, as well as humans, ill with anthrax. In one experiment he injected twenty-four sheep with the virulent form of anthrax as well as an additional twenty-four sheep that had been immunized with his attenuated bacilli. All twenty-four nonimmunized sheep died; all of the immunized sheep lived. His most famous accomplishment was his treatment of a young boy who had contracted rabies from the bite of a rabid dog. Pasteur developed the antirabies attenuated virus to inoculate the boy who, after several treatments, survived. The concept of vaccination was improved and expanded by others to include inoculations for diphtheria, typhoid, cholera, plague, poliomyelitis, smallpox, measles, and other diseases of humans and animals. Louis Pasteur is known as the father of microbiology.

*See also* Koch; Lister; Redi

**PAULING'S THEORY OF CHEMICAL BONDING:** Chemistry: *Linus Carl Pauling* (1901–1994), United States. Linus Pauling was one of the few people to be awarded two Nobel Prizes, the 1954 Nobel Prize for Chemistry and the 1962 Nobel Peace Prize.

> *The nature of chemical bonding of elements and molecular structure can be determined by the application of quantum mechanics.*

Linus Pauling's career covered not only his early work in determining the nature of chemical bonding of atoms, the complex structure of organic molecules and crystals, the nature of oxygen binding to hemoglobin (sickle cell anemia), and vitamin C therapy, but also for activities to halt nuclear testing, for which he received the 1962 Nobel Peace Prize. Pauling used many techniques to study the structure and bonding properties of atoms, including electron and X-ray diffraction of large protein molecules and electromagnetic instruments to assist in determining molecular structure. His pioneering approach was the use of quantum mechanics to describe how electrons are arranged in orbits, how they bond, at what angles they bond, their bonding energies, and the distances between electrons in different atoms that are combined (See Figure S2 under Sidgwick). This was important because, up to this time, quantum mechanics was usually limited to explanations of phenomena at the larger atomic level rather than the subatomic and energy levels. Pauling found that some elements and compounds do not follow the classical valence bonding of single electrons but rather exist in two or more forms through the process of resonance. He made two unique observations. One was the concept of hybrid molecules, which accounted for various shapes of molecules. The other was the concept of resonance for molecules, which explained how some molecules could appear to be somewhat like similar molecules yet have slightly different structures and thus different characteristics. This is how he conceived the idea that oxygen molecules do not bond in a normal way with some types of hemoglobin cells that are "sickle" shaped. The American physician James Bryan Herrick (1861–1954) first discovered sickle cell disease in 1904; Pauling identified its cause as a genetic malformation of blood cells. People who inherit this *molecular disease* often die at an early age because it affects their blood's capacity to carry oxygen to tissue cells. In 1950 Pauling described the structure of the complex protein molecules involved in the chromosomes of cells as an alpha-helix structure, which almost described the double-helix structure of DNA announced in 1954 by James Watson and Francis Crick. Pauling was the first to use quantum mechanics to explain how atoms bond to form molecules, including the concept of *electronegativity*. (Note: Electronegativity is the power of an atom in a molecule to attract electrons to itself, which is dependent on its valence state.) Pauling's concepts that describe the types of atomic bonding are depicted as complex three-dimensional figures.

*See also* Bohr; Crick; Ingram; Pauli; Watson (James)

**PAULI'S EXCLUSION PRINCIPLE:** Physics: *Wolfgang Pauli* (1900–1958), Switzerland. Wolfgang Pauli was awarded the 1945 Nobel Prize for Physics.

> *Only two electrons of opposite spin can occupy the same energy level (same quantum number) simultaneously in an atom's orbit.*

Wolfgang Pauli was aware of Niels Bohr's application of quantum theory to the electrons orbiting the nuclei of atoms. At the time, Niels Bohr based his "planetary" structure of electrons orbiting the nuclei on the new concept of quantum mechanics. In cooperation with the German theoretical physicist Arnold Sommerfield (1868–1951), Bohr expanded his "solar system" atom model to include the energy associated with the electrons in different orbits. The quantum aspect stated that each orbiting electron could have only one of three quantum numbers, referred to as $n$, $l$, and $m$. Pauli proposed the concept that required each electron to be one member of a pair of electrons, one of which "spun" around its axis in one direction, while the other spun in the opposite direction. For this situation to exist, he introduced a fourth quantum number for electrons, $s$, which has the value of $+1/2$ or $-1/2$. The numbers correspond to the spin of each member of the pair. This is the point at which his famous 1924 *exclusionary principle* entered the picture (also referred to as the *Pauli principle of exclusion*). The principle states that no two electrons in an atom may have the same four quantum numbers—*n*, *l*, *m*, or *s*. The spin and exclusion principle, which explained many mysteries related to the structure of the atom, was confirmed a few years later. In other words, if an electron is at a specific energy level in a particular orbit, no other electron can be at that same exact energy level; thus, other electrons are "excluded." In 1930, Pauli identified a particle in the atom's nucleus that he believed was another type of neutron that had no charge and was emitted along with an electron during beta decay of a neutron. The problem was that the new particle was too light to be a neutron. Some years later Sir James Chadwick discovered the actual neutron that exists in the nucleus. Pauli's "neutron" was later verified and renamed by Enrico Fermi as the *neutrino*.

*See also* Bohr; Chadwick; Fermi; Steinberger

**PAVLOV'S THEORY OF ASSOCIATIVE LEARNING BY RESPONDENT CONDITIONING:** Physiology: *Ivan Petrovich Pavlov* (1849–1936), Russia. Ivan Pavlov was awarded the 1904 Nobel Prize for Physiology or Medicine.

*An unconditioned response is an automatic response resulting from an unconditioned stimulus.*

After graduating from St. Petersburg University, Russia, in 1875 with a degree in natural science, Ivan Pavlov attended the Military Medical Academy (also in St. Petersburg). He received his medical degree in 1879 and another postdoctoral degree in physiology in 1883. Pavlov was interested in mammalian physiology and digestion and continued this work in Germany. However, he returned to St. Petersburg in 1890 where he built a physiology research center. Pavlov focused on the physiology of digestion and glandular secretions. He developed surgical skills that involved the separation of a small part of a dog's stomach from the rest of the stomach. This became known as a "Pavlov pouch" from which he could collect samples of the dog's gastric juices. During his research, he observed that the dogs salivated when they became stimulated during feeding—even before the food arrived. This observation led to his lifelong work in the field of the body's reflex systems. His first research involved the study of involuntary reactions to stress and pain that often led to the body's natural response to "shut down" during excessive stress or pain.

Ivan Pavlov's experiments with dogs included a variety of stimuli to the dogs in addition to the well-known ringing of a bell during their feeding. He used a variety of methods of stimulation, including auditory sounds such as whistles, metronomes, tuning forks, and bells. He also used various visual stimuli. The dogs were soon conditioned to salivate at sounds they connected to the sight of their food. Pavlov's research and theories have had a huge influence on psychological theories related to conditioning of human and animal behavior, as well as learning and neurosis. Today, his followers believe that conditioned reflexes are responsible for most of human behavior and psychological disorders. In the 1940s B.F. Skinner (1904–1990), a behavioral psychologist, proposed that human behavior is a set of responses to a person's environment, and through therapy involving behavior modification and training that includes rewards, one's behavior could be "reconditioned." Unfortunately, the theory has also been used as the more controversial "aversion-therapy" where pain is induced by electrical or other stimulation to alter what is perceived as undesirable behavior.

During this research on involuntary response to external stimuli,

PEANO'S CURVE

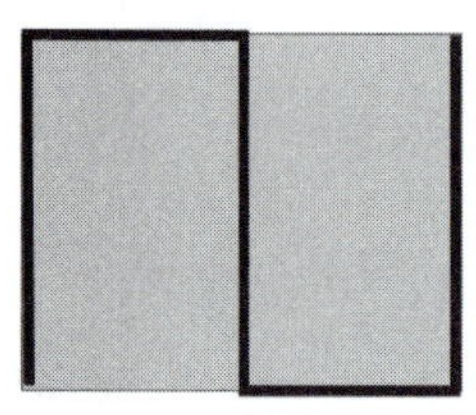

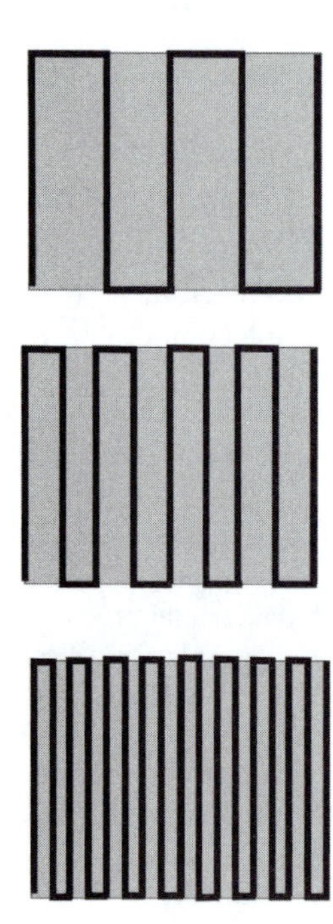

Figure P2. Peano's curve is related to Hilbert's curve and Koch's curve. All versions of continuous curves are based on the concept that many different types of spaces can be completely filled by continuous curves (of many kinds) if space is contained, the curves are connected, and the curve (lines) does not cross itself.

Pavlov performed experiments that showed that the secretion of digestive juices does not require the presence of food in the dog's stomach. The experiment involved allowing the dogs to see, smell, and actually swallow their food, but he prevented the food from going to their stomachs by surgically removing the food from their throats. He then noted that gastric juices were secreted in the stomach just as if the food had entered the stomach. His work on the nerves and gastric glands involved won him the 1904 Nobel Prize in Physiology or Medicine. His papers reporting his research in the early 1900s led to the concept and terms "conditioned reflex" and "Pavlov's dog."

**PEANO'S AXIOMS AND CURVE THEOREM:** Mathematics: *Giuseppe Peano* (1858–1932), Italy.

***Peano's axioms:*** *The use of symbolic logic and the axiomatic method provide rigor to the theorems of mathematics.*

Mathematical logic is the use of symbols instead of words to explain mathematical statements. This form of logic is sometimes referred to as a *universal language* because universal symbols make working with equations easy for anyone to understand, no matter what language is spoken. Giuseppe Peano conceived nine different logical mathematical axioms. Five of his most famous axioms involved the logic of developing numbers:

1. The figure 1 is a number.
2. Any number that follows 1 in sequence is also a number.
3. No two numbers can have the same successor number.
4. One (1) cannot be a successor to any other number.
5. If 1 has a property and any successor also has that property, so do all numbers.

These axioms are in the form of one of Peano's **syllogisms**, which describe numbers in terms of a set of elements. They are a logical series of statements used to define what a "number" is and is not. Peano claimed that the natural (real) number system can be derived from his axioms. Another of Peano's axioms defined natural numbers in terms of (sets.)

***Peano's curve theorem:*** *There are continuous curves that completely fill all the space inside squares or cubes.*

Peano's continuous curve theorem is based on his idea of the cluster point of a function, which is the basic element for geometric calculus. He defined a curved surface as the length of an arc as compared to the area within a curved surface. He named this geometric function a *space-filled curve*. Peano's curve theorem was revised to include a great variety of types of space that may be completely filled up by a continuously drawn curve, if and only if that space is connected, compact, has a continuous border, and is measurable. For the theorem to apply to enclosed circles and squares, the "curve" or continuous line must not cross itself. For a square, the space-filling curve must have at least three points of multiplicity (*see* Figure P2).

Peano's curve theory introduced the basic elements of geometric calculus and provided a new definition for the length of an arc. It can be used to measure space (areas) within a curve (*see also* Leibniz).

***Peano's theory of differential equations:*** *If f (x,y) is continuous, then the first order differential equation dy/dx = f(x,y) has a solution.*

The solution for the *f* above was first proffered by others. However, just a few years later, Peano showed that it was not unique. As an example, he stated that the differential equation: $dy/dx = 3y^{2/3}$, with $y(0) = 0$.

Peano's interest in mathematics led him to consider the relationships between mathematics and language. He was a pioneer in the field of auxiliary artificial languages, and he created the language called "interlingua" in the early 1900s. He made up his own vocabulary by using words from other languages, including English, French, German, Latin, as well as others. Others carried out further research in this area, including the development of what is known today as the universal language Esperanto.

Two of Peano's greatest achievements are 1) he devised a very efficient notation system for mathematical logic which was clear and easy to understand as well as use and 2) he indicated how simple arithmetic is derived from using a simple logical basis (*see* the above five axioms). Peano is known as the father of mathematical logic.

Esperanto is a constructed international language that was created in 1887 by an eye doctor, Dr. Ludovic Lazarus Zamenhof (1859–1914), a Polish Jew who lived in what was Germany at the time and is now Poland. Dr. Zamenhof described Esperanto's structure and morphology as based on Romanic and Indo-European languages with contributions from Germanic, Polish, French, and some English words. Although it is not the official language of any country, it is offered as an elective subject in some schools and is learned by many who find it an interesting alternative language. Today it is estimated to have somewhere between one and two million fluent speakers. After the 1911 revolution in China, Esperanto was considered as a replacement for Chinese languages as a possible means to help that country enter the twentieth century. In 1924 the American Radio Relay League used Esperanto as its official language in the hope that it might be adopted by some nations and lead to world peace. Several nonprofit international organizations used it as their official language. Today the largest group to use the language is the World Esperanto Association connected with the United Nations UNESCO organization. Speakers of Esperanto are found mostly in northern and eastern Europe, urban areas of China, as well as regions of Korea, Japan, and Iran. It is spoken also in South America in Brazil, Argentina, and parts of Mexico, as well as on the island of Madagascar in Africa. Its structure is basically pragmatic with new words "invented" by the use of many prefixes and suffixes. The core vocabulary contains about nine hundred roots, which can be expanded by use of the many prefixes and suffixes. It uses just five vowels and twenty-three consonants. It is a relatively easy language to learn and use, and many people and organizations who support its use see it as a possible future world language that will help unite differing nations with a common language to assist them in solving their differences in a more peaceful manner.

**PEARSON'S STATISTICAL THEORIES:** Mathematics: *Karl Pearson* (1857–1936), England.

*It is possible to measure statistically the continuous variations responsible for natural selection.*

Karl Pearson developed several important statistical methods for treating data related to evolution. The concept of natural selection as a series of continuous variations was, according to Pearson, a problem dealing with biometrics (using information from living systems to develop synthetic systems). His statistical approach was in direct conflict with that of other scientists, who believed that evolution (natural selection) was a process of discontinuous variations based on breeding as the main mechanism. Among the important statistical concepts that grew out of Pearson's work with evolution are:

- *regression analysis,* used to describe the relationship between two or more variables. In controlled experiments, it is used to estimate the value of the dependent variable (the experimental factor) on the basis of independent variables (controlled factors).
- *correlation coefficient,* the statistical measurement between two variables that is quantitative or qualitative in nature. The measurements are unchanged by the addition or multiplication of random variables. This statistical method expresses measurable (and probable) differences between two events.
- *chi-square test of statistical significance,* which is a means for determining the "goodness of fit" between two binomial populations (two different but related groups), and where each population has a normal distribution. It is a test for determining how far the event is from the statistical mean or how factors for two different events "match."
- *standard deviation* (a term originated by Pearson), the statistical treatment used to determine the difference between a random variable and its mean. It is expressed as the square root of the expected value of the square of this difference.

Many of Pearson's concepts deal with averages. "Mean" is the term we usually associate with "average." It is determined by adding all the values or a set of numbers and dividing the sum by the total number of values—for example, $2 + 4 + 9 + 3 = 18 \div 4 = 4.5$ as the mean (average).

"Median" is the central point in a series (set) of numbers when arranged in order of numbers or value. For the median, there is an equal number of greater (larger) and lesser (smaller numbers above and below the median value). For example, in the sequence 1, 2, 3, 4, 5, 7, 9, 4 is the number halfway between those numbers above and below it, and thus the median. The median may equal the mean, or it may not.

The "mode" is the number most frequently occurring in a set of numbers or values. In the sequence, 1, 1, 2, 2, 5, 5, 5, 6, 8, 9, 12, 16, 25, the mode is 5; it occurs most often in the set. Using the mode instead of the mean can influence the meaning of the data. For instance, if the salaries of one hundred workers and five executives in a company are listed and the mode is used to express a "pseudo-average" salary for company employees, the figure will certainly be lower than if the mean (average) salary is used for the calculation.

*See also* Darwin; Galton; Mendel

**PEIERLS' CONCEPT FOR SEPARATING U-235 FROM U-238:** Physics: *Sir Rudolph Ernst Peierls* (1907–1995), England.

*Rare uranium-235, which is an isotope of uranium-238, can be separated from uranium-238 by the process of gaseous diffusion.*

Rudolph Peierls was familiar with Henri Becquerel's accidental discovery of radioactivity, as well as the Curies' work in separating radium from uranium ore, and Otto Frisch and Lise Meitner's theory of fission of unstable nuclei of uranium. In 1940 Peierls and Frisch collaborated on forming the rare isotope of uranium 235 (U-235) into a small mass that would spontaneously fission, resulting in the production of tremendous energy that might be used to construct a giant bomb. In 1913, Frederick Soddy theorized that when a radioactive element gives off alpha particles (helium nuclei), it changes from one element into a different element with a different atomic number. With the loss of a beta particle (electron), there is also a loss of a negative charge. Thus, the nucleus gains one positive charge, which also makes it a different element. Soddy realized that if neutrons were added to or removed from the nuclei of atoms, the charge would not change (the atomic number would not change), thus the element would still fit its same place in the **Periodic Table of the Chemical Elements**, although it would have a different atomic weight. Soddy called these variations of elements by weight *isotopes*.

Based on the information that uranium contained at least two different isotopes (common U-238 has 92 protons and 146 neutrons, while U-235 consists of 92 protons and 143 neutrons), Peierls and Frisch attempted to separate the two isotopes of uranium. This proved to be difficult because only 1 out of 140 atoms of uranium ore (U-238) is the U-235 isotope. Even so, they made some calculations for the energy output that could result from the fission of U-235 to sustain a chain reaction. For example, once a few atoms started to fission, they would, in very rapid geometric progression, cause all the other nuclei also to fission almost instantly, thus causing a huge explosion. The next problem was how to separate the fissionable U-235 from the stable U-238. This was accomplished by a massive effort using gaseous diffusion. This process converted purified uranium ore into a gas that allows the heavier U-238 to be separated through diffusion filters from the lighter U-235. The first atomic bombs were produced by this method based on the slight difference of atomic weights of U-238 and U-235.

A much simpler and less expensive method of producing a fissionable material was to force uranium-238 nuclei to absorb neutrons inside a nuclear reactor. This nuclear reactor formed neptunium-239 (93 protons and 146 neutrons), which decays into plutonium-239 (94 protons and 145 neutrons). Although plutonium is stable, if it is forced to absorb slow neutrons, it gains atomic weight and thus is radioactive, unstable, and is fissionable. This procedure has become the basis for producing modern nuclear (atomic) weapons.

*See also* Becquerel; the Curies; Frisch; Meitner; Soddy

## PENROSE'S THEORIES FOR BLACK HOLES, "TWISTORS," AND "TILING": Physics: *Roger Penrose* (1931–), England.

***Penrose's hypothesis:*** *Black holes are singularities with "event horizons."*

Roger Penrose, along with Stephen Hawking, applied Einstein's theory of general relativity to prove that black holes are *space-time singularities* (a single event in space-time within a trapped surface). They proposed this concept even though such phenomena have no volume, are infinitely dense, and evolve as space-time events. Penrose's hypothesis is based on the fact that the singularities are not "naked" but rather have an "event horizon," which is the outer limit boundary where all mass, including light once it enters this border area, will be sucked into the black hole by tremendous gravity. The outer rims of black holes (accretion disks) emit visible light but much less than do neutron stars. Penrose stated that things do happen in a black hole. Only a massive, dense object in deep space with an event horizon creates the energy adequate to cause energy (light) to disappear. Particles break down into new particles, one of which would be trapped in the black hole while the other, containing more mass and energy than the original particle, might escape out the bottom. Such an arrangement of matter being changed and compressed on entering one hole and then exiting might be responsible for the birth of new galaxies or universes. If this proves to be correct, the physical concept of conservation of energy will be preserved. Orbiting spacecrafts that carry X-ray instruments continue to explore black holes to learn more about them.

***Penrose's twistor theory:*** *There are massless objects in space existing in "twistor space."*

Penrose's "twistor" theory is a new, complex geometric construct that he developed to explain a synthesis of quantum theory and relativity in space-time. Penrose refers to his theory as "twistor space," where particles have no mass at rest but do exhibit properties of linear and angular momentum (movement) when the particles change position from their point of origin. The particle in space-time may have spin as well as a vector of four dimensions. He developed his theory to replace the Einsteinian theory of relativity and a four-dimensional space-time construct. Twistor theory is a complicated and a not-well-understood or accepted area of research in mathematics and theoretical physics. Penrose's concept of twistors was a result of his investigations into how space-time might be structured. He proposed a wide range of applications for this particular form of mathematics. Even though it is an elegant mathematical formulation, it still remains a mystery and is now considered an "unfashionable" area of mathematical research.

***Penrose's tile theory:*** *"Tiles" can intersect at their boundaries while never overlapping.*

Tilings, also known, as *tessellations*, occur when geometric figures continue to repeat themselves. In other words, tessellations happen when an arbitrarily large plane surface is arranged with nonoverlapping tiles (i.e., the tiles connect only at their boundaries). The tile figures can be constructed by triangles, squares, and hexagons (with three, four, and six possible symmetries) (*see* Figure P3).

A "tiling" figure can never have five sides, such as a pentagon, because it cannot be "folded" so that the edges meet perfectly (i.e., it is not symmetrical). Penrose's tiling also explains the structure of crystals that can have three-, four-, or six-folded rotation symmetry, but not a five-fold rotational symmetry. In 1984 this belief about a five-fold crystal structure being impossible seemed to be disproved when a crystal of an alloy composed of aluminum and manganese was rapidly cooled to form such a crystal. The possible distinction could be that the crystal is three-dimensional while the three, four, and six symmetries were two-dimensional. Tiling may be thought of as a periodic pattern that carries the design into itself. **Fractals** are similar to tiling in that they are self-contained repeating patterns of decreasing size (See Figure W1 under Wolfram). The generating of Penrose's tilings and fractals on personal computers has become a popular exercise for creating geometric designs.

*See also* Wolfram

Figure P3. Penrose tiles use the same graphic figure repeatedly, but not overlapping, to cover an arbitrarily large two-dimensional area with a continuous design. The figure may be constructed by using triangles, squares, and hexagons, but may never have 5-sided pentagons. Tessellation is the placing of congruent polygons in a plane.

**PENZIAS' THEORY FOR THE BIG BANG:** Astronomy: *Arno Allan Penzias* (1933–), United States. Arno Penzias shared the 1978 Nobel Prize for Physics with Pyotr L. Kapitsa and Robert W. Wilson.

*Background radiation received on Earth from all directions in space is the leftover microwave radiation following the big bang.*

In the early 1960s Penzias and his colleague at Bell Laboratories, Robert W. Wilson, were working with a special radio antenna they designed to detect signals from communication satellites, when they received some unexplained background noise. They continually picked up all kinds of signals—some of this "noise" was generated by the internal electronics of their instruments, some from Earth sources, and some were unexplained. By 1964 they had eliminated much of the other noises but still continued to receive signals near the wavelength of $10^{-3}$ meters, indicating that the signal was at about 3.6 kelvin. This type of signal proved to be on a timescale much older than Earth itself. Therefore, they concluded it was leftover microwave emission from the big bang origin of the universe about thirteen to fifteen billion years ago. This theory concerning cosmic background microwave radiation is considered an important breakthrough in modern astrophysics.

*See also* Gamow; Hale; Hawking

**PERL'S THEORY FOR A NEW LEPTON:** Physics: *Martin Lewis Perl* (1927–), United States. Martin Perl shared the 1995 Nobel Prize for Physics with Fredrick Reines.

> *A new lepton, the tau particle, generated by a particle accelerator will decay into an electron plus a neutrino/antineutrino, or a muon.*

There are two classes of subatomic particles: 1) the *hadrons*, which include the proton and neutron (protons and neutrons along with their "binding" quarks are also considered **fermions**), which have half-integer spin and are quantum individualistic in the sense that they obey the Pauli exclusion principle, and 2) *leptons*, which include four types of particles: the *electron* and the *muon*, along with their two related *neutrinos*. Perl proceeded to use a powerful accelerator to generate a new lepton. It is necessary to use some form of detection device to record the existence of the subatomic particles generated in a particle accelerator before they decay. He generated a record of over ten thousand events, of which only a few of the new particles were detected as he predicted. This new particle turned out to be a new, heavy lepton with more mass than a proton. This caused some problems because the four known leptons are strongly related by symmetry to their four quarks. Therefore, to maintain symmetry, one of the basic constants of physics, a new quark was predicted to match the new lepton. In 1977 this new quark was discovered and named the upsilon particle by Leon Lederman. Thus, both a new lepton and a new quark joined the growing multitude of subatomic particles.

*See also* Lederman

**PERRIN'S THEORY OF MOLECULAR MOTION:** Physics: *Jean Baptiste Perrin* (1870–1942), France. Jean Perrin was awarded the 1926 Nobel Prize for Physics.

> *Einstein's formula for molecular motion can be confirmed by determining the size of molecules.*

Jean Perrin was familiar with Scottish botanist Robert Brown's concept that the motion of molecules of water caused the motion of tiny pollen grains suspended in water (Brownian motion) and Einstein's theory that the average distance the pollen particle traveled in the water increased with the square of the time elapsed for the motion. After controlling for conditions of temperature and the type of liquid in which the pollen was suspended, Einstein was able to predict, on the average, how far a particle would travel. However, at the time Einstein made this prediction, there was no way to confirm it. Perrin, who was known for his theories concerning the discontinuous structure of matter and how sediments obtain equilibrium, related these ideas to the importance of knowing the size and energy of molecules. In 1908 Perrin, who had access to a "super" microscope, experimentally controlled and measured the size of molecules as related to molecular movement, thus was able to confirm Einstein's theory of how far a particle should travel in a specific time period. This formula for the size of water molecules was important for the confirmation of the kinetic theory of matter (motion) and Avogadro's number.

Earlier in 1895 Perrin worked on the development of the cathode-ray and X-ray tubes. He was the first to determine that cathode rays must have a negative charge because they were deflected by a magnetic field. He was working on how to determine the ratio of the charge of an electron to its mass, but was superseded in this accomplishment by J. J. Thomson. Perrin is also remembered for his 1913 publication of *Les Atomes*, a comprehensive work that brought together what was then known about atoms and molecules, as well as knowledge about the chemistry of radioactivity, "blackbody" radiation, and other information about atoms and their behavior. This book established the field of atomism.

*See also* Avogadro; Einstein

**PERUTZ'S THEORY OF MOLECULAR STRUCTURE OF HEMOGLOBIN:** Biology (Molecular Biology): *Max Ferdinard Perutz* (1914–2002), Austria and England. Max Perutz shared the 1962 Nobel Prize in Chemistry with the British biochemist and crystallographer John Kendrew.

> *The hemoglobin molecule is composed of four separate globin (polypeptide chains) molecules (two alpha and two beta), while the alpha globin consists of 141 amino acids and the beta globin has 146 amino acids that exist near the molecule's surface.*

Max Perutz received his undergraduate education at the University of Vienna. Later in 1940 he received his PhD from Cambridge University in England. He and his first graduate student John Kendrew (1917–1997), the renowned British biochemist and crystallographer, founded a medical research group to continue their research to determine the molecular structure of protein hemoglobin. They used the powerful tool of X-ray diffraction to study this unique hemoglobin molecule. The only limitation of their early procedure was that they were unable to "see" the molecules in three dimensions, thus their model was limited because the actual molecule has over twelve-thousand atoms. To overcome this limitation, they used atoms from several heavy metals (gold and mercury) to replace some of the hemoglobin's atoms, making the X-ray diffraction pattern clearer as to the positions of the atoms in the hemoglobin's molecules. In addition to determining that the structure of the molecule was composed of a tetrahedron consisting of four chains, they determined that four heme groups of atoms existed just beneath the surface of the molecule. (Note: A heme is a prosthetic group consisting of iron—protoporphyrin complex that is associated with each polypeptide unit of hemoglobin.) Perutz also discovered that by exposing hemoglobin to oxygen, it was possible to rearrange the four chains. In time, this discovery was used to explain the mechanism in the hemoglobin molecule that transports oxygen in the blood throughout the body.

Perutz's laboratory was combined with another group at Cambridge to create in 1947 the Medical Research Council Unit for Molecular Biology (MRC), which under his chairmanship grew to over four hundred people. MRC's research led to numerous discoveries and inventions in the area of molecular biology. Max Perutz wrote many books promoting science, including *Is Science Necessary* and *Science is Not a Quiet Life: Unraveling the Atomic Mechanism of Hemoglobin* among others. He is considered one of the founders of the field of molecular biology.

**PFEIFFER'S PHENOMENON: THE THEORY OF BACTERIOLYSIS:** Biology (Bacteriology): *Richard Friedrich Johannes Pfeiffer* (1858–1945), Germany.

> *Disease-causing bacteria can be destroyed (lysis) by heating them to just above 60°C, which causes them to swell up and burst.*

Richard Pfeiffer was a military surgeon in the German army in the late 1880s. After his discharge, he joined the German biologist and germ-disease theorist Robert Koch at the Institute of Hygiene, which was followed by a professorship at the University of Konigsberg, Germany, then onto Breslau where he did most of his research in bacteriology.

His first discovery, made in 1892, was the bacteria known as *Haemophilus influenzae* (also known as *Pfeiffer's bacillus*). Because the bacillus was found in the throats of flu patients, he at first thought that it also caused the disease. Even in the world's deadliest influenza pandemic that began in 1918 scientists mistakenly believed that Pfeiffer's bacillus caused the disease. This pandemic, commonly called the Spanish flu, killed as many as one hundred million people. It was later determined that the Pfeiffer bacillus was not found in all influenza (flu) patients. Two medical researchers at the Rockefeller Institute in New York City used a filter that contained pores small enough to block bacteria but large enough to allow viruses to pass through. Peter Olitsky (1886–1964) and Frederick Gates (dates unknown) used a Berkefeld filter to filter the nasal secretions of influenza patients. They believed that what passed through was an "atypical" type of bacteria. However, what they really discovered was the influenza virus.

Pfeiffer's theory of bacteriolysis (the destruction of bacteria) resulted from his research involving guinea pigs. He injected live viruses that are responsible for the disease cholera into the peritoneal cavity (the space between the outer wall of the stomach and intestines and the abdominal wall) of guinea pigs that had previously been immunized against the disease. This caused the virus to lose its motility and disintegrate, which he observed under a microscope. He also determined that he could produce the same results if his "bacterioloytic" serum was injected into the peritoneal cavity of nonimmunized guinea pigs. In addition, he discovered the specific bacteria-destroying immune bodies in the bacteria that cause typhus as well as cholera. At the time, several other bacteriologists were working on a vaccination theory for typhus. Another scientist, the British pathologist Almroth Wright (1861–1947) was at first given credit for the discovery of the vaccination for typhus, but after reviewing the literature, it was determined that the credit should go to Richard Pfeiffer.

*See also* Koch

**PLANCK'S FORMULA AND QUANTUM THEORY:** Physics: *Max Karl Ernst Ludwig Planck* (1858–1947), Germany. Max Planck was awarded the 1918 Nobel Prize for Physics.

***Planck's formula:*** $E = \hbar v$; *where* $E$ = *the energy involved,* $\hbar$ = *Planck's constant of proportionality, and* $v$ = *the frequency of the radiation.*

For centuries, physicists were puzzled by the two theories of light. During the nineteenth century, some thought the corpuscular (particle) theory and the wave (electromagnetic radiation) theory were inconsistent with the then current theory of molecules and thermodynamics. In the 1860s, Gustav Robert Kirchhoff and other scientists experimented with blackbody radiation, an ideal surface, such as a hollow metal ball with a small hole that absorbs all light, and that does not reflect back any light but rather emits radiant energy of all wavelengths. They found that a body at "red heat" emitted radiation at low frequencies for the spectrum of light waves (infrared and deep-red), and that "white heat" emitted radiation at the higher frequencies at the yellow, green, blue end of the light spectrum. From these data, scientists projected curves on a graph to explain their theories. Max Planck plotted a new set of curves representing these data and advanced a different formula. His formula $E = \hbar v$ explained that energy radiated from the blackbody was specifically in quanta (small bits) of energy, not continuously as previously believed. These quanta were represented by the $\hbar v$ in the formula, where $v$ is the frequency of the radiation and the $\hbar$ is the action of the quanta of energy, which is a proportionality that can only assume integral multiples of specific quantities (quantum theory). The $\hbar$ is now known as *Planck's constant* and is one of the major constants in physics. In other words, the energy of a quantum of light is equal to the frequency of the light multiplied by Planck's constant.

***Planck's elementary quantum action theory:*** *Energy does not flow in an unbroken stream but rather proceeds or jumps in discrete packets or quanta.*

The science of quantum mechanics is based on Planck's theory that energy can only be emitted or absorbed by substances in small, discrete packets he called quanta. This theory has been used and expanded by many scientists. After the 1950s, Planck's quantum theory was used extensively in producing and identifying numerous subatomic and subnuclear particles and energy quanta. There are dozens of other examples in science that make use of Planck's quantum theory. Today many people mistakenly consider a "quantum leap" to be a great stride or large advancement of events or accomplishments. Originally it referred to a very small packet of energy or mass, which may be thought of as a tiny "particle wave" of light or an electron's tiny gain or loss of energy when it moves from one orbit to another (*see* Figure D6 under Dehmelt).

*See also* Chadwick; Compton; Einstein; Heisenberg; Kirchhoff; Pauli; Rutherford; Schrödinger

**POGSON'S THEORY FOR STAR BRIGHTNESS:** Astronomy: *Norman Robert Pogson* (1829–1891), England.

> *Pogson's ratio is the interval of star magnitudes that might be represented by a multiple of five magnitudes.*

From ancient times, the magnitude (brightness) of stars was based on what could be judged from Earth. The brightness of stars was ranked in just six magnitudes, the first being the brightest stars (excluding our sun) and the sixth faintest were those just barely visible. Norman Pogson, an Englishman who spent his life in India as an official astronomer, proposed a more rational and useful system to determine the magnitude of stars. He realized the first-magnitude stars were about one hundred times brighter than those in the sixth-magnitude category. From these data he devised a *ratio of brightness* of 2.512. This means a fifth-magnitude star is 2.512 (about two-and-a-half) times as bright as is a sixth-magnitude star. It was soon evident his "ratio" was not adequate for the actual range of brightness to cover all luminosities, so negative magnitudes were introduced. For instance, the sun is a −26.7 magnitude star (as viewed from Earth), and the brightest star beyond the sun is the −1.5 magnitude star, Sirius. The moon has a luminosity of −11.0. Pogson's ratio is still used today, but it is augmented by using the spectrum and colors of stars as recorded on photographic plates, which, through timed exposures, can record stars beyond the twenty magnitude level (*see* Figure H6 under Hertzsprung).

**PONNAMPERUMA'S CHEMICAL THEORY FOR THE ORIGIN OF LIFE:** Chemistry: *Cyril Andrew Ponnamperuma* (1923–1994), Sri Lanka and United States.

> *It was possible for chemicals and energy existing in the primordial atmosphere to synthesize protein molecules and nucleic acids required for life.*

Cyril Ponnamperuma's theory is based on three processes that must proceed in sequence for life to form from chemical elements and energy. First, the necessary atoms must form into the required molecules. Second, these molecules must combine into self-replicating polymers. And, third, these polymers (large organic molecules) must unite into living cells, tissues, organs, systems, and finally organisms. Ponnamperuma and several other scientists attempted to achieve this process actinically in the laboratory. One attempt exposed a mixture of water, methane, and ammonia to beta radiation, expecting to produce adenine (a purine found in RNA), but in the second and third stages of this process no success was achieved. Another attempt exposed formaldehyde to ultraviolet light to produce a polymer, again without success. Several other scientists have synthetically produced a variety of organic molecules, but none of these experiments met the three stages required in the process of producing life as described by Ponnamperuma.

*See also* Chambers; Miller

**PORTER'S THEORY FOR THE STRUCTURE OF HUMAN GAMMA GLOBULIN:** Biochemistry: *Rodney Robert Porter* (1917–1985), England. Rodney Porter shared the 1972 Nobel Prize for Physiology or Medicine with Gerald Maurice Edelman.

> *The antibody known as gamma globulin (IgF) is composed of two identical halves with each half having one long and one short chain.*

After serving in the military from 1940 to 1946 Rodney Porter continued his education and received his PhD from the University of Cambridge in 1948. He worked at

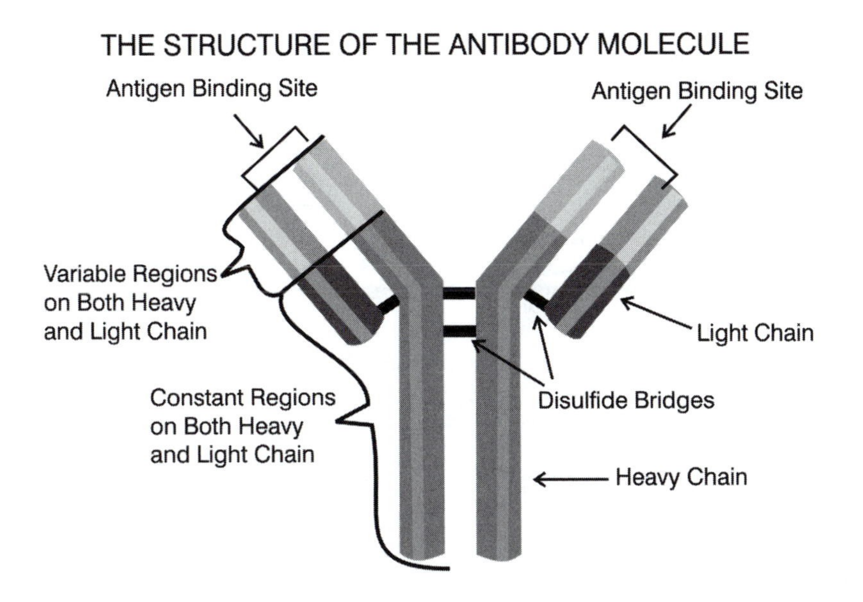


Figure P4. Porter's conception of antibody molecules confirmed what was then currently known (circa. 1960) about antibodies.

the National Institute of Medical Research (NIMR) on the outskirts of London from 1949 to 1960. He then went onto St. Mary's Medical Hospital, London, as the first Pfizer Professor of Immunology. In 1967 he accepted a professorship in biochemistry at the University of Oxford where he developed an interest in antibodies. He was aware of the data resulting from microscopic work related to antibody molecules using electron microscopes. In 1950 while at the NIMR he showed that some antibody molecules could be broken down and still maintain their antigen-binding characteristics. After a decade of research, in 1960 he showed that antibodies contained both "heavy" and "light" protein chains, and that each chain has three distinct regions. Two of these regions on the chains are alike and serve the function of binding antibodies. The two "light" chains form the branches of the "Y" structure in the chain, while the two "heavy" ones form the trunk of the "Y" (*see* Figure P4).

This insight was more-or-less an informed guess as to the actual structure of the molecule of antibodies. However, his scheme confirmed then-known information about antibodies and inspired other biochemists, including Landsteiner and Pauling, to continue his work in the field of immunology.

*See also* Landsteiner; Pauling

**POSEIDONIUS' CONCEPT OF THE EARTH'S CIRCUMFERENCE:** Astronomy: *Poseidonius of Apamea* (c.35–51 BCE), Greece.

*The circumference of Earth can be calculated by measuring distances between two locations, both on the same meridian circle.*

About two hundred years before Poseidonius conceived his method for measuring the circumference of Earth, Eratosthenes of Cyrene measured, simultaneously at the time of the summer solstice, the distance between two distant cities and then used this figure to calculate Earth's circumference as 25,054 miles (the current average circumference of Earth is 24,857 miles). Poseidonius used the figure of 5,000 stadia as the distance between two cities located on the same meridian (*stadia* is an ancient Greek measurement of distance based on the length of the course in a stadium. It is equal to approximately 607 feet, or 185 meters). This meridian encompasses 1/48 of the circle of Earth's circumference. In other words, he projected that the distance between the cities equaled about 1/48 of the distance around the globe at that particular meridian. From this he multiplied 48 times 5,000 to arrive at a circumference of 240,000 stadia, which compared favorably with Eratosthenes' figure of 250,000 stadia. We now assume there are 8.75 stadia to the mile, so the circumference comes out to about 27,000 miles. Poseidonius thought 240,000 stadia was much too large, so he reduced his figure to only 180,000 stadia. About one thousand years later this had unexpected consequences when Christopher Columbus used Poseidonius' figure for a much smaller Earth rather than the one provided by the ancient astronomer Eratosthenes. Therefore, Columbus believed Asia was only 3,000 miles west of the European coast, which made his trip to the New World, which he thought was India, much longer than expected.

*See also* Eratosthenes

**POYNTING'S THEORIES:** Physics: *John Henry Poynting* (1852–1914), England.

***Poynting–Robertson effect:*** *Solar radiation causes dust grains within the solar system to spiral slowly inward while increasing their orbital speed.*

Poynting theorized that the sun's radiation caused small particles to become suspended in space and, in time, to spiral closer and closer to the sun, eventually crashing into the sun. He proposed that the radiation pressure upon the particles acted tangentially to the motion of this system of particles. Early in his work in 1903 Poynting considered this phenomenon to be related to the luminiferous nature of the aether that at the time was still believed to be an aethereal substance that filled space and was the medium that carried electromagnetic waves, such as light and radio waves. In 1907 the American physicist, Howard Robertson (1903–1961) correctly described Poynting's theory by ignoring the concept of aether and correctly stating it in terms of special relativity. Since then it has become known as the "Poynting–Robertson Effect."

***Poynting vector:*** *The Poynting vector points in the direction of traveling electromagnetic waves.*

In 1884 Poynting wrote a paper titled "Transfer of Energy in the Electromagnetic Field" that introduced his point theory as the flow of energy through the surface of conductors in terms of electric and magnetic properties. His theory showed that the flow of energy (power) at any point in a conductor may be expressed by a simple formula in terms of vectors indicating the electric and magnetic forces at that point (vectors are scalar quantities that give direction and magnitude). He is best known for this electrical and magnetic phenomena known as the "Poynting vector."

In 1891 Poynting determined the mean density of Earth as well as its gravitational constant by using a well-made instrument known as a torsion balance to arrive at his conclusions that were published in *The Mean Density of Earth* in 1894 and later in 1913 in *The Earth: Its Shape, Size, Weight, and Spin.*

*See also* Cavendish

**PRÉVOST'S THEORY FOR THE EXCHANGE OF HEAT RADIATION:** Physics: *Pierre Prévost* (1751–1839), Switzerland.

> *A body that radiates heat energy is independent of the body's environment; and if the body's temperature increases or decreases, it will be dependent on whether radiation is gained or lost.*

Pierre Prévost was originally a professor of philosophy who studied political economy and the fine arts, including poetry. He became friends with the notable French mathematician Joseph-Louis Lagrange, which led him to the study of physics—in particular magnetism and heat. At this time in history, it was believed that heat was a fluid called "caloric," thought to be responsible for heat always flowing from hot substances to cooler bodies. This concept made it reasonable to conceive that cold was also a fluid called "frigoric" and was responsible for the flow of cold to warmer bodies. Some scientists used the example that if a piece of ice is held near a thermometer, the temperature recorded in the thermometer would drop, thus explaining the "frigoric" concept. Prévost's book, *Traite de Physique*, published in 1791, explained his "law of exchange of radiation" and helped clarify the nature of heat. He still believed that heat was a fluid, but a single fluid that flowed from hot objects to cooler ones—never in the other direction. He is the first to describe the concept of equilibrium that explains that if a body is colder than another, it will absorb radiation until its temperature is in equilibrium with its environment's temperature. This means that the cold body does not stop radiating heat but rather radiates just enough heat to reach the point where the radiation is in equilibrium. This became known as "Prévost theory of exchanges" and influenced future generations of physicists to arrive at the kinetic theory of heat based on atomic/molecular motion about a century later.

*See also* Carnot; Helmholtz; Kelvin; Mayer; Maxwell

**PRIESTLEY'S THEORIES OF ELECTRICAL FORCE AND DEPHLOGISTICATED AIR:** Chemistry: *Joseph Priestley* (1733–1804), England.

***Priestley's theory of electrical force:*** *The force between two charged bodies decreases as the square of the distance separates the charged bodies.*

Joseph Priestley's friend, Benjamin Franklin, encouraged him to investigate the new phenomenon of electricity. Priestley was the first to measure the electrical force between two charged bodies as related to the distance between them. He calculated that if the distance between the two bodies is increased by a factor of two, the electrical force is decreased by a factor of four. This follows the well-documented general physics principle of the general square law, which was confirmed by other scientists. He also was the first to determine that charcoal (carbon) could conduct electricity. This became an important concept when applied to the new uses of electricity, such as the carbon arc light, the arch furnace, electric motors, and the telephone.

***Priestley's dephlogisticated air:*** *When the oxides of certain metals are heated, they produce an air that has lost its phlogiston.*

Although not the first to experiment with the heating of materials to drive off gases, Joseph Priestley was one of the first chemists to make careful observations and measurements of what happened to the materials he used. He was also the first to try different experiments to help him understand respiration and combustion. In 1771, he hypothesized that when a candle is burned in a closed jar, it consumes much of the "pure" air. Therefore, there must be some way for nature to replenish the air dissipated by burning objects, or else it would all be used up in the atmosphere and none left for respiration. Then he placed a small green plant in the same jar with the candle and found that after several days, the air again would support combustion. Using similar techniques, Priestley isolated several other gases by heating different substances. He produced sulfur dioxide, ammonia, nitrous oxide, hydrogen chloride, and carbon monoxide. He collected a gas, which was known as "fixed air" ($CO_2$), given off from the vats in a brewery. He then bubbled it through water; the result was carbonated water. Priestley then heated a small amount of mercury in a closed container and noticed that it formed a red "calx" on the surface similar to rust. He proceeded to place a candle, and then a mouse, in this air given off by the heating of this red calx which is the

In addition to being a successful untrained scientist, Joseph Priestley was also somewhat of a religious iconoclast. Priestley's father, Jonas Priestley, was a finisher of cloth and his mother Mary Smith was raised on a farm. He was born in the parish of Birstal near Leeds, England, in 1733. The oldest of six children, he was raised on his grandfather's farm after his mother died in 1741, only to be adopted by his Aunt Sarah who was childless. It was in his aunt's household among many local people with differing views of religion that he was exposed to liberal religion and political beliefs. In elementary school Joseph learned not only Greek and Latin but also Hebrew. As a teenager, he contracted tuberculosis and during his illness considered entering the ministry. As he recovered, he taught himself several other languages as well as geometry and algebra. As his health improved, he began questioning what he had learned earlier about orthodox Calvinism. Instead of attending a religious school, he entered the liberal Daventry Academy in Northamptonshire where dissention in the form of liberal education was taught to followers of natural philosophy and other nonconformist doctrines. Priestley accepted a position as a minister in a church with a poor congregation, staying only three years because his slight speech impediment, along with his acceptance of Unitarianism, made him an unpopular preacher. His next job not only paid more but the congregation was more tolerant of his different theology. Also, with extra income from tutoring in languages, he was now able to buy a number of instruments to aid him in his scientific research. After marriage to Mary Wilkinson, he was ordained in what was known as the "Dissenting Ministry." This was the period when he indoctrinated his students and others into his liberal political theories of law and other fields. With a growing family he moved to a congregation near his birthplace and continued his work with gases and electricity, while at the same time he continued visiting London where he met Benjamin Franklin and other scientists. At this time Priestley turned his interests to politics. He wrote *The First Principles of Government and the Nature of Political, Civil, and Religious Liberty* in 1768. Later in 1774, at the encouragement of his friends, he wrote *The State of Public Liberty in General and of American Affairs in Particular* which was a pamphlet attacking the British role in America. In 1782 he wrote *The History of the Corruptions of Christianity,* followed by *History of Early Opinions Concerning Jesus Christ* in 1786. These books did not make him popular with the public or the ruling politicians because they proposed Unitarian ideas. King George III was convinced that Priestley was an atheist. In 1791 he wrote *A Political Dialogue on the General Principles of Government* which was similar to the *Rights of Man,* written by the American revolutionary Thomas Paine (1737–1809). Many people attacked Priestley for his political and religious views that soon exploded into a mob breaking into his house and destroying books and papers, and even his scientific equipment. He and some members of his family emigrated to the United States where they settled in the northern Pennsylvania town of Northumberland on the Susquehanna River. He is remembered not only for his scientific achievements but also for the establishment of the first Unitarian Church in America.

component mercury oxide. The candle burned much brighter and the mouse lived much longer in this new air that he called "dephlogisticated" air because he believed it lost its "phlogiston." Although Joseph Priestley is credited with the discovery of oxygen, it was Antoine Lavoisier who named *oxygen* from the Greek word meaning "sharp" because, at one time, scientists mistakenly thought that all acids contained oxygen.

*See also* Franklin (Benjamin); Lavoisier; Scheele

**PRIGOGINE'S THEORIES OF DISSIPATIVE STRUCTURES AND COMPLEX SYSTEMS:** Chemistry: *Ilya Prigogine* (1917–2003), Russia. Ilya Prigogine received the 1977 Nobel Prize for Chemistry.

***Prigogine's dissipative structures:*** *States of thermodynamic equilibrium for systems are rare. More common states exist where there is a flow or exchange of energy between systems.*

One example Ilya Prigogine used to explain his "dissipative structures" was the solar system. Without the sun's continual bathing of Earth with energy, Earth's atmosphere would soon reach thermal equilibrium, meaning it would reach a sustained very cold temperature because heat always flows to cold, not the reverse. Because the sun provides a steady flow of energy to Earth, this might be thought of as *negative entropy* or, as Prigogine believed, a process that reverses irreversible equilibrium states. His work on irreversible processes is credited with forming a bond between the physical sciences and biology that deals with systems that over time have not obtained equilibrium—life and growth. An example is what happens in living cells as they constantly exchange substances and energy with their surroundings in tissues. The process of thermodynamic entropy is irreversible only if there is no exchange of energy between or among complex systems. Theoretically, some billions of years in the future of the universe, entropy (the complete disorganization of matter) and the irreversible attainment of thermal equilibrium will win out unless a new source of universal energy is forthcoming.

***Prigogine's theory of complex systems:*** *Simple molecules can spontaneously self-organize themselves into more complex structures.*

Ilya Prigogine is known as the grandfather of *chaos theory*, which in the related science of complex structures has a more specialized meaning than the concept of chaos used in ancient as well as modern times. Before the development of Prigogine's theory dealing with dissipative and irreversible processes, chaos theory was thought of as a mathematical curiosity. More recently, chaos theory, as related to complex systems, has had a widespread impact on several science disciplines, particularly biology, but also economics. Chaos deals with initial conditions and how these conditions alter the causes that create problems when trying to predict effects. A classic example is that the knowledge of initial conditions of a weather system does not provide adequate information, down the line, to be able to predict weather with any degree of accuracy (*see* Figure W1 under Wolfram). Weather is a classic complex system where the chaos theory is applicable (to some extent, chaos theory also applies to climate change that might be thought of as weather changes over large geographic [or worldwide] regions that occur over long periods of time).

Prigogine believed that very simple inanimate and inorganic molecules, at least at one time, had the ability to organize themselves spontaneously into higher, more complex organic molecules and organisms. This process must have involved some exchange of energy for the self-organizing molecules to reverse entropy (or, as Prigogine would say, "nonequilibrium thermodynamics"). This is not exactly the same as the old idea of spontaneous generation, but it might be thought of as a modern version of that idea, which he also related to evolution. Ilya Prigogine received a Nobel Prize for his work in nonequilibrium thermodynamics (dissipative structures), which relates to concepts in chemistry, physics, and biology.

*See also* Margulis

**PROUST'S LAW OF DEFINITE PROPORTIONS:** Chemistry. *Joseph-Louis Proust* (1754–1826), France.

*Elements in a compound always combine in definite proportions by mass.*

Based on research that Proust conducted on two tin oxides, two iron sulfides, and several other metals, he found that each compound had a definite proportion of weights between the elements of the compound's molecules. An example is the molecule of water ($H_2O$). The two elements, hydrogen and oxygen, in the molecules of the compound water always exist in the ration of two atoms of (H) to one atom of (O). (Note: There is a compound of $H_2O_2$ known as hydrogen peroxide which is not exactly the same as water, but as the oxygen escapes this saturated molecule, it will again be $H_2O$.) From these experiments he deduced his principle in 1806 stated as "I have established that . . . iron like many other metals is subject to the law of nature, which presides at every true combination . . . that unites with two constant proportions of oxygen." He went on to say: "In this respect it does not differ from tin, mercury, and lead, and in a word, almost every known combustible." Up until this time the French chemist Claude-Louis Berthollet (1748–1822) stated that elements could combine to form compounds in a large range of proportions. The distinction between Proust's concepts and Berthollet's idea is that Proust described the proportional mixtures of elements that make up compounds and Berthollet referred to mixtures and solutions, not compounds. Somewhat later Berthollet admitted that he was not correct and that Proust's law was sound. Also at one time it was thought that a chemical reaction depended on the amount of the original mass of all the reactants—not proportionally. It was the work of the great chemist John Dalton, who many consider the father of modern chemistry, which validated that Proust's law of definite proportions was based on a definite number of atoms of elements joining together to form molecules.

Today, in the field of nanostoichiometric chemistry (viewing reactions at the atomic/molecular levels) chemists have found some minor differences in the proportions of elements in some compounds. For instance, a form of iron oxide known as wüsite can contain between 0.83 and 0.95 atoms of iron for every oxygen atom. Thus, this compound of iron may contain between 23% and 25% oxygen. The small variations are largely due to the various isotopes (molecules of the same element with different atomic weights) that compose the compounds. Also in the field of polymer chemistry the proportion of elements forming a polymer molecule may vary. Some examples are proteins, such as DNA, as well as carbohydrates. Some chemists do not even consider polymers to be absolutely pure chemical compounds, except when their molecular weights are uniform. Proust's research was not accurate enough to detect these slight variations in proportions. Proust was also interested in studying the types of sugars found in some vegetables and fruits. During his research with grapes he discovered that they had the same type of sugar as honey. This discovery later became known as glucose.

*See also* Dalton

**PTOLEMY'S THEORY OF A GEOCENTRIC UNIVERSE:** Astronomy: *Claudius Ptolemaeus (Ptolemy of Alexandria)* (c.90–170), Egypt.

> *Earth being the heaviest of all bodies in the universe finds its natural place at the center of all the cosmos.*

Ptolemy collected and compiled a great deal of information from other astronomers. From Aristotle, he gleaned there were two parts to the universe—Earth and the heavens and that Earth's natural place is at the center of the entire universe. He considered Earth the sublunary region where all things are born, grow, and die, whereas the heavens are composed of compact concentric crystal spheres surrounding Earth (*see* Figure P5).

Each shell was the home of a heavenly body arranged in the order of Moon, Mercury, Venus, Sun, Mars, Jupiter, and Saturn, followed by the fixed stars and the "prime

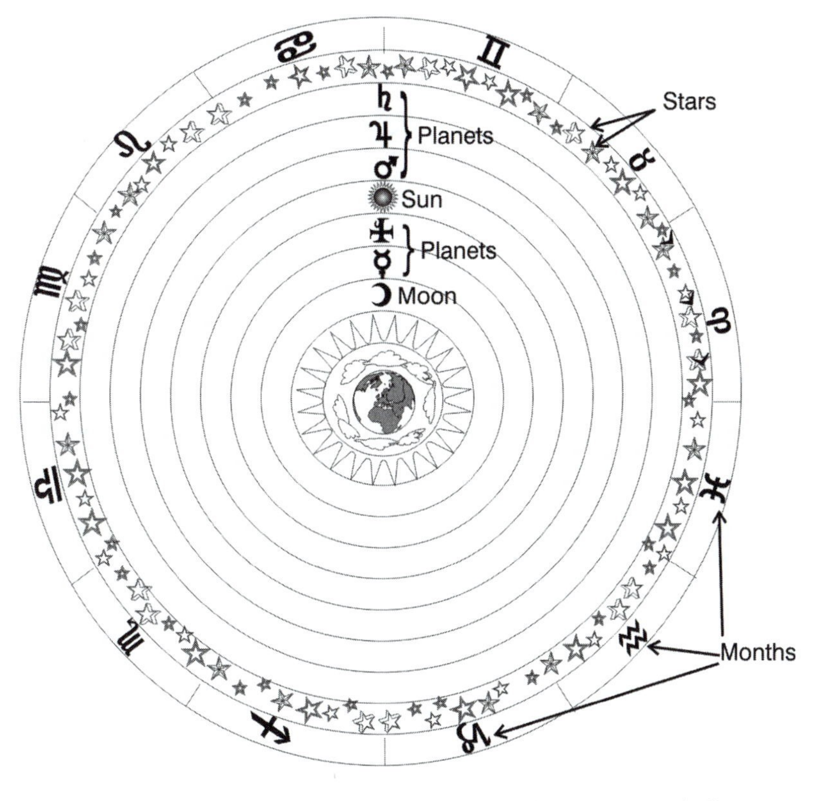


Figure P5. Ptolemy's Earth-centered universe consisted of concentric "shells" starting with the moon's orbit, followed by the inner, then the outer planets, then the stars and finally the "prime mover."

mover," who kept the whole system moving. Hipparchus of Nicaea (c.190–120 BCE), who formulated positions and motions for the planets and moon, was another early Greek astronomer and mathematician who influenced Ptolemy. From this background Ptolemy claimed not only the universe was geocentric, but that all bodies that revolve in orbits do so in perfect circles and at constant velocities, whereas the stars move in elliptical orbits at inconsistent velocities. This required the application of complicated geometry, which Ptolemy used to describe these motions. These three kinds of motions traced the following geometric paths: eccentric, epicycle, and the equant. Ptolemy combined these to form his *Ptolemaic system* for planetary motion. His system was not accurate enough to determine all the motions of heavenly objects, but his system was used by other astronomers for over thirteen hundred years, until in 1514 Nicholas Copernicus developed his first heliocentric model of the universe, which he continued to refine for the next thirty years. It might be mentioned that much of the pseudoscience of today's astrology can be traced back to Ptolemy's planetary motion and the idea that there is some form of physical "rays" emanating from the "heavens" that affects the lives of humans.

Ptolemy's book titled *Optics* was not only his final book but also his best work in which he described a variety of elementary physical principles. He demonstrated that he understood the principles of reflection and **incidence** and to some extent the refraction of light. He even developed tables from his studies of the refraction of rays of light passing from a source of light into water from different angles of incidence.

*See also* Aristotle; Copernicus; Galileo

**PURCELL'S THEORY OF NUCLEAR MAGNETIC RESONANCE (NMR):** Physics: *Edward Mills Purcell* (1912–1997), United States. Edward Purcell shared the 1952 Nobel Prize for Physics with Felix Bloch.

> *When the nuclei of atoms are affected by a magnetic field, they absorb energy in a particular radiofrequency range of the electromagnetic spectrum, and then re-emit this energy as the nuclei revert to their original energy state.*

In the 1930s the American physicist Isidor Rabi developed a method of observing a specific atomic spectrum by focusing an electromagnetic beam on the nuclei of atoms and molecules. Rabi thought this phenomenon, later called "nuclear magnetic resonance" (NMR), was caused by his equipment and thus did not recognize its importance. Ten years later Edward Purcell, a young PhD from Harvard, was the leader of a group of researchers at Massachusetts Institute of Technology (MIT), Cambridge, Mssachusetts. His research covered many fields, including radio astronomy, radar, astrophysics, biophysics, and in particular nuclear magnetism. In 1945 while spending time in his laboratory after his regular workday, Purcell observed the phenomenon of NMR. His discovery was based on the work of the Irish physicist Sir Joseph Larmor who determined that the angular frequency of the "spins" of nuclei are proportional to the strength of the magnetic field.

"Nuclear" refers to atomic nuclei of certain elements—not to nuclear energy or the atomic bomb as many people mistakenly believe when they hear the term "NMR."

"Magnetic" as used in this process referred to a magnetic field that is applied to the nuclei.

"Resonance" refers to the oscillating motion of the nuclei caused by the electromagnetic frequency applied to the system.

Edward Purcell of MIT and Felix Bloch of Stanford University expanded this research and found that certain nuclei, when placed in a magnetic field, absorbed energy in the electromagnetic spectrum that relates to the frequencies of radio waves. In addition to absorbing this energy, which caused precession "spins" of the nuclei, the oscillating nuclei gave up this energy and returned to their natural state. With the discovery of NMR it was soon used as a method to study and analyze the exact quantities and quality of various chemical compounds as well as to determine the structure of various materials.

Not long after, others recognized that the single dimension images produced by NMR spectroscopy, although an improvement over other forms of chemical/physical analysis, might be further improved to produce three-dimensional images. First, computed axial tomography (CAT scan) technology was a step in the right direction as others developed what became known as MR and later MRI (magnetic resonance imaging). (Note the "N" for "nuclear" was removed because many people, due to their misunderstanding, did not want to subject themselves to anything related to nuclear energy.) MRI has become a successful diagnostic instrument for viewing the human body in three dimensions. Presently it is also being developed for use as a "real-time" fluoroscope that would enable body processes, such as the flow of blood through veins and arteries, to be viewed as it surges through tissue.

*See also* Ernst; Kusch; Mansfield; Rabi; Ramsey

**PYTHAGORAS' THEOREM:** Mathematics: *Pythagoras of Samos* (c.580–500 BC), Greece.

*The square of the hypotenuse of a right triangle is equal to the sum of the squares of the other two sides of the triangle.*

Pythagoras believed that whole numbers, as well as fractions expressed as ratios of whole numbers, were not only "rational numbers" but also explained the basis of the universe. However, when he compared the sides of a right triangle with the ratio of 1 to 2, the opposite side of the 90° angle (hypotenuse) was an "irrational number." In other words, the diagonal of a square cannot be related to the sides expressed in whole numbers and thus the ambiguity of the square root of 2. The original concept for the Pythagorean theorem goes back one thousand years before Pythagoras to Babylon when the idea was first conceived that any three-sided figure with sides containing the ratio of 3:4:5 would form a 90°right-angle triangle. Proof for the theorem was derived by the Pythagoreans ($a^2 + b^2 = c^2$) with the credit for the proof given to Pythagoras who was the leader of an academic "cult" of mathematicians who believed their work was sacred and should be kept secret. They believed all events and all things can be reduced to mathematical relationships. Their motto was stated as: "All things are numbers," and their secrecy is one reason it is difficult to determine which writings were by Pythagoras and which by his fellow Pythagoreans.

# Q

**QUANTUM THEORIES: FROM 1900 TO 2008:** Over the last century many theoretical physicists from several countries contributed to the development of quantum theories involving particles and waves.

Note: Because there are many scientists involved in the story of the development of quantum theory, this special summary of their contributions is presented. The term *quantum* is derived from the Greek word meaning "how great," or "how much," or "how far." In its modern use it is conceived as referring to the submicroscopic, subatomic, subnuclear phenomena involving unimaginably small particles and energies or waves. Unfortunately, some people and the media confuse the modern use of "quanta" with something very large—they could not be more wrong—even by the ancient Greek concepts. For instance, when an electron is at a specific energy level (orbit) of an atom and either receives or loses energy, the electron will "jump" to a "higher" or "lower" energy level (orbit). This slight jump is referred to as a "quantum leap" and is extremely tiny and does not mean a large shift of movement. There are three crucial dates related to quantum theory. They are as follows:

**1900**: The concept that atoms are elementary particles of all matter is very old—as far back in time to the Greek philosophers Leucippus of Miletus (c.490–430 BCE) and his student Democritus who are credited with originating the atomic theory that stated nothing could be separated or divided further than the minute atom. From the age of Newtonian classical physics of motion and the development of a more modern concept of the atom based on experiments by Thomson, Rutherford, Niels Bohr, as well as others, it became obvious that the atom was not just a solid ball of matter but rather much more complicated in structure. In 1900 many, but not all, scientists accepted the atomic theory. Even though atoms were defined as indiscrete point particles, it was not understood why and how atoms of different elements differed. For instance, the electron was not discovered until just three years before 1900. No one knew where they were located on or within the atoms, or what their function(s) were.

December 14, 1900, is considered by some the birthdate of the quantum concept of matter and energy. There were many observations of experiments involving the different colors of light emitted by various degrees of heat. For example, a low level of heat produces infrared or red colored light while a higher degree of heat produces bluish colors. The explanation for this is that heat is just the "jiggling" of atoms while remaining in their place. This is explained as the greater the temperature, the faster the jiggling, and thus the greater the heat and the shorter the radiation's wavelengths resulting in the faster jiggling that produces different colors. This also explains the relationship between heat and molecular motion—the faster the molecules move in a substance (solid, liquid, or gas) the hotter the substance—and vice versa. While considering these observation and theories, the German physicist Max Planck arrived at a formula that explained this phenomenon, yet he could not explain why. It just did—and he also stated that jiggling atoms could not assume any of the possible energy levels, but only a level at specific "permissible" values. His formula was announced on December 14, 1900, but his theory did not take the science community by storm. One reason was that his formula did not explain how the energy of a group of jiggling atoms existing at one allowed energy level could change to another energy level if each particle has a specific level of jiggling. What happens in the transition for one level to the next level of energy?

**1905:** Up to this time Planck's ideas did not seem to fit in with classical physics, but his theories were thought by some to be an extension of "classical mechanics" when only specific values of energy were involved. Some physicists considered that energy consisted of these small quantities, that is, bits of energy that were referred to as being "quantized." After Planck's ideas became more accepted and more was learned about matter and energy, the classical concepts of physics became known as the "old quantum theory." At the age of twenty-six Albert Einstein, drawing on the work of others regarding the nature of heat and the structure of atoms, theorized that a beam of light is quantized. Based on Planck's ideas Einstein proposed that light was quantized and traveled in little packages or bits of electromagnetic energy (light). Gilbert N. Lewis, the famous American physical chemist, suggested the name "photon" for these bits of light quantum. This was accepted by Einstein who went on to propose the nature of these photon packets of light as a duality of the particle-wave theory of light. Another related achievement was Einstein's theory explaining the equivalent nature of mass and energy as in his famous equation $E = mc^2$.

**1927**: This was the year in which Werner Heisenberg presented the results of his observations of particles, often called his "indeterminacy principle" but more commonly known as the "uncertainty principle." In essence, it states that the more you learn about a particle's energy (momentum) the less you would know about its position in time. And, the reverse is true—using high-powered microscopes, the more you observe a particle's position at a specific time, the less is known about its energy.

Since the discoveries of the electron (1897), the proton (1919), and the neutron (1931), over thirty-five subnuclear particles have been identified in the atom and the atom's nucleus. With the construction of larger, improved, and more powerful circular cyclotrons and linear particle accelerators, they are capable of slamming accelerated "bullet particles" into "target particles" at great forces. This results in the "target particles" breaking or separating into many smaller bits (particles) of matter or bits of energy. It is something like striking a bowling ball with a revolver shot—it may split the ball in a few pieces, but if you hit it with a high-speed artillery shell, it would pulverize the ball into thousands of smaller pieces—all jiggling with specific energies.

Back in the 1920s there were several theories explaining the quantum nature of matter. One was called "matrix mechanics" theory that was based on a complicated form of "mathematical matrices." Another theory was based on the "wave function" of quantum mechanics. And, still another is called "amplitude formation" of quantum mechanics, also known as the "path to integral formulation," or as it was called by Richard Feynman, "the path of least action" that is based on a paper by Paul Dirac. This theory seems to be related to a principle of nature, which says that all actions follow the course of least energy, required for completing that particular process. Based on these several different quantum theories, research into minute bits of particles and energy progressed rapidly. It was not long before it was realized that, mathematically, many of these theories were related and could be applied to solving problems with the very small (atoms and molecules and their particles), as well as explaining very large phenomena including electromagnetism, gravity, and the composition of stars. These theories for quantum mechanics were extended to include relativistic theories, field theories, chemical bonding, and many more problems in various fields of physics. The modern field of quantum mechanics has become a successful theory for science and is based on the research of many physicists. See the index for names of individual scientists who have contributed to quantum theory and who are contained in these two volumes.

# R

**RABI'S THEORY OF MAGNETIC MOMENT OF PARTICLES:** Physics: *Isidor Isaac Rabi* (1898–1988), United States. Isidor Rabi was awarded the 1944 Nobel Prize for Physics.

> *Neutron beams can be used to determine the magnetic moments of fundamental particles such as the electron.*

Isidor Rabi advanced the work of Otto Stern who in 1922 used a beam of molecules to determine the spacing of atomic particles referred to as *space quantization*. Rabi

Isidor I. Rabi's parents were from Eastern Europe, where his father, David, could not earn enough to maintain a family, which led to his emigration to the United States. Because he had no education or skills, he could only find work in New York City's garment district. When he saved enough money, he sent for his family the year after Isidor's birth. With a loan from other Yiddish speaking friends, his father opened a small grocery store. The Rabis soon moved from the Lower East Side to Brooklyn where Isidor discovered science books in the local library. He finally entered Cornell University and graduated with a BA in chemistry in 1919 and later attended Columbia University where he received a PhD in 1927. He spent his entire career at Columbia, with the exception of a two-year tour of Europe. In 1937 he was appointed the first professor of physics at Columbia and held that position until his retirement in 1964. While in Germany in the late 1920s, he worked with Otto Stern and Walter Gerlach whose experiments with molecular beams led to space quantization. When he returned to Columbia in 1929, he invented an atomic/molecular beam system for identifying the magnetic moments of subatomic particles. His work led to theories of quantum electrodynamics, the atomic clock, nuclear magnetic resonance, and the laser. He later was responsible for the planning of CERN—the international physics laboratory and the site of high-energy accelerators and equipment in Europe for the study of subatomic particles.

proceeded to develop a beam composed of various atoms and molecules that he used to produce magnetic resonance (oscillations), which could accurately determine the magnetic moments of fundamental particles. His theory and experiments resulted in the development of nuclear magnetic resonance (NMR), making it possible to measure the energies absorbed and the energies given off by the resonating atoms and molecules, which then can be used to identify substances. The process was revised and improved to produce a better image of human tissue than X-rays, and its name was changed from NMR to magnetic resonance imaging (MRI) because of the mistaken belief that "nuclear" referred to nuclear radiation, rather than the oscillating nuclei of the atoms in the tissue cells of human bodies.

*See also* Purcell; Ramsey; Stern; Tyndall

**RAMAN'S THEORY OF LIGHT SCATTERING:** Physics: *Sir Chandrasekhara Venkata Raman* (1888–1970), India. Chandrasekhara Raman was awarded the 1930 Nobel Prize in Physics.

*A small amount of light of specific frequencies will be reflected from a substance exposed to a direct beam of light of a single frequency.*

Chandrasekhara Raman determined that a beam of light of a single frequency, when striking a substance at right angles, would produce some frequencies different from the original single frequency beam. He further discovered that these new frequencies were specific to the type of material from which the beam was reflected. This became known as the *Raman effect*, which is the exchange of infrared frequency of the light and the material reflecting the light. Although the Raman effect is very weak—about 1/100,000 times less intense than the light of the incident beam—this scattered light of different frequencies can be used to measure the exchange of energy between the light and the substance being examined. The characteristics of the molecules of the examined substance exhibit intensities proportional to the number of scattering molecules that happen to be in the beam of light. This technique can identify specific gases, liquids, and solids. Gases have a low molecular concentration and thus produce a very weak Raman effect. Even so, the Raman effect is a very accurate and effective tool for quantitative and qualitative analysis.

*See also* Tyndall

**RAMÓN Y CAJAL'S NEURON THEORY:** Biology: *Santiago Ramón y Cajal* (1852–1934), Spain. Santiago Ramón y Cajal shared the 1906 Nobel Prize for Physiology or Medicine with Camillo Golgi.

*Neurons are discrete basic cells separate from a network and are the basic structures that function as units of the nervous system.*

Up until about 1800 it was believed that nerve cells were not individual cells as are most other cells of the human body but rather were connected in a mesh or network of cells. In addition, there was the reticularist theory that presumed that nerves continuously communicated. Cajal's theory stated that nerve cells are distinct units consisting of cell bodies, axons, and dendrites, and signals are transferred from cell to cell by the proximity of one cell to the next cell. In other words, they communicate by contiguity. He also proved that neural transmission of signals goes only one way, from dendrites toward axons.

The Italian cytologist Camillo Golgi (1843–1926) developed a silver nitrate stain that could identify specific nerve tissue. Up to this time the stains used for microscopic examinations of body tissues were unable to determine one type of body cell from another. Ramón y Cajal improved Golgi's stain so that it could be "cell specific" as well as being able to identify parts of the cell. He used this technique to examine the structure of the cells, as well as how they connected to each other in the brain. This led to Ramón's neuron theory. Later his research was used to help detect brain tumors. Using his discovery that the neuron was the fundamental unit of the nervous system, coupled with Golgi's staining techniques, Ramón found that nerve cells are discrete and not part of a "mass" of fused cells. Therefore, the Karolinska Institute for the Nobel Prize awarded the 1906 Nobel Prize for Physiology or Medicine to Ramón y Cajal and Camillo Golgi.

**RAMSAY'S HYPOTHESIS FOR INERT GASES:** Chemistry: *Sir William Ramsay* (1852–1916), England. William Ramsay was awarded the 1904 Nobel Prize for Chemistry.

> *The placement of the inert gas argon in the Periodic Table of the Chemical Elements indicates there will be other similar inert gases, still to be detected, with greater atomic weights.*

Sir William Ramsay followed the work of Lord Rayleigh and Henry Cavendish, both of whom experimented with air and discovered that, after removal of all the nitrogen and oxygen, there appeared to be some "leftover" gas. Rayleigh and Cavendish believed this small amount (1/20 of the original sample) of unidentified gas was the result of contamination by a lighter gas. Ramsay collected this small amount of gas and had William Crookes examine its properties by using spectroanalysis. Ramsay's sample was identified as being the same gas that had been previously detected by the French astronomer Pierre Janssen during a 1868 solar eclipse. In 1898 Ramsay determined this gas was a new gaseous element that was heavier, not lighter than predicted by Rayleigh and Cavendish. He named it *argon* from the Greek word *argos*, meaning "inert." Based on his theory of the placement of argon in the **Periodic Table of the Chemical Elements**, Ramsay predicted there were at least three other heavier inert gases yet to be found. He and several colleagues proceeded to identify helium as the gas emitted from the radioactive decay of uranium. Other inert gases and their meanings are neon (new), krypton (hidden), and xenon (stranger). In 1900 the German physicist Friedrich Ernst Dorn (1848–1916) found that the radioactive element radium gave off a radioactive gas that proved to be the sixth noble gas he called "radon" which is a variation of the word "radium." All six of these gases, also called noble gases, have many uses, including the gas in light bulbs, neon tubing, lasers, photographic speed lights, the decarbonizing of iron during smelting, and as a nonoxidizing gas for welding. Radon, which is the only radioactive inert gas, has limited uses in the treatment of cancer, as well as for detecting leaks and flow rates in liquid pipelines.

*See also* Cavendish; Rayleigh

**RAMSEY'S CHEMICAL SHIFT THEORY FOR IMPROVED MRI:** Physics: *Norman Foster Ramsey* (1915–), United States. Norman Ramsey shared the 1989 Nobel Prize for Physics with Hans G. Dehmelt and the German physicist Wolfgang Paul (1913–1993).

*Nuclear magnetic resonance using two different radiofrequency fields can identify the chemical shift of molecules by using magnetic shielding.*

Norman Ramsey improved Isidor Rabi's nuclear magnetic resonance (NMR) technique by using two different radio frequencies, which result in more accurate measurement of the magnetic effects on atoms. Magnetic shielding encloses the magnetic field within a specified area, preventing external static charges from interfering with the process. The NMR process causes nuclei in atoms and molecules in cell tissues to resonate (vibrate in position) thus revealing the magnetic properties of their atoms and molecules. It can be used to analyze the structure of molecules and their interactions with other nuclei in close proximity. The modern NMR, now called magnetic resonance imaging (MRI), can detect a variety of conditions in the human body. Ramsey also utilized his concept of separate oscillating fields to produce molecular beams in a **maser** to run a very accurate atomic clock. In addition, he worked out a statistical model for negative thermodynamic temperature systems, theorizing the possibility of temperatures below absolute zero (below 0 kelvin or −273.16°C).

*See also* Dehmelt; Rabi

**RAOULT'S LAW:** Physics: *François-Marie Raoult* (1830–1901), France.

*The amount of decrease in the freezing point of a dissolved substance (as compared to just the solvent) is related to the amount of the dissolved solute as well as to the molecular mass of the solute.*

Raoult's law was based on Jacobus Van't Hoff's work on solutions previously done with the optical activity of organic compounds when in solution. Raoult's observations were founded on how and why the freezing point of salt changed when dissolved in water as compared to its being dissolved in an organic solvent. Many organic compounds are optically active in the sense that they rotate the plane of polarized light. Raoult's law is important for understanding the structure and determining the molecular weights of organic compounds.

*See also* Van't Hoff

**RAUP'S THEORY OF CYCLIC EXTINCTION OF ANIMALS:** Biology: *David Malcolm Raup* (1933–), United States.

*The cyclic extinction rates for individual species of animals peaks every twenty-six million years.*

David Raup based his theory of a twenty-six-million-year cycle for the mass extinction of life on Earth on data gathered in cooperation with his collaborator, the paleontologist from the University of Chicago, J. John Sepkoski (1948–1999). Raup believed that fossil evidence of a smooth, evolutionary transition from one species to another, as Darwin claimed, is not convincing. Darwin assumed these gaps in the fossil records would be filled in time and with more exploration. More than a hundred years after Darwin, Raup maintained that the fossil record was still too incomplete to account for a gradual evolution and suggested that general extinctions were caused by extraterrestrial catastrophic phenomena (e.g., asteroids, comets, meteors), not terrestrial disasters, such as earthquakes and volcanic eruptions. Raup's theory is somewhat related to the catastrophic theory proposed by Eldredge and Gould for the extinction of the dinosaurs sixty-five million years ago by an asteroid. When this huge asteroid crashed into Earth, it sent massive clouds of dust and debris into the atmosphere blocking out the sun and resulting in the death of plants, thus depriving animals, such as dinosaurs, of food, ultimately leading to extinction. Raup proposed that Earth has a "companion" star with a twenty-six-million-year orbital period, meaning it returns to the region of the solar system on a periodic basis, bringing with it showers of asteroids that impact Earth. He called this scenario with the companion sun the Nemesis theory. He explained his theory in more detail in his book titled *Nemesis Affair*. Most astronomers reject the Nemesis theory.

*See also* Agassiz; Cuvier; Darwin; Eldredge; Gould

**RAYLEIGH'S LIGHT SCATTERING LAW:** Physics: *Third Baron Rayleigh (Rayleigh was born John William Strutt)* (1842–1919), England.

> *When energy is removed from a beam of electromagnetic radiation (light), the change in the direction (angle) and wavelength of the emitted radiation is dependent on the scattering nature of the medium through which it passes (i.e., gases or liquids).*

Third Baron John Rayleigh confirmed John Tyndall's theory that light passing through the atmosphere is scattered by small particles suspended in the air. The *Tyndall effect* explains that because water droplets in clouds are larger than wavelengths of light, the clouds appear white. Rayleigh applied mathematics to this concept of scattering to explain why the sky is blue in color. He claimed that light from overhead (midday) is more direct and thus is less scattered (fewer particles for light to travel through) than the light coming from the sun when it is near the horizon. Because there is less scattering of overhead light, the wavelengths of this visible light of the electromagnetic spectrum are shorter and thus appear blue. The same reasoning can explain red sunsets, where sunlight is scattered by more particles and thus the light waves are longer and more toward the red end of the spectrum. Rayleigh accomplished this by determining that the amount of scattering was dependent on the wavelength of light. There are two kinds of scattering: instantaneous, considered "true" scattering, which occurs rapidly when electromagnetic energy is absorbed from the incident beams and then re-radiated; and "delayed" scattering, during which a time lapse between the absorption of the energy and its re-radiation takes place. Delayed scattering causes luminescence. This led to an expansion of the scattering law to include how longer AM radio (amplitude modulation) waves are scattered by the atmosphere and thus can travel around corners and around mountains and buildings, whereas shorter radio wavelengths, such as FM radio (frequency modulation) waves and TV (television) waves, cannot go around buildings and mountains but travel in rather straight lines. Scattering experiments that cause beams of electrons, alpha particles, or other subatomic particles to collide with atomic nuclei have uncovered a great deal about atomic structure and the fundamental nature of matter. These experiments use high-energy particle accelerators designed to scatter the particles and record the resulting paths of collisions. Rayleigh also explained there was another type of wave that followed along a surface whose motion decayed exponentially with the depth of the source from the surface. This type of surface wave is now called a *Rayleigh wave* and is the basis for the development of the science of earthquake detection.

*See also* Maxwell; Raman; Tyndall

**RAY'S THEORIES OF FOSSILS AND PLANT CLASSIFICATION:** Biology: *John Ray* (1627–1705), England.

***Ray's theories for the origin of fossils:*** *Fossils were formed by natural processes, not by God.*

John Ray's religion was based on his concept of "natural theology" in which he claimed that if one wants to understand God, one must study His creations of the natural world. Ray proposed several theories about fossils that were considered controversial at that time. There are several other theories for the existence of fossils: 1) some scientists claimed fossils were formed by a creative force on Earth such as earthquakes, floods, and so forth; 2) God was making "models" of different kinds of life; and 3) Satanic forces placed fossils on Earth just to confuse people. Ray's proposed theory was that some organisms possibly washed into big cracks in Earth during the biblical flood. However, he did not believe this was a major cause because this would expose them, and the flood most likely would have washed the fossils away. In addition, most fossils are found in sedimentary beds or stratified rocks. His major theory stated that these organisms were created in the oceans that covered Earth at the time of creation. As the oceans receded, the living organisms were deposited on dry land and then covered with mud and silt, later to become fossilized. His theories that fossils were at one time "natural" living organisms laid the groundwork for future scientists, including Charles Darwin, to explore evolutionary adaptation (*see also* Darwin; Wallace).

***Ray's classification system:*** *Plants and animals can be classified by differences in structure of species rather than by individuals.*

Ray is best known for his classification systems of plants and his later attempts to classify animals by structural similarities and differences. A major contribution was his division of the plant world by distinguishing between *monocotyledons* (based on seeds with a single opening leaf, e.g., grass, corn) and *dicotyledons* (based on seeds with two opening leaves, e.g., trees, beans). Ray also established the basis of classification systems on species, not individuals, and used this system to classify about nineteen thousand different plant species. Ray's classification system influenced Carolus Linnaeus and other taxonomists for several centuries and led others to explore the concept of biological evolution.

*See also* Agassiz; Darwin; Linnaeus; Lyell; Theophrastus

**REDI'S THEORY OF SPONTANEOUS GENERATION:** Biology: *Francesco Redi* (1626–1697), Italy.

> *Flies do not generate spontaneously but rather develop from eggs, whereas some other worms and types of insects may appear by spontaneous generation.*

From ancient times through the Renaissance period, it was accepted that some forms of lower life formed spontaneously from nonliving matter. It seemed obvious to most people that garbage generated rats, and food and manure sooner or later spontaneously generated flies. William Harvey was one of the first to contend that vermin, such as flies and rats, do not appear spontaneously but rather come from such vermin breeding and laying eggs. Francesco Redi decided to investigate Harvey's idea and conducted one of the first examples of a controlled experiment. First, he placed cooked meat in eight jars, covering four of them while leaving four uncovered. Maggots and flies developed in the uncovered jars but not the covered ones. He wondered if the air had something to do with the appearance of flies. Next, he placed more meat in another eight jars, covering four with gauze but otherwise leaving them open to air. He left the other four jars uncovered and exposed to air. Redi concluded that maggots do not develop in covered jars that allow air in but keep out flies; therefore spontaneous generation is not a reality, at least for flies. He also concluded that flies must lay eggs too small to be seen in the open jars, and these eggs develop into maggots, which hatch into flies. However, Redi still believed that spontaneous generation was possible for some living organisms, but his controlled experiments did encourage others to perform more definitive experiments.

*See also* Harvey; Pasteur; Spallanzani

**REED'S THEORY OF THE TRANSMISSION OF YELLOW FEVER:** Biology: *Walter Reed* (1851–1902), United States.

*Yellow fever, also known as Yellow Jack, is carried and transmitted by the Stegomyia fasciata (Aedes aegypti) mosquitoes and is not transmitted by contact with sick patients or their clothing.*

The tropical disease of yellow fever was known since ancient times and along with other insect borne diseases, such as malaria and plague, has killed millions of people. In the 1880s the French attempted to dig a canal across the Isthmus of Panama in Central America to facilitate passage between the Atlantic and Pacific Oceans. During the period of 1881 to 1889 one-third of the workers (about twenty-thousand men) digging the canal died from the acute viral disease called Yellow Jack. This and financial troubles drove the French from the area. Following the 113-day Spanish-American War in 1898, troops from the United States that occupied Cuba were devastated by Yellow Jack. The disease takes a few days to develop after being bitten by an infected mosquito. Damage to the liver occurs. The skin turns yellow, followed by high fever, usually ending in fatal coma. The few survivors are often damaged for life. No cure was available, and in 1900 the exact nature of the disease was unknown.

In 1899 Major Walter Reed traveled to Cuba to study the outbreak of disease in the Army's encampments. The following year the U.S. Surgeon General George Miller Sternberg (1838–1915) created a small committee headed by Reed to examine yellow fever as well as other tropical diseases. The committee was known as the U.S. Army Yellow Fever Commission in Cuba. A Cuban doctor Carlos Finlay (1835–1915) was one of the first to theorize that yellow fever was spread by the bite of a mosquito. However, this theory was not accepted by most of the world's doctors. Even so, Reed's Yellow Fever Commission and a contingent of army/soldier volunteers set out to test

Finlay's theories. The first "guinea pig" was Reed's friend and fellow Commission member Dr. James Carroll (1854–1907) who allowed himself to be bitten by a mosquito thought to carry Yellow Jack fever. He became very ill but survived. Other volunteers were not so fortunate. Team member Dr. Jesse William Lazear (1886–1900) died, and Reed himself became very ill. Although he survived, he sustained ill health for the rest of his life. All the army volunteers refused special pay to engage in a definitive test that kept one group in open tents and another group in screened tents with mosquito netting over their bunks. This and other experiments indicated the mosquito's life cycle. It picks up the disease in the first three days that a patient has yellow fever. It then takes twelve days for the disease to incubate in the mosquito's body, followed by the insect's ability to infect other people.

Discovering the cause of the disease was only the first step. Much later a vaccine was developed, and some forms of medication helped the patients, but the main issue for solving the problem was how to get rid of the insect. Major William Crawford Gorgas (1854–1920), 22nd surgeon general of the U.S. Army, eliminated the Yellow Jack mosquito in Cuba by cleaning out all the low areas containing standing freshwater that served as the breeding grounds for the insects. In an attempt to avoid the same disaster that the French encountered in their effort to build a canal, that is, the deaths of thousands of workers from yellow fever, Gorgas was brought to Panama to do the same job he had done in Cuba. This was a challenge because the isthmus region was larger with more breeding grounds for insects and other vermin. In time he succeeded in not only eradicating the Aedes aegypti mosquito, but also the species that causes malaria, as well as the many rats infested with fleas that carried bubonic plague. By the time the canal was finished in 1914 (the U.S. efforts to build the canal began in 1904), the death rate in the area was about half that of the death rate from mosquito-borne diseases in the southern United States. Most of the mosquito-borne diseases in the United States, as well as malaria, yellow fever, plague, and other insect-borne diseases in the third world countries, were eliminated by the use of the insecticide DDT. It proved to be the most effective means for the eradication of disease-causing insects until it was banned in 1972. DDT's ban was based on data contained in the book, *Silent Spring,* by Rachel Carson (1907–1964) who claimed that it was an environmental disaster. Although this reaction resulted in eliminating the use of DDT, it did not eliminate insect-borne diseases that are now responsible for millions of deaths in undeveloped countries.

*See also* Koch; Pasteur

**REGIOMONTANUS' THEORY FOR TRIGONOMETRY:** Mathematics: *Johannes Müller von Königsberg (Regiomontanus)* (1436–1476), Germany.

> *In trigonometry, the use of tables of 1) sines for minutes and 2) tangents for degrees are more useful than using chords.*

Johannes Müller, the son of a miller, was a well-known fifteenth-century mathematician, astronomer, writer, and translator of Arabic and Greek science and mathematics. Regiomontanus, known as a young prodigy, was admitted to the University of Leipzig at the age of eleven where he studied for three years. He then entered the University of Vienna in 1450 to study mathematics and astronomy under the Austrian astronomer and mathematician Georg von Peurbach (1423–1461), who became his mentor for life.

Regiomontanus was awarded a baccalaureate degree in 1452 at the age of sixteen. He continued his education at the University of Vienna despite the school's regulation that to receive a master's degree, a student must be twenty-one years of age. When he turned twenty-one, he was awarded his MA degree. In 1457 he was appointed to a position on the faculty at the University of Vienna where he continued working with his mentor, Peurbach. He taught courses on Euclid, perspectives, mathematics, and astronomy, while at the same time he constructed his own astronomical instruments including astrolabes. He became interested in reading and interpreting old science, mathematics, and astronomy books, and making copies for his own use—some of which are still in existence. He had a successful career in several other countries, including Italy and Hungary, in addition to writing several important books. One was *Epitome of the Almagest*, that was begun by Peurbach, in which Regiomontanus not only translated some ancient works, but revised some computations dealing with Ptolemy's lunar theory related to the measurement of the apparent diameter of the moon. While writing *Epitome*, he became aware of the need to revise trigonometry as related to astronomy. In his five-volume book *De triangulis omnomodis libri quinque* he demonstrated a new method of solving triangles as used in astronomical observations. The first and second books of the series were most likely the most important where he, in essence, modernized trigonometry by presenting the definitions for quantity, ratio, equality, circles, arcs, chords, and the sine functions. This was followed by axioms and fifty-six theorems on geometry. The last three books were related to spherical trigonometry as related to astronomy. Later, he calculated two different tables of sines. In his *Tables of directions* he first constructed sine tables in sexagesimal numbers (the number system using the base of 60). Later tables of sines were computed using the decimal base.

In the 1470s Regiomontanus made several important astronomical observations including a lunar eclipse and comets. During this period he also built an observatory and workshop in which to construct his instruments, including dials, quadrants, astrolabes, armillary astrolabes, torqueta, parallactic rulers, Jacob's staffs, among many others that were accurate enough for him to identify what later became known as Halley's comet.

Although instruments of that day were not accurate enough to measure precise positions of the moon that could be used for navigation, Regiomontanus described how the moon's exact positions could be used to determine longitude—a problem for which astronomers throughout the ages had sought an answer. He published these finding in *Ephemerides* using his own printing press. Christopher Columbus and Italian explorer and cartographer Amerigo Vespucci (1454–1512) used this publication to measure longitudes in their travels. Regiomontanus was called to Rome in 1475 by Pope Sixtus IV (1414–1484) to assist in revising the calendar. He died in Rome of the plague that broke out after the River Tiber flooded the region in 1476, although some maintained that he was poisoned.

**REICHENBACH'S THEORY OF PROBABILITY BASED ON LOGICAL EMPIRICISM (AKA LOGICAL POSITIVISM):** Mathematics and Philosophy: *Hans Reichenbach* (1891–1953), United States.

***Reichenbach's theory of probability:*** *Probability statements are only about measurable frequencies based on three foundations of probability: 1) the law of consistency—only a new probability can be derived from an existing probability, 2) probability rules are given for* *situations where no probability is present, and 3) 1 and 2 are based on the meaning of probability and how it is applied to problems in mathematics and science.*

Reichenbach's theory of *logical empiricism* is an alternative name for *logical positivism*—both are versions of rationalism that itself is a belief that human knowledge includes some knowledge that is not derived directly from empirical observations. His theory is based on the principle of verification, which means that a statement can only have meaning if it is verified by its own methods. Many theological and ethical statements are, in essence, meaningless and only give credence to the beliefs of the person making such statements. He stressed that only scientific, logical, and mathematical concepts can be meaningful and valid.

Reichenbach completed his early schooling in Hamburg, Germany, and later attended several universities in Germany to study engineering, physics, mathematics, and philosophy. Among his teachers were Max Planck, Arnold Sommerfeld, and Max Born. He held several positions, mainly in philosophy, in several universities and was considered a "philosopher of science" and a "scientific philosopher." After Hitler came to power, Reichenbach emigrated to Turkey and accepted a professorship just before papers arrived that expelled him from Germany because of his Jewish background. While in Turkey, he introduced the concept of interdisciplinary courses in science and in 1935 wrote a definitive paper "The Theory of Probability." In 1938 he came to the United States, accepting a position at the University of California, Los Angeles. While there, he published work on the philosophical foundations of quantum mechanics and on space and time. He published *Philosophic Foundations of Quantum Mechanics* in 1944, in which he claims, "[T]here is not any exhaustive interpretation of quantum mechanics which is free from causal anomalies." Hans Reichenbach was a popular teacher because he encouraged his students to ask questions. He also held discussions on a variety of related topics, which students also enjoyed. He was also a prolific writer during these years. In 1947 he wrote *Elements of Symbolic Logic*, and in 1951 the *Rise of Scientific Philosophy*. Two books on which he was working at the time of his death were published in 1954 and 1956: *Nomological Statements and Admissible Operations*, and *The Direction of Time* which distinguishes between the *order* of time (where "A" events occur before "B" events), and the *directions* of time (a process that is irreversible).

**REICHSTEIN'S THEORY OF THE CHEMICAL ROLE OF THE ADRENAL GLAND:** Biology and Chemistry: *Tadeus Reichstein* (1897–1996), Switzerland. Tadeus Reichstein shared the 1950 Nobel Prize for Physiology or Medicine with Philip Hench and Edward Kendall.

*Six of the twenty-nine identified chemical steroids are essential to prolong life in animals with damaged adrenal glands.*

In 1946 Tadeus Reichstein isolated and identified twenty-nine **steroid** hormones in adrenal glands. He synthesized aldosterone, corticosterone, and hydrocortisone that he synthetically produced on an industrial scale. He also synthesized the steroid deoxycorticosterone that is used to treat Addison's disease. Earlier, he isolated what is known as adrenocorticotropic hormone (ACTH) or, more commonly known as cortisone that is used in the treatment of arthritis, skin rashes, and joint diseases where inflammation is a major symptom. In 1933 Reichstein also artificially synthesized ascorbic acid (vitamin C), the first vitamin that could be mass-produced.

**REINES' THEORY OF NATURAL NEUTRINOS:** Physics: *Frederick Reines* (1918–1988), United States. Frederick Reines shared the 1995 Nobel Prize for Physics with Martin L. Perl.

*If neutrinos exist in the high levels of radiation found inside nuclear reactors, they should exist in the cosmic radiation.*

In 1930 Wolfgang Pauli proposed the theoretical existence of what was called the **neutrino,** a fundamental physical particle that seemed to have no charge and much less mass than the neutron. Pauli claimed that such a particle was necessary to comply with the law of conservation of matter (*see* Figure F2 under Fermi). The problem was that it existed for only a very short period and was no longer detectable when it weakly interacted with other particles. Frederick Reines and his colleague, the American chemical engineer Clyde Cowan (1919–1974), were the first to investigate the neutrino's properties, interactions, and role. First, they confirmed the neutrino's existence as being produced by the high levels of radiation in nuclear reactors. The neutrino is difficult to detect because it travels only a very short distance before weakly interacting with matter and then disappearing. Confirming the neutrino's existence was an extremely difficult task, which Reines and Cowan accomplished in a deep pit near a nuclear reactor in Augusta, Georgia. It was necessary to shield out other high-energy particles and to use tanks of water to slow the neutrinos produced by the reactor so their instruments could record the neutrinos' interactions with other particles. At first, they detected only about three or four events per hour, but this was adequate to prove the existence of neutrinos. Reines and other collaborators were the first to discover neutrinos being emitted from the stellar supernova SN1987A, confirming his theory that neutrinos can be generated from outer space, most likely from the collapse of stars. Reines also found that neutrinos from outer space enter the ground (Earth), from which **muons** are produced. Neutrinos scatter electrons, which produce **antineutrinos**; and oscillating neutrinos can be transformed into different types. Reines' theory of cosmic neutrinos was the forerunner to neutrino physics and neutrino astronomy, which study the interactions of cosmic neutrinos with particles in the atmosphere and their sources in the cosmos. The research related to neutrinos continues as a study of particle physics, which may someday lead to a better understanding of the fundamental laws of conservation of energy and matter.

*See also* Fermi; Pauli

**REVELLE'S THEORY OF GLOBAL WARMING:** Chemistry: *Roger Randall Dougan Revelle* (1909–1991), United States.

*Energy from sunlight arriving at the surface of Earth as ultraviolet and visible light frequencies is absorbed by rocks, soil, and water at Earth's surface, but not by the small amount of carbon dioxide in the atmosphere at Earth's surface. Thus, this absorbed energy from the surface of Earth is radiated back into space from the surface of Earth as infrared radiation (heat) that is absorbed by carbon dioxide in the atmosphere where it acts somewhat like the glass in a greenhouse and is radiated back to Earth as heat.*

Roger Revelle received his PhD degree in oceanography in 1936 from the University of California, Berkeley, while he was employed at the Scripps Institute of

Oceanography (SIO). He became its director from 1950 to 1964, then he moved to Harvard University until his retirement in 1976. While at SIO, he was one of the first scientists to study and verify the magnetic reversals of Earth's magnetic field that led to a better understanding of the tectonics responsible for the spreading of the seafloor and the Great Atlantic Riff. Both are a result of the drifting of continental landmasses. During the 1950s he turned his interest to the new concern of global warming that was just being explored.

In 1896 the Swedish chemist Svante August Arrhenius was the first to suggest that carbon dioxide ($CO_2$) gas had the capability of absorbing heat and that this could possibly present a problem of warming Earth. He was also the first to propose that an increase in the release of $CO_2$ into the atmosphere could possibly cause an increase in global temperatures. He proposed that the percentage of $CO_2$in the upper atmosphere regulated Earth's temperature. Arrhenius believed that this factor was responsible for the past heating and cooling effects on Earth, including the possible cause of the last ice age. Arrhenius also calculated that doubling the amount of $CO_2$ in the atmosphere could possibly raise Earth's temperature by 10°C. His research led to more studies related to the "greenhouse effect" that led to the current global warming debate and increased political and environmental speculation for pending global disasters. Doubling the amount of $CO_2$ (a 100% increase) of carbon dioxide as proposed by Arrhenius is not realistic. At the beginning of the twentieth century $CO_2$ in the atmosphere was about 0.03% of at least eighteen different gases that make up the total atmosphere. This amount of $CO_2$ is equal to about 325 parts per million volume (ppmv) of the total atmosphere. Today it has risen to about 0.035 percent and may rise to about 0.040 percent or about 400 ppmv, which may raise global temperatures about one degree Celsius by the year 2100. (Note: Some climate experts estimate that the amount of $CO_2$ in the atmosphere will be greater by the year 2100 at the present rate of release into the atmosphere.) The term "greenhouse effect" is somewhat confusing because the glass roof in a greenhouse does not "radiate" the heat back inside the greenhouse but rather "traps" the heat inside, while $CO_2$ and other gases in the atmosphere actually "radiate" the heat back to Earth (this might be considered a distinction without a difference). Revelle was partially responsible for the establishment of sites at Mauna Loa in Hawaii and at the South Pole to measure the real extent to which $CO_2$ and other gases were involved in global warming. The National Science Foundation supported a Greenland Ice Sheet Project that used new techniques to glean information from ice cores form Greenland's ice sheets that gives a climate record of over one hundred thousand years. These cores and others from Antarctica and the Himalayas provide the chemistry of past atmospheres that indicated climate changes as often as a few decades to over periods of thousands of years. Most of Earth's climate changes occurred long before human civilization and the more recent release of greenhouse gases.

There are many causes for the warming effects of Earth's atmosphere in addition to $CO_2$. First of all, many people do not recognize that Earth is an evolving planet. It is dynamic and undergoes constant physical and geological changes, usually slowly but not always, as in earthquakes. Second, there are other gases that make up the atmosphere, including so-called pollution gases, such as several oxides of nitrogen, and especially the hydrocarbon gas methane that is expelled constantly from vents in the bottoms of oceans as well from gas and oil wells. Methane ($CH_4$), although not as concentrated as $CO_2$ in the atmosphere, is many times more effective as a greenhouse gas. There are many factors that account for changes over the centuries of Earth surface temperatures. A few examples are the precession of Earth on its axis; the variations in the sun resulting in the fluxing output of the sun's energy; internal geological changes in the deep oceans and geological structures; and many more, including the results of human activities such as forest destruction, industrial and automobile emissions, and other activities—human and natural.

There is no argument that Earth's surface temperature has increased about 1°C over the last century—a span that has included periods of warming as well as cooling. One problem is the use of advanced computers to "model" the causes and extent of climate change related to global warming. There are a multitude of factors involved in global warming that are not included, or just roughly estimated, that are included in the computer climate modeling programs. These models are improving, but even using weather satellites scientists cannot accurately predict weather for more than a few days, and current methods are still unable to predict accurately the global climate changes for future years or centuries. There are a few solutions to solving the problem of excess $CO_2$ expelled into the atmosphere. One is that because plants use $CO_2$ in the process of photosynthesis to make food for all animals, including us, exposing crops to concentrated $CO_2$ will increase growth and production. This includes trees that act as a huge sink for $CO_2$, so are the vast oceans that absorb not only $CO_2$, but other greenhouse gases as well. The oceans sequester a significant amount of excess $CO_2$ in the forms of many carbon compounds in sea life such as shells and coral. Another method used by some Scandinavian industries is to pump their excess $CO_2$ deep into the oceans where it is sequestered on the ocean bottom. This process can also be used on land by pumping $CO_2$ into deep wells or empty gas wells. And another is the market-based proposal of establishing "caps" on industrial pollution with the possibility of selling unused allowances to other entities. Still another possible solution is to develop more machinery and automobiles that use alternative fuels that do not produce $CO_2$. One example is to extensively use nuclear power to generate electricity. Over the last fifty years or so it has become obvious that even though nuclear power is safer than coal- or oil-produced electricity, nuclear generated electricity is still not accepted by many people.

*See also* Arrhenius; Rowland

**RICCIOLO'S THEORY OF FALLING BODIES:** Physics: *Giovanni Battista Ricciolo* (1598–1671), Italy.

> *A pendulum that beats once per second can be used to confirm Galileo's theory of falling bodies.*

Giovanni Ricciolo, an observational astronomer of the seventeenth century, disagreed with many of Copernicus' theories. Even so, he mapped mountains and craters on the moon and was the first to identify Mizar as a double star. He attempted to confirm Galileo's theory that the period of a swinging pendulum is the square of its length. He, and others who assisted him, tried to count the number of swings each day. If the number of swings per day could be adjusted to 86,400 (60 sec./min. × 60 min./hr. × 24 hr./day = 86,400 sec. per day), they would succeed in developing a pendulum that could count seconds accurately. They tired of counting all day and night and so abandoned the project. However, Ricciolo used his pendulum to measure falling bodies. It is assumed that it was either Ricciolo or Simon Stevin, not Galileo, who dropped two balls of different sizes (weights) from the Tower of Pisa. Galileo used inclined planes to slow the descent of the balls and thus was able to time them with his heart pulse beat. Ricciolo used a pendulum as an accurate timekeeper. They both came up with the figure for g (the gravity constant) of approximately 9.144 meters per sec. squared that compares with today's figure of 9.807 (about 30 feet per sec. squared).

*See also* Galileo

**RICHARDSON'S LAW OF THERMIONIC EMISSION:** Physics: *Sir Owen Willans Richardson* (1879–1959), England. Sir Owen Richardson was awarded the 1928 Nobel Prize for Physics.

> *The kinetic energy of electrons emitted from the surface of a solid is exponentially related to the increase in the emitter's temperature.*

Sir Owen Richardson proposed an explanation for Thomas Edison's observation of the emission of electrons from hot surfaces (Edison effect). The electrons came from inside the solid, which was heated, and escaped from this material when the electrons achieved enough kinetic energy to overcome the "grasp" of the surface of the solid. Richardson's law states: *The electron's temperature (kinetic energy) increases exponentially with the increase of the emitter's temperature.* He claimed that the electrons came from within the solid and eventually escaped if they had achieved adequate kinetic energy (heat) to overcome the energy (heat) barrier existing at the surface of the metal. This is what he meant by **thermionic emission** of metals, which he related to the thermal activity of molecules within a liquid that achieve adequate kinetic energy to pass the surface tension of the liquid resulting in the escape of the molecules from the surface of a liquid during the process of evaporation and boiling. This law became important in the development of electron tubes used in early radio, TV, and radar prior to the days of **transistors** and computer chips.

*See also* Edison; Shockley

**RICHTER'S THEORY OF EARTHQUAKE MAGNITUDE:** Geology: *Charles Francis Richter* (1900–1985), United States.

> *Earthquakes can be measured on an absolute scale based on the amplitude of the waves produced.*

Several earthquake scales existed before Charles Richter developed his absolute log scale. In 1902 the Italian volcanologist Giuseppe Mercalli (1850–1914) devised a descriptive scale based on the extent of devastation caused by an earthquake, as well as descriptions of the aftereffects. This was a very subjective means for determining the actual strength of earthquakes because it depended on the nonstandard observations of humans and the types of structures in the region of the earthquake. In 1935 Charles Richter created a scale based on the maximum magnitude of the waves as $\log_{10}$ (logarithm base 10 or a tenfold increase in power for each numerical increase in the scale), as measured in microns. His scale has values of 1 to 9, where 1 is the least damaging and 9 the most damaging. Using logarithms for this scale can be confusing because each increase in number represents a tenfold increase in the power and severity of the earthquake. In other words, an earthquake measured at 5 on the Richter scale is ten times stronger than one with a 4 reading, and a 6 is ten times stronger than a 5 and so on. However, the same 1 unit increase in magnitude corresponds to an increase of approximately thirty-two times the earthquake's energy. It is estimated that for every fifty thousand earthquakes of magnitude 3 or 4, only one of a magnitude 8 or 9 will occur. The United States Geological Survey (USGS) National Earthquake Information Center lists the 9.5 magnitude earthquake in Chile in 1960 as the most powerful earthquake since 1900. The December 2004 earthquake off the west coast of Northern Sumatra that caused the devastating tsunami in that region of the world registered at 9.1.

**RIEMANN'S THEORY FOR DIFFERENTIAL GEOMETRY:** Mathematics: *Georg Friedrich Bernhard Riemann* (1826–1866), Germany.

> *The curvature for the tensor (multilinear function) of surfaces can be reduced to a scalar number that is either positive, negative, or zero, while the nonzero as well constant cases are models of non-Euclidean geometries.*

Riemann geometry (also known as elliptical geometry) is a form of non-Euclidean geometry. Non-Euclidean geometry differs from two of Euclid's original postulates. Euclid's 5th Postulate deals with parallel lines, whereas Riemannian geometry states that there are no parallel lines. Euclid's 2nd Postulate states that a straight line can be extended to infinity, whereas Riemannian geometry states that all straight lines are of the same length. Another example that most mathematics students learned in geometry classes was that Euclid's geometry states that the sum of the three angles for any triangle will always total 180°. Conversely, Riemann's non-Euclidean geometry states that the sum of the angles in a triangle may be greater than 180°. Thus, Riemann's version was used to explain Einstein's theory of general relativity that includes the concept of curved space.

Riemann made contributions to pure mathematics and analytical mathematics that are related to various areas of physics. Several examples are given below. (Note: A more detailed explanation of Riemann's mathematical concepts is beyond the scope of this book.)

1. *Riemann's zeta function* is an important concept in number theory, primarily because of its relation to the distribution of prime numbers. It also has applications in physics, probability theory, as well as applied statistics. The following are the most commonly used values in the Riemann zeta function:

   $(x) = 1 + (1/2)\hat{\ }x + (1/3)\hat{\ }\,x + (1/4)\hat{\ }\,x + \ldots$. etc.,
   Only where $x = 1$; and,
   if $x > 1$, then it is a finite number; and,
   if $x <$ *than 1* then it is an infinite number.

2. *Riemann integral* is a branch of mathematics that involves real analysis by using approximations for areas. In other words, taking better and better approximations, in time, will result in the exact area under a curve. In essence, the

Riemann integral is the sum of the partitions under the curve as they get finer and finer and the sections under the curve become smaller and smaller where the limit is zero (this is something akin to Archimedes' "squaring the circle" by using polygons to arrive at a better approximation for pi).

3. *The Riemann hypothesis* is related to the *Riemann zeta function* that is explained in 1) above that was first introduced by Euler. However, it was Riemann who generalized its use. Hence, it is known as the Riemann hypothesis which is the most famous problem in mathematics that has never been solved. Many young mathematicians have tackled aspects of this hypothesis either by evaluations of the Riemann zeta functions or calculating the distributions of zeros in the zeta functions The current record of calculation of the amount of zeros has been verified to $10^{13}$ zeros.

*See also* Euler; Gauss

**ROBBINS' THEORY FOR THE POLIOVIRUS:** Biology: *Frederick Chapman Robbins* (1916–2003), United States. Frederick Robbins shared the 1954 Nobel Prize for Physiology or Medicine with American virologists John Enders and Thomas Weller (1915–).

*Because the poliovirus can multiply outside nerve tissue, it can exist in other tissue as well.*

Frederick Robbins' medical background included collaborating with the U.S. Army to find cures for diseases caused by viruses and parasitic microorganisms. In 1952, Robbins and his colleagues grew the virus that caused poliomyelitis in cultures produced outside a living organism. Up to this time, it was thought the poliovirus could exist only in nerve cells of the central nervous system. They proved this particular virus could live in tissue other than nerve tissue, leading to the theory that the virus survives in body tissue and later attacks the central nervous system. This research resulted in the development of vaccines and new techniques for culturing and detecting the poliovirus that may be dormant in body tissue and later attack the central nervous system.

*See also* Delbruck; Enders; Sabin

**ROBERTS' THEORY OF SPLIT GENES:** Biology: *Richard J. Roberts* (1943–), England. Richard Roberts shared the 1993 Nobel Prize for Physiology or Medicine with American microbiologist Phillip A. Sharp.

*The DNA of prokaryotic cells becomes messenger RNA, which acts as a template to assemble amino acids into proteins.*

Prokaryotic cells have very primitive, poorly defined nuclei, and their DNA has no membrane surrounding them. Some examples are blue-green algae and some primitive bacteria, such as *Escherichia coli* whose primitive nuclei have a single chromosome with only about three million DNA base pairs. Because amino acids require about nine hundred DNA base pairs to form proteins, this prokaryotic type cell should be able to produce about three thousand different proteins. This is in comparison with cells of mammals, called *eukaryotic cells*, which have a well-defined nucleus and contain about four billion DNA base pairs and can produce over 3 million proteins—many more than mammals need. Roberts found that part of the DNA of the prokaryotic cells with no nuclei can split into separate messenger RNA capable of producing proteins. Roberts' simple explanation of the structure of primitive prokaryotic nuclei made the study of the formation of RNA as genetic messengers for the DNA much easier than simply studying the very complex RNA and DNA of mammals. Roberts' theory advanced a better understanding of how amino acids form proteins within the human body.

*See also* Crick; Sharp; Watson (James)

**ROCHE'S "LIMIT" THEORY:** Astronomy: *Edouard Albert Roche* (1820–1883), France.

> *A satellite of a planet cannot be closer than 2.44 radii of the larger body without disintegrating.*

In 1850 Edouard Roche proposed what is now known as the *Roche limit*, based on the concept that if a satellite and the planet that it is orbiting are the same density, there is a limit to their proximity to each other without the satellite, or both, breaking up under the force of gravity. The Roche limit explains the existence of the rings of the planet Saturn. Because the outer ring of Saturn is only 2.3 times the radius of Saturn, it might have been a solid satellite that came too close and broke into fragments. The Roche limit also explains why these many small chunks of matter did not re-form into a solid body orbiting Saturn. The orbit of Earth's satellite (moon) is many times the distance of 2.44 radii (the Roche limit) of Earth, thus there is little chance for it to be affected by gravity to the extent it would break into fragments.

*See also* Cassini; Schiaparelli

**RÖENTGEN'S THEORY OF X-RAYS:** Physics: *Wilhelm Conrad Röentgen* (1845–1923), Germany. Wilhelm Röentgen received the first Nobel Prize for Physics in 1901.

> *Cathode tubes are capable of sending unknown and undetectable rays to screens, thus causing fluorescence.*

In 1895 Wilhelm Röentgen experimented with a Crookes' tube (a high-voltage gaseous-discharge tube), which produces cathode rays, which produce fluorescence when focused onto a sensitive screen, and detected an unknown form of radiation (*see* Figure C6 under Crookes). He noticed that a cardboard coated with a yellow-green crystal fluorescent material, $BaPt(CN)_4$ (barium cyanoplatinite), located in another part of the room, also was fluorescing when the tube was in operation, even though no rays were directed toward it. He concluded that because **cathode** rays can travel only a very short distance, they must originate from some unknown radiation. Thus, he called them X-rays (also known as Röentgen rays in his honor). He continued to study X-rays, recording accurate descriptions of their characteristics as listed below:

- X-rays had a much greater range than cathode rays.
- X-rays traveled in straight lines but may also be scattered in straight lines from their source.

- X -rays were not affected by magnetic fields or electrical charges.
- X-rays could pass through cardboard and thin metal sheets. Most materials, except lead, are transparent to some degree.
- X-rays could expose photographic materials.
- X-rays passed through the human hand and outlined the bone structure.
- X-rays are longitudinal vibrations (waves), whereas light consists of transverse vibrations.

The discovery of X-rays did not solve the issue of the particle–wave duality nature of light, which was being explored at that time. Rather, it complicated the dilemma because some characteristics of X-rays are similar to light rays and some are not. The use of X-rays became important in the study of crystal structures, as well as in medical diagnosis, and later led to the discovery of radioactivity. Röentgen and his assistant were subjected to excessive exposure to X-rays; both died from radiation poisoning.

*See also* Becquerel; Curies

**ROMER'S THEORY FOR THE SPEED OF LIGHT:** Physics: *Olaus Christensen Romer* (1644–1710), Denmark.

*The motion of Earth to or away from Jupiter can be used to establish the speed of light.*

In the 1670s while examining the records of Giovanni Cassini who had determined Jupiter's rotational period and its distance from Earth, Olaus Romer noticed the figures varied depending on whether Earth and Jupiter were approaching each other in their orbits or receding from each other. There was a difference of ten minutes from the time Jupiter's four then-known moons went behind the planet (were eclipsed) while at the same time Earth's orbital path was *proceeding* in the direction of Jupiter, and when Jupiter's major moons were next eclipsed as Earth's orbital path was *receding* from Jupiter. This ten-minute difference was the amount of time it took the light from Jupiter to reach Earth from these two different distances between Jupiter and Earth. This provided the necessary data for Romer to calculate the speed of light since Cassini had previously determined the distance of Jupiter from Earth. In 1676, Romer announced his theory for establishing the fundamental constant of the speed of light as 140,000 miles per second, which is only about 75% of today's figure of 25,000 kilometers per second (about 186,000 miles/sec.) This was the first proof that light has a finite speed.

*See also* Cassini; Michelson

**ROSSI'S THEORY FOR COSMIC RADIATION:** Physics: *Bruno Benedetti Rossi* (1905–1994), Italy.

*The charge on cosmic rays can be detected by the influence of Earth's magnetic field.*

Cosmic rays were first detected in the early 1900s, but little was known about them except they were a form of high-energy, penetrating radiation. In 1930 Bruno Rossi tested his cosmic ray theory using the east–west symmetry concept. Earth's eastward and westward magnetic fields would act differently on incoming cosmic rays due to the direction of the fields' motions. Rossi set up several Geiger counters (radiation detectors) on a high mountain, facing some eastward and facing several westward, so they could detect and count the cosmic rays coming from outer space from different directions as Earth rotated on its axis. He found an excess of 26% of cosmic rays coming eastward toward Earth. Thus, he concluded they were mainly composed of positive protons and other positive particles, along with some electrons. These were all high-energy particles coming from the sun and possibly supernovae (stars). Later, Rossi believed X-rays must also originate from astronomical bodies in outer space but were not detectible on Earth because they were all absorbed by the atmosphere. In the 1960s, Rossi was one of the pioneers in the use of rockets that carried instruments to detect cosmic X-rays above Earth's atmosphere. He found some X-rays originating from the Crab Nebula, the Scorpio constellation, and many other sources beyond the solar system. Currently a special telescope is orbiting Earth, detecting X-rays leftover from the big bang.

**ROWLAND'S THEORY OF CHLOROFLUOROCARBONS' EFFECTS ON THE OZONE:** Chemistry: *F. Sherwood (Sherry) Rowland* (1927–), United States. Sherry Rowland shared the 1995 Nobel Prize in Chemistry with two other atmospheric chemists Paul J. Crutzen from the Netherlands and Mario J. Molina from Mexico for their efforts in identifying the threat of chlorofluorocarbons on Earth's atmosphere.

Rowland's theory for CFCs effects on ozone: *Chlorofluorocarbons will decompose in the upper atmosphere, releasing chlorine, which reacts with and breaks down ozone molecules:* $Cl + O_3 \rightarrow ClO + O_2$*, and* $ClO + O \rightarrow Cl + O_2$ *under the influence of UV.*

Rowland's theory is based on one type of reaction of oxygen atoms proposed in 1930 by the British astronomer and geophysicist Sydney Chapman (1888–1970) as a sequence of oxygen to ozone and then reversed as: $O + O_2 \rightarrow O_3$; then $O_3 + O \rightarrow 2O_2$. In the early 1970s, Sherry Rowland and his postdoctoral student, Mario J. Molina, began investigating the possible effects of chlorofluorocarbons (CFCs) on the ozone. In the laboratory setting they were able to work out the reaction as $Cl + O_3 \rightarrow ClO + O_2$, and $ClO + O \rightarrow Cl + O_2$, with a net result of $O_3 + O \rightarrow 2O_2$. In this reaction the *Cl* is a free chlorine atom, $O_3$ is a form of oxygen molecule called ozone, and O is a nascent oxygen atom, while $O_2$ is a regular oxygen molecule. This reaction is driven by ultraviolet radiation (UV) striking the CFC molecules and separating out the chlorine atoms that collide with the ozone molecules that then bond with one of the ozone's oxygen atoms forming chlorine monoxide (ClO). In the second reaction, the chlorine is regenerated (freed) to start the process all over again, but some of the oxygen atoms in the second reaction can also combine with the oxygen molecule to re-form the ozone molecule. It takes only a relatively small amount of the CFCs to start the reaction because the chlorine (Cl) in the CFCs can be used over and over again. The question that has not yet been settled is whether this laboratory reaction is the same as what actually happens in the 15- to 30-mile-high **ozone layer**. As the amount of CFCs entering the atmosphere increased after the 1970s, there was a detectable decrease in the ozone layer over Antarctica, but not much of a "hole" over the North Pole. The thickness of the ozone layer has always been cyclic and is always thinner over the equator because this is the area where it is generated and then spreads out to the polar regions. Because refrigeration and air-conditioning used most of the CFCs, these industries in the United States and most of Europe have eliminated their use and are substituting less reactive substances, such as hydrochlorofluorocarbons (HCFCs) and hydrofluorocarbons (HFCs). (However, many underdeveloped countries continue to use CFCs.) Experts claim that these and about eighteen other possible HFC substitutes will not cause **global warming** and ozone depletion because they do not contain free chlorine and will decompose in the lower atmosphere. Research continues for even better substitute fluids for refrigeration and air-conditioning use.

*See also* Arrhenius

**RUBBIA'S THEORY OF INTERMEDIATE VECTOR BOSONS:** Physics: *Carlo Rubbia* (1934–), Italy. Carlo Rubbia shared the 1984 Nobel Prize for Physics with Simon van der Meer, the Dutch accelerator physicist at CERN.

> *Intermediated vector bosons might be produced in super energy accelerators by using colliding beams of protons and antiprotons that hit each other head on.*

The history of the discovery of elementary particles found in atoms begins with J.J. Thomson's discovery of the electron in 1897, followed by the discovery of the nucleus of the atom in 1911 by Ernest Rutherford and the realization that the nucleus of the hydrogen atom consisted of a single proton. The discovery of the neutron in 1932 was somewhat more difficult because it has no electrical charge. The model of the atom (except for the element hydrogen) was soon determined to contain a nucleus consisting of three types of particles: the positive protons, neutral neutrons, and negative electrons. This model did not last long as additional elementary particles were found in the nuclei of atoms. They became so numerous that they soon were described by a single theoretical name called the Standard Model that includes *quarks* and *leptons* and how they interact within three types of forces known as the *strong*, *weak*, and *electromagnetic* forces (so far the force of gravity is not included in the Standard Model). Gauge bosons are force-related particles that are not the same as quarks and leptons. The Standard Model, although useful, does not identify the mass of elementary particles. So far, the lightest of elementary particles that have mass are electrons and the heaviest so far discovered is the top quark that weighs about two hundred thousand times that of the electron.

Two classes of elementary particles are described by the statistics that are based on two cases. First, is the Fermi–Dirac statistics that apply to fermions (lepton and quarks). According to the Pauli exclusion principle, no two fermions are allowed to occupy the same quantum state at the same time. Second, is the Bose–Einstein statistical rule that states they are the particles known as bosons, which are elementary particles not affected by the Pauli exclusion principle (the boson is named after the Indian [Bengali] mathematical physicist Satyendra Nath Bose [1894–1975], who provided the foundation for the Bose–Einstein condensate). That means that there is no restriction on the number of particles that can exist in the same quantum state. In essence, fermions compose the structures of atoms and nuclei of atoms, whereas bosons are related to forces that interact with fermions. Some examples of bosons that provide these forces are: photons, gluons, and the W and Z particles. Several other particles have been described by unique and somewhat exotic statistical behaviors. Some of these exotic particles are mesons that consist of pairs of quark–antiquark and baryons as quark triplets. Some of the more elusive particles cannot be classified due to their very short lifetime existence, thus they leave no tracks in bubble or cloud chambers, and they cannot yet be detected.

Carlo Rubbia was involved with a new type of particle accelerator at CERN that used intersecting storage rings that cause beams of protons accelerated in one direction and anti-protons beams accelerated in the opposite direction to collide with each other with tremendous force. Using this colliding accelerator enabled Rubbia to discover *intermediate bosons*, which are particles approximately one hundred times heavier than regular protons.

*See also* Dirac; Einstein; Fermi; Higgs; Rutherford; Thomson

**RUBIN'S THEORY OF DARK MATTER:** Astronomy: *Vera Cooper Rubin* (1928–), United States.

*Galactic rotation indicates there is more mass in galaxies than is visible from Earth.*

Vera Rubin studied spiral galaxies by measuring the rotational velocity of their arms by using the Doppler shift that indicates that the light from a body moving away from the viewer will appear redder, and when moving toward the viewer, it appears bluer. In addition, Kepler's law of rotation of bodies in space states that the velocity of a rotating body decreases with the distance. When the gravitational constant is applied to a revolving mass, the following equation should apply: $v^2 = GM/r^2$, where $v$ is the velocity, $r$ is the radius of the orbiting mass, M is the mass, and G is the gravitational constant. Rubin found this equation did not apply to some spiral galaxies because they increased their speed with distance and their mass seemed much too low. She interpreted this to mean the mass had to be there but that it was not visible from Earth. She called this unseen mass **dark matter.** Rubin also concluded that over 90% of all the matter in the universe does not emit much radiation and thus is dark and relatively "cold" in the sense that no light or infrared (heat) radiation is detected. Finally, she concluded there are more dark galaxies than luminous ones. In addition, she believes there has to be much more matter than can be seen because it is required to provide the gravity to hold galaxies together so they do not "fly apart." Solving the puzzle of dark matter may lead to an understanding of the fundamental nature of the universe. Most astronomers accept her concept of dark matter although more research is still required to completely understand our universe.

*See also* Doppler; Kepler

**RUMFORD'S THEORY OF RELATING WORK TO HEAT:** Physics: *Count Benjamin Thomson Rumford* (1753–1814), England.

*A specific amount of work can be converted into a measurable amount of heat.*

Benjamin Rumford was impressed by the amount of heat generated by the process of boring out holes in metal cannon barrels even when water was used to cool the operation. Rumford was familiar with the old concept of "caloric" as being the property within substances that was released by friction or by forcing it out of solids in some way. Some scientists said the boring process "wrung out" the caloric from the metal; others said all those fine shavings created the heat. Rumford had a different theory. He believed heat was generated by the mechanical work performed and proposed there was a conservation of work (friction) and heat (motion or energy). This was one of the first concepts of the conservation of matter and energy and that heat involves motion of some sort (kinetic energy). Several other developments furthered Rumford's theory. These and other theories led to the laws of conservation of mass, energy, and momentum. Rumford invented the **calorimeter**, which measures the amount of heat generated by mechanical work.

*See also* Joule; Lavoisier

**RUSSELL'S THEORY OF STELLAR EVOLUTION:** Astronomy: *Henry Norris Russell* (1877–1957), United States.

*Based on the correlation of the magnitude of stars to their types, stars evolve through stages of contraction from hot giants, to smaller stars, and finally into cold dwarfs.*

In 1913 Henry Russell published the results of his research relating the classification of stars by type to their brightness (magnitude). At about the same time another astronomer, Ejnar Hertzsprung, produced similar data. Their combined data were placed in graph form, known as the Hertzsprung–Russell diagram, which depicts a main sequence of stars as distinct from supergiants, giants, and white dwarfs (*see* Figure H6 under Hertzsprung for a depiction of their graph). Russell was the first to use the terms "giant" and "dwarf" to describe groups of stars. He was also the first to use photographic plates to record stellar **parallax** and to measure a star's luminosity. The diagram depicts the concentration of the supergiants and giants (located in the upper right of the graph) that in time become hot stars in the main sequence, followed by their collapse under the force of gravity to form cool, white dwarfs (located in the lower left of the diagram). Russell developed a method for measuring the size and orbital period for stars as well as their spectra. His work enabled other astronomers to determine galactic distances for stars that were beyond the parallax technique for making measurements. His work also led to new theories for stellar evolution.

*See also* Hertzsprung

**RUTHERFORD'S THEORIES OF RADIOACTIVITY/TRANSMUTATION AND ATOMIC STRUCTURE:** Physics: *Baron Ernest Rutherford* (1871–1937), New Zealand.

***Rutherford's theory of radioactivity and transmutation:*** *Radioactive substances emit three different types of radiation by which one element is changed into a different element.*

Baron Ernest Rutherford was one of the first to explore the emissions of polonium and thorium, in addition to radium. In 1899 he discovered there were two different types of radiation emissions from these mineral elements that he referred to as *radioactivity*. He named one type of radiation *alpha*, which would cause ionization but could be stopped by a piece of paper (helium nuclei). The other he named *beta*, later known as high-energy electron emission, which was not ionizing but somewhat more penetrating than beta radiation. He then determined there was a third type of radioactivity, which was characterized by high energy, deep penetration, and highly ionizing, but was not affected by a magnetic field, which he named *gamma rays*. Rutherford used this knowledge to devise an unusual theory, called *atomic transmutation*, which almost sounded like the

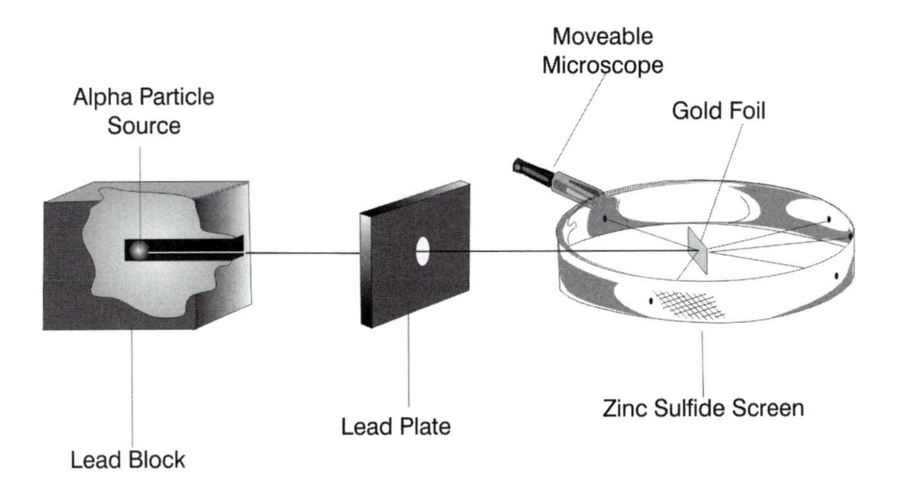


Figure R1. Rutherford's apparatus' arrangement to shoot alpha particles into a small sheet of gold foil that scattered a few particles but allowed most of them to go straight through without hitting any gold nuclei.

old alchemists' dream of the philosophers' stone that was able to change lead into gold. Rutherford's idea stated that as some of these radioactive particles were emitted from their source element, the mass and charge (number of protons) of the original atoms were changed to become a different element. Rutherford and Frederick Soddy confirmed this theory with experiments using radioactive thorium that decayed into another active form, which they called *thorium-X*, which resulted when a series of chemical and physical changes converted one type of atom into another. This was known as *transmutation*. As a result, Rutherford became most interested in the alpha particles and their effect on substances.

***Rutherford's theory of atomic structure:*** *The atom, which is composed mostly of "empty space," has a mass that is concentrated in a very small, dense, central particle that contains a charge.*

Ernest Rutherford knew alpha particles (hydrogen nuclei) could expose photographic plates and could be beamed through very thin pieces of material to produce a fuzzy image. Two of his students conducted an experiment where alpha particles were beamed through a very thin piece of gold foil (about 0.00004 centimeter, which is only a few atoms thick) to determine what type of pattern the particles would form on the other side of the foil.

Rutherford noticed that most of the alpha particles went straight through the foil and were recorded by the detecting instrument directly behind the foil. But a detector off to the side at about 45 degrees also picked up some signals, indicating that something in the foil was deflecting a few of the alpha particles. Rutherford noticed that although most particles went through the foil as if nothing was there, a few were deflected to the side, and a few actually seemed to bounce backwards toward the source. He said, "It was almost as incredible as if you fired a 15-inch shell at a piece of

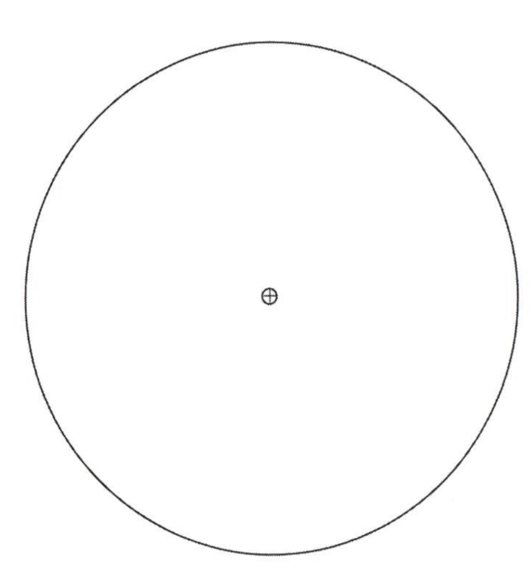


Figure R2. Rutherford's experiment indicated that the positively charged atom consisted almost entirely of a massive but very small positive nucleus. The atom consisted of a vast area around this tiny nucleus. Negatively charged electrons that weigh much less than the positively charged central nucleus occupy the outer reaches of this area. His conclusion was that the atom is mostly empty space.

tissue paper, and it came back to hit you." After making some calculations, he concluded that this backward scattering of the alpha particles is evidence of a few collisions with something where almost all the mass is concentrated in a central, very small "nucleus." It was at this point that he realized this central nucleus had a positive charge (*see* Figure R2).

*See also* Bohr; Curies; Mosely; Soddy

**RYDBERG'S THEORY OF PERIODICITY FOR ATOMIC STRUCTURE:** Physics: *Johannes Robert Rydberg* (1854–1919), Sweden.

> *Elements can be organized by the structure of their atoms based on their spectra rather than according to their mass.*

In the 1880s Johannes Rydberg was aware that Swiss mathematician and physicist Johann Jakob Balmer first discovered the relevance of the spectral lines of the hydrogen atom. Balmer found there was a simple relationship between the wavelength of the lines and the spaces between them when expressed on a graph. Rydberg's theory and experiments provided the explanation for this relationship. He examined the spectra of hydrogen atoms and discovered that the frequencies of the excited atoms produced a spectrum that can be stated as a constant, relating the wavelength to a series of lines in the spectrum. The Rydberg equation can be stated as: $\lambda = R(1/m^2 - 1/n^2)$, where $\lambda$ is the wavelength, $R$ is the Rydberg constant, and $m$ and $n$ are whole numbers squared. It can also be expressed as $1/\lambda\ R(1/1^2 - 1/m^2)$, where $m$ must be an integer larger than 1. The Balmer spectra series for atoms represented only the shorter ultraviolet range. Rydberg proceeded to reorganize Dmitri Mendeleev's **Periodic Table of the Chemical Elements** according to the structure of atoms based on their spectral lines (*see* Figure M5 under Mendeleev). After applying his equations to the patterns of atomic structure, Rydberg developed a spiral form of the periodic table. Soon after Rydberg developed his equations, Henry Moseley determined that the nuclei of atoms had positive charges, which confirmed Rydberg's and Rutherford's theories.

*See also* Balmer; Bohr; Mendeleev; Moseley; Newlands; Rutherford

**RYLE'S THEORY OF USING RADIO ASTRONOMY FOR OBSERVING DISTANT GALAXIES:** Astronomy: *Martin Ryle* (1918–1984), England. Martin Ryle shared the 1974 Nobel Prize in Physics with Antony Hewish.

> *By using two smaller astronomical interferometer type radio telescopes located in a 1.5 kilometer diameter area, a synthesis of their apertures can be analyzed and thus improve the computation of data received from radio signals from distant quasi-stellar objects called "quasars."*

Sir Martin Ryle earned a degree in physics from the University of Oxford in England in 1939. During World War II he helped design radar equipment that eventually helped save England from German air attacks. After the war, he became director of the Mullard Radio Astronomy Observatory, which is located at the University of Cambridge, and later a professor of radio astronomy. He was elected to the Royal Society in 1952, knighted in 1966, and later became the Astronomer Royal from 1972 to 1982.

Early in his career he led the effort of the Cambridge Radio Astronomy Group in their production of several radio astronomy catalogues that, in time, led to the discovery of the first quasar. His major achievement was the development of the theory that greater definition and depth could be achieved by connecting two smaller radio telescopes into one large one, thus enlarging the viewing aperture of the whole system. He led the construction of such a system. His technique is known as "aperture synthesis," which means by using two smaller telescopes on rails with adjustable positions within a 1.5 kilometer diameter area, he could obtain the same results as one huge radio telescope. While using this system, he discovered the location of the first pulsar. Later, his theory of aperture synthesis was expanded by constructing several smaller radio telescopes on rails in an area 5 kilometers in diameter. By adjusting the positions of the entire telescope system and synthesizing their individual signals, he was able to make observations crucial to the study of the physical characteristics of stars and systems of the cosmos. It also led to a better understanding of the universe as a whole. One example is the discovery of a unique type of distant signal that sent pulsating radio waves that were repeated on a very regular basis several seconds apart. His aperture-synthesis system of radio telescopes also established the presence of neutron stars that were predicted by astronomers for some time but not previously discovered. Neutron stars are relatively small—only about 10 kilometers in diameter. However, merely one cubic centimeter of its "stuff" is estimated to weigh millions of tons. It seems pulsars consist of matter similar to the neutron stars and have a magnetic field of great energy, stronger than any magnet ever produced in the laboratory. Both the pulsar stars and neutron stars are surrounded by gas-like plasma. When viewed from Earth, the quasar-neutron star appears as a radio beacon. The best-known pulsar star was first viewed by the Chinese in the eleventh century and is found in the Crab Nebula. It is a glowing gas cloud, the remains of a giant stellar explosion. At the center of the Crab Nebula is an expanding neutron star that sends several different frequencies of electromagnetic radiation, including light pulses, X-ray pulses, as well as the detected radio pulses.

*See also* Hewish

One of the great debates in the history of astronomy took place in the 1950s between two British astronomers, Sir Martin Ryle and Sir Fred Hoyle. It seems that Fred Hoyle came up with the term "big bang" as a pejorative term for those who believed in the cosmic theory of an expanding universe starting from a singularity point and rapidly expanding, and which is still expanding. Hoyle was a proponent of the steady-state universe, while Martin Ryle accepted the expanding universe theory. Their arguments became not only a scientific debate but also a personal feud. One thing this dispute did was awaken an interest in the two theories and spurred much more research into the origins of the universe, as well as cosmology in general. Recent research indicates that the big bang is not only the correct theory, but it is now assumed that it all began with one giant "explosion" of matter and energy.

# S

**SABIN'S THEORY FOR ATTENUATED LIVE POLIO VACCINE:** Biology: *Albert Bruce Sabin* (1906–1993), United States.

*If live poliovirus can be grown in tissue cultures, it can be attenuated (weakened) and used to vaccinate humans against poliomyelitis (infantile paralysis).*

During World War II Albert Sabin developed vaccines for diseases such as dengue fever and encephalitis and was familiar with the work of other microbiologists who experimented with the growth of viruses in the brains of mice. In 1954 Jonas Salk (1914–1995) used poliovirus "killed" by formaldehyde, which was then injected to stimulate the human immune system, thus developing antibodies against the disease. Outbreaks of polio were common, particularly in the summer months. In 1952 and 1953 the United States alone had outbreaks of the disease with fifty-eight thousand and thirty-five thousand cases, respectively. Prior to this, the usual number of cases reported was around twenty thousand. Thus, immunization was not completely successful, and the vaccine had to be injected several times over a period of years. Also, the dead virus's effects on the immune system did not last a lifetime. Albert Sabin developed a live but weakened version of the poliovirus in the kidney tissue of monkeys. His version for providing immunity to the virus could be taken orally and had a lasting effect for producing antibodies as a preventative against the virus. The oral-attenuated virus is still being improved. After testing his attenuated live virus on animals, Sabin tested it on himself and several prisoners who volunteered to test its efficacy. The U.S. public was skeptical of Sabin's vaccine due to the problems experienced with the earlier Salk vaccine. Finally, after successful use in Russia and England, it was accepted in the early 1960s and extensively used in the United States as an oral vaccine that prevented the outbreak of polio epidemics. Currently, there are two types of polio vaccine available. One is called inactivated polio vaccine (IPV) that is administered via an inoculation with a sterile syringe. The other is called a live oral polio vaccine (OPV) that is a liquid that is swallowed. As of the year 2000 the United States uses the IPV form of the vaccine almost exclusively. (OPV is administered in special circumstances.) However, OPV is used in parts of the world where polio remains a threat to the population because it is more effective in preventing the spread of the disease. In rare cases, OPV can cause polio. IPV does not. The incidence of polio has drastically decreased. As recently as 1988, it was estimated that three hundred and fifty thousand people were stricken with the virus. According to the Center for Global Development, in 2006 only four countries were endemic for the poliovirus with less than seven hundred cases reported worldwide. The last outbreak of endemically transmitted poliovirus in the Unites States occurred in 1979 among the Amish population in several midwestern states.

*See also* Jenner; Pasteur

**SACHS' THEORY OF PHOTOSYNTHESIS:** Biology: *Julius von Sachs* (1832–1897), Germany.

> *Photosynthesis and starch formation occurs in the green pigment in plant cells which absorb energy from light and that are found in discrete bodies called chloroplasts.*

Over the centuries a number of scientists theorized about how plants grow, as well as the nature of the green material in their leaves. Aristotle believed that plants received all their food from soil. In the 1600s Johann Baptista Van Helmont conducted one of the first controlled experiments with plants. He planted a willow tree in a given amount of soil that he had carefully weighed. After five years of natural growth, he weighed the tree and the soil in which it was grown. The tree gained 160 pounds while the soil lost only a couple of ounces. He concluded that the willow tree gained not only its food but also its increase in mass from water. Until 1862 scientists believed the green material in plants was distributed more or less evenly throughout individual plants. Julius von Sachs was the first to theorize that the green matter was contained in small, discrete bodies he named *chromoplasts* (a colored cell, later given the name ***chloroplast***). He coated several leaves of a plant with wax and left others unwaxed. After exposure to sunlight, the unwaxed leaves produced starch, while the waxed leaves did not. Sachs concluded that the unwaxed leaves were able to absorb carbon dioxide, while the coated ones could not let this gas enter, even in sunlight. Thus, photosynthesis (from the Greek *photo*, which means "light," and *synthesis*, which means "put together") is the process whereby in the presence of light, chlorophyll in green plants converts carbon dioxide and water into starch: $6CO_2 + 6H_2O + \text{light energy} \rightarrow C_6H_{12}O_6 + 6O_2 \uparrow$ (the $\uparrow$ represents heat energy given off by the reaction).

*See also* Calvin; Cohn; Ingenhousz

**SAGAN'S THEORIES OF NUCLEAR WINTER AND THE COSMOS:** Astronomy: *Carl Edward Sagan* (1934–1996), United States.

***Sagan's theory of nuclear winter:*** *A large-scale nuclear war can cause the ejection into the upper atmosphere of large amounts of smoke, ash, soot, and dust from burning cities and* *forests resulting in the blockage of sunlight and solar heat, the aftermath of which will be an extended period of winter on Earth.*

Scientists have known for many years that natural events, such as giant volcanic eruptions, great desert dust storms, and the collision of asteroids with Earth, can spew hundreds of millions of tons of particulate debris into the upper atmosphere, effectively blocking sunlight from reaching Earth's surface for months or years at a time. A massive blockage of sunlight would create extremely cold temperatures on Earth and greatly reduce plant growth resulting in the destruction of the supply of basic food for animals including humans. In the early 1980s a study called "TTAPS" (the initials stand for the names of the five researchers involved in the study: R. P. Turco, O. B. Toon, T. P Ackerman, J. B. Pollack, and more famously, Carl Sagan) developed a limited model based on a specific latitude plus a two-dimensional archetype of the atmosphere of the flat Earth. Other studies expanded their model to assume that at least half of the stockpiles of nuclear weapons of all nations would need to be used to destroy at least one thousand cities and most forests along with the production of tremendous amounts of fine dust in the atmosphere that would be opaque to solar radiation, resulting in the cooling of Earth's surface between 20° to 40°C for at least several weeks after the war. The results would be the opposite of global warming. In addition, much of the ozone layer would be destroyed, increasing the amount of ultraviolet radiation reaching Earth's surface by about 200%. A more recent 2006 study found that even a small-scale regional nuclear war could produce as many casualties as did World War II, and that global climate would be disrupted for many years. The study concluded that even a limited nuclear war could send millions of tons of soot into the stratosphere producing a cooling of several degrees over much of Earth, including the regions where most of the food plants are grown. As a continuum to his nuclear winter theory, in 1991 Carl Sagan predicted on ABC's *Nightline* TV program that the smoky oil fires resulting from the burning of the hundreds of oil wells in Kuwait during the Persian Gulf War could cause worldwide ecological disaster resulting in global cooling. The atmospheric scientist Fred Singer, also a guest on *Nightline* appearing with Sagan, was a skeptic and stated that Sagan's prediction was nonsense because the clouds from the burning oil wells would dissipate in a few days. Richard Feynman and Freeman Dyson, well-known scientists, are quoted as responding to Sagan's paper on nuclear winter as an "absolutely atrocious piece of science." Sagan later admitted in one of his books that his prediction related to the smoke from the burning oil wells was one of his mistakes. Since the days of "nuclear winter" and the end of the Cold War, the emphasis and interest of climate scientists, politicians, and the media has shifted to global warming.

***Sagan's planetary and cosmos theories:*** *Life, either present or past, may have existed on other planets or satellites of planets (moons) in the solar system, or even beyond in deep space.*

Carl Sagan based many of his planetary theories on laboratory research experiments and data collected from artificial satellites that provided information concerning the various types of organic molecules that make up the atmospheres on Earth, as well as other planets. His main goal related to this research was to discover and understand the origins of life on Earth, as well as possible life elsewhere in the universe. He postulated that there are "billions and billions" of stars in billions of galaxies, and thus billions of these stars are of average size with planets orbiting them. He further estimated that at least one of these planets among the billions of stars is at an optimum distance from their sun and has similar chemical and physical conditions that could amicably create and support life akin to Earth's life forms. Early in 2007 NASA announced the discovery of a new planet they named Gliese 581 c that is about five times heavier than Earth and is about 1½ times larger in diameter. It is revolving around a Red Dwarf Star. The planet has a temperature ranging from 34°F to 124°F, which is not too hot for life (temperatures on Earth have ranged from −131°F to 136°F). There is still much unknown about this new planet as there is unknown about the other 220 planets so far discovered outside our solar system.

During his time as a graduate student at the University of Chicago, Carl Sagan studied astrophysics and became interested in the atmospheres of planets. His 1960 PhD thesis proposed that the planet Venus in its past had undergone some drastic physical change that caused a massive greenhouse effect that was still evident on the planet. He was invited to join the team working on the first NASA satellite expedition to Venus in 1962 called *Mariner II*. The signals sent back to Earth from this satellite gave proof to Sagan's speculations concerning the greenhouse nature of the atmosphere of Venus. In the 1970s his research interest was the physical and chemical aspects of the solar system's planets, particularly the planet Mars. During his time as an adviser to NASA, Sagan and his colleagues, the exobiologist Cyril Ponnamperuna and NASA lab technician Ruth Mariner, demonstrated that particular organic molecules, including amino acids that are required to build proteins, could be produced by the energy of ultraviolet light. Included among these amino acids was adenosine triphosphate (ATP) that is a universal source of energy that would be needed for the origin of life. (Note: Exobiology is branch of biology that studies and searches for extraterrestrial living organisms.)

*See also* Miller; Ponnamperuma; Urey

Carl Sagan was somewhat of an enigma in life and as a scientist. From early childhood he marveled at the stars and the vastness of the universe. He followed this interest for the rest of his life. After attending several universities and being involved with NASA's early space program, his interests expanded to include concepts of religion and scientific beliefs, origins of life, other life in the universe, and democratic humanism. He became a national spokesperson for the wonders of the universe as he wrote several well-known books, as well as a prominent TV commentator and speaker at many conferences. Although accused of being an atheist, he insisted that an atheist is someone who is certain that God does not exist and has proven evidence for that belief. He said that he had no evidence that a god does not exist and that people with open questioning minds should challenge dogmatic claims about all aspects of the ultimate reality, both scientific and religious. He refuted "postmodernism" claims that 1) scientists idolize science and 2) that the scientific method destroys the philosophies that ponder the mysteries of nature and religious beliefs. He claimed that science is not a form of idolatry but that the scientific method is a way that humans can distinguish the false idols from reality and is the "best" method yet devised for humans to use as a way of life. He also believed that neither scientists nor theologians have all the answers and comprehend the vastness of the universe. Sagan gently chided traditional religions for persisting in their assertions about the natural world and cosmos that were contradicted by answers to questions of nature. He distinguished between mysticism (magic and the occult) and spirituality, which he believed is compatible with science—science itself is a source of spirituality. Not all scientists agreed with this last statement, but, Sagan, although a skeptic, was a great humanist who believed that science was a source of spirituality. He deplored scientific illiteracy and advocated that skepticism should be integrated into early education programs. Carl Sagan was a great popularizer of science who died much too young at the age of sixty-two.

**SAHA'S THEORY OF THERMAL IONIZATION:** Astronomy: *Meghnad N. Saha* (1894–1956), India.

*The composition of a star's spectrum varies with the temperature of the light source.*

Meghnad Saha theorized that the degree of ionization (electrons stripped from atoms to form ions) was dependent on the temperature of the atoms. He applied his concept to the spectrum of the light from stars. It was known that the light spectrum from some stars pointed to the presence of only hydrogen or helium, the two lightest elements. His examination of the spectra of stars indicated there were heavier elements (metals) that were being ionized in some stars. He developed a system that suggested the degree of ionization, and thus the stars' temperatures could represent spectral lines of stars. In other words, as the temperatures of stars increase, so does the degree of the ionization of the nuclei of the stars' atoms. Thus, atoms that have two or three ionization states will absorb sunlight at different wavelengths that produce different stellar spectra, whereas each light spectrum becomes stronger according to their proportions of the atoms' ionization increases. Saha's theory led to the linking of gas thermodynamics (heat) with the kinetics (molecular motion) of plasmas, which aided in interpreting the spectral lines of stars. His theory also enabled astronomers to determine the chemical makeup of different stars and confirmed the idea that heavy elements originated in stars, including the sun's gases.

*See also* Sakharov; Teller; Ulam

**SAKHAROV'S NUCLEAR FUSION THEORY:** Physics: *Andrei Dmitriyevich Sakharov* (1921–1989), Russia. Andrei Sakharov was awarded the 1975 Nobel Peace Prize.

*Controlled nuclear fusion can be achieved by containing the plasma in a magnetic "bottle."*

Andrei Sakharov's theory described how by confining a deuterium plasma (a highly ionized gas) within a strong magnetic field, the temperature could be raised to the point where the heavy hydrogen (deuterium) gas molecules would be forced to fuse to form helium molecules. The resulting reaction would release a tremendous amount of energy that could be used to produce electricity, much like the controlled nuclear fission reaction in nuclear power plants. In the early 1940s the United States developed the first nuclear fission bombs. In 1954 Sakharov was involved in the explosion of Russia's first atomic (**fission**) bomb, as well as its first nuclear (**fusion**) H-bomb. After realizing the tremendous destruction that would result from a nuclear war, Sakharov became an advocate for nuclear disarmament, which led to his demotion and exile, and eventually the Nobel Peace Prize.

*See also* Teller

**SALAM'S THEORY FOR THE PROPERTIES OF ELEMENTARY PARTICLES:** Physics: *Abdus Salam* (1926–1996), Pakistan. Abdus Salam shared the 1979 Nobel Prize for Physics with Sheldon Glashow and Steven Weinberg.

*At very high temperatures both the electromagnetic and weak interacting forces act as a single interacting force for elementary particles.*

Four basic forces account for the interactions of elementary particles: 1) electromagnetic forces and 2) gravity are observed to interact over long distances throughout the universe. The electromagnetic force both attracts and repels, thus counteracting its strength. Gravity is a very weak force that only attracts and never repels (at least there is no known *negative* gravity force). Even so, it is the most dominant force in the universe. 3) The strong force interacts with hadrons (nuclei of atoms), while the 4) weak interacting force, which is much less strong, interacts with leptons (similar to electrons and neutrinos). The forces that interact with hadrons and leptons are evident only at the very small atomic and subnuclear distances of matter (*see* Figure G11 under Glashow for a chart of the Standard Model for the unification of the particles of physics). Albert Einstein attempted to devise a mathematical solution to combine the electromagnetic and gravity interacting forces. He was unsuccessful, and at that time, the other two interactions with elementary particles were unknown. In 1968 Abdus Salam and his colleagues successfully combined the electromagnetic and weak forces, which behaved as a single interacting force, but only at high temperatures.

The analogy used is that at high temperatures, water turns into steam. As the temperature drops, it again becomes liquid, and if it drops further, it becomes solid (ice). In other words, for electromagnetic and weak interactions, there are two different physical states of being, depending on the temperatures involved (combined at high temperatures and separate at low temperatures). Abdus Salam and the other Nobel Prize winners referred to the high-temperature combined state as the electroweak interaction. Salam's theory is a great step in the development of a grand unification theory (GUT) and Einstein's unified field theory.

*See also* Einstein; Glashow; Weinberg

## SANDAGE'S THEORIES OF QUASARS AND THE AGE OF THE UNIVERSE:

Astronomy: *Allan Rex Sandage* (1926–), United States.

***Sandage's theory of quasars:*** *Quasars can be identified by their emitted radio signals, ultraviolet radiation, and blue light.*

In the early 1960s Allan Sandage detected radio signals from a small area in the distant universe. These radio signals seemed too strong to be originating from such a distant dim star. He and other astronomers referred to these objects that produced ultraviolet and blue light radiation as blue star objects (BSOs). Sandage determined they were not really radio stars. Therefore, he called them quasi-stellar or *star-like* bodies (*quasi* means "apparently" and *stellar* means "star"). This term was changed to *quasar* and used ever since. Sandage realized these objects exhibited a great Doppler redshift, which overcame the ultraviolet and blue light. He concluded this could only mean that the quasars were located at tremendous distances within the universe and that what could be seen from Earth was really the center of a huge galaxy. Later Allan Sandage and his team estimated the distance of such quasar galaxies to be over 12 billion light-years away. The first quasar that Sandage discovered was named 3C 48, which had the brightness of a sixteenth-magnitude star (*see* Pogson).

***Sandage's theory for the age of the universe:*** *The universe has an eighty-billion-year cycle of growth: forty billion years of expansion and forty billion years of contraction.*

Allan Sandage does not believe the universe is static, regenerating itself, or expanding indefinitely. His contention is that the universe oscillates in cycles of expansion (birth and growth) and contraction (shrinking, death, and followed by rebirth). His theory states that after a forty-billion-year period of expansion (it still has about twenty-five billion years left to reach maximum growth because the current universe is about thirteen or fourteen billion years old), the universe will cease to expand, reverse itself, and start contracting for another forty billion years of the cycle. At the end of this contraction period, he contends it will form back into its original, tiny, very dense "point" particle for a new singularity that will again create another big bang that will start the process all over again. It might be mentioned that not all cosmologists agree with Sandage's theory, which is just one of many dealing with the dynamic nature and age of the universe and is partly based on the estimation of the density of matter in the universe, which has yet to be accurately determined. More recently, it has been estimated that the vast majority of matter and energy in the universe is **dark matter** and dark energy that cannot be seen but implies an infinite universe.

*See also* Einstein, Glashow; Rubin; Schmidt; Weinberg

**SANGER'S THEORIES OF THE STRUCTURE OF PROTEINS AND GENE SPLITTING:** Chemistry: *Frederick Sanger* (1918–), England. Frederick Sanger was one of the few people to be awarded two Nobel Prizes. The first one in Chemistry in 1958 for determining the structure of proteins, and the second prize, also in chemistry, that he shared with Paul Berg and Walter Gilbert, in 1980 for determining the base sequences of nucleic acids.

***Sanger's theory of protein structure:*** *The four amino acids that make up proteins connect in groupings and can be identified in sequences inside the protein molecule.*

Frederick Sanger used the process called paper **chromatography** to separate and count the number of amino acids in specific protein molecules. Once these groupings were broken into segments of two, three, or four amino acids, he determined the structure of large and complete protein molecules. In 1953 he used this procedure to outline the complete structure of the molecule for the protein hormone insulin. It consists of fifty amino acids combined in two connecting chains. Since his identification of the exact order of the amino acid groups in the chains, it became possible for other scientists to produce synthetic insulin, used in the treatment of diabetes.

***Sanger's theory of gene splitting:*** *DNA can be split into fragments of various sizes, isolating a few cases of genes within genes.*

Frederick Sanger developed a new technique of splitting DNA into fragments to determine the base sequences of the nucleotides. In 1977 he was the first to describe the entire sequence of nucleotides in the DNA of a bacteriophage (a virus that infects a bacteria cell, also called **phage**) called Phi-X 174. To accomplish this, he needed to ascertain the order of about fifty-five hundred nucleotides in just one strand of the phage's DNA. While examining his results, he unexpectedly discovered several situations where genes were located within other genes. Today this phenomenon is used to explain traits of genetic expression. Sanger's theory and research contributed to the foundations for the science of genetic engineering.

*See also* Sharp

**SARICH'S THEORY OF UTILIZING PROTEIN TO GENETICALLY DATE MAN/APE DIVERGENCE:** Biology: *Vincent Sarich* (1934–), United States.

*When species split into two branches, future mutations for each branch are accumulated in a linear manner, and the greater the number of mutations, the greater the time divergence.*

Vincent Sarich used the albumin found in blood protein as a determinant for the divergence of humans and apes from a common ancestor. The concept is based on the facts that there is only about 1% difference between humans and apes in the DNA protein molecules and that mutations of individual genes not only differ for individual genes but also mutate on a random basis at a measurable rate over time. Thus, it should be possible to determine the rate of changes in the albumin of humans and apes over a long time span. He called this technique the *molecular clock*. Sarich and his colleague Allan Wilson began with the base of thirty million years ago as the estimated time the species of humans and Old World apes evolved separately from a common ancestor (the "missing link"). After analyzing the data that compared antigens from humans and other anthropoids, which have a common genetic base, they proposed a time factor of about five to seven million years ago when the two species evolved in their own directions. Other scientists disputed this short time period and claimed that fossil evidence places the division of **hominoids** and **hominids** from a common ancestor at about fifteen million years. Sarich responded that his molecular data were more accurate than the estimations of the age of the oldest human fossils. The accuracy of his work led to agreement for his five-million-year figure and is now accepted, as recently as 2006. Also there is now some evidence suggesting that chimpanzees are more genetically similar to humans (over 98% of the same DNA) than they are to gorillas and that chimps and humans should be classed in the same genetic family.

*See also* Wilson (Allan)

**SCHEELE'S THEORY OF THE CHEMICAL COMPOSITION OF AIR:** Chemistry: *Karl Wilhelm Scheele* (1742–1786), Sweden.

*Air is composed of two gases, one of which supports combustion, while the other does not.*

One of the gases that Karl Scheele isolated in air he named *fire air* (oxygen) because it supported combustion. The other, which prevented combustion, he named *vitiated air* (nitrogen). In 1772, about two years before Joseph Priestley produced oxygen (which Lavoisier named), Scheele actually isolated oxygen and described it in a paper. However, his findings were not published until after Priestley's discovery of the same gas. Priestley published his discovery first, thus he was given the credit.

*See also* Lavoisier; Priestley

**SCHIAPARELLI'S THEORY OF REGULARITY IN THE SOLAR SYSTEM:** Astronomy: *Giovanni Virginio Schiaparelli* (1835–1910), Italy.

*Meteors, planets, and the rings of Saturn follow regular patterns within the solar system.*

In 1877 the planet Mars was in conjunction with Earth—only about 35 million miles away. This proximity provided Giovanni Schiaparelli and other astronomers an opportunity to view Mars' surface for details. Despite the fact the atmospheres of Earth and Mars hindered a clear view, others confirmed Schiaparelli's record of his observations. He reported narrow and larger dark markings on the surface of Mars, which he concluded were bodies of water connected by narrow "channels" he called *canali*. He claimed these markings represented geometric patterns, which indicated some degree of regularity. It was speculated these patterns were the result of "constructions" by Martians. Others later expanded his concepts of regular structures to propose that Mars was a dying planet and these channels were the work of a desperate race attempting to bring water from the Martian ice caps to the tropical areas, where it could be used to grow plants. Myths about life on Mars existed for many decades until modern exploratory spacecrafts were sent to Mars to examine its atmosphere and surface.

Another example of Schiaparelli's theory of regularity is his claim that some natural physical process formed the rings of Saturn. Another is that meteor showers are caused by the breakup of comets. Thus, meteors must follow regular orbits similar to comets. One of his major theories of regularity states that the rotations of Venus and Mercury on their axes are synchronized with their sidereal periods. Therefore, these two planets always keep their same side facing Earth. He based this concept on his viewing of the same markings on these planets' surfaces when they were in a specific position. In the 1960s Schiaparelli's theory that these two planets keep their same side facing the sun was disproved when radar signals bouncing off their surfaces indicated that Venus rotates on its axis every 243 days, while its sidereal period is 225 days; Mercury rotates on its axis about once every 59 days, while its sidereal period is 88 days.

*See also* Cassini; Huygens; Lowell; Roche

**SCHLEIDEN'S CELL THEORY FOR PLANTS:** Biology: *Matthias Jakob Schleiden* (1804–1881), Germany.

*Plant structures are composed of small, distinct "walled" units known as "cells."*

In 1838 Matthias Schleiden first recognized and reported on the "cellular" basis of plants, which he referred to as "units" of plant life. (Note: Robert Hooke was the first to use the term "cell" for the minute structures he observed in a slice of cork. Using his compound microscope, the gaps in the cork's texture reminded him of the tiny monks' rooms in monasteries that were called *cells*.) Schleiden was first to note the importance of the nuclei in the reproduction of plant cells. He mistakenly thought that cells reproduced by the "budding" of new cells from the "mother" cell's nuclei. After Schleiden's discovery of plant cells, Theodor Schwann, also in about 1838, announced that animal tissues were also composed of cells, but with much less well defined cell walls. This led to the biological concept that cells are a basic unit of all living organic things. From this fundamental idea and the research of several other biologists, Schleiden and Schwann have been credited with the formulation of the cell theory, which states:

- All plants and animals are composed of cells or substances derived from cells.
- Cells are living matter, with membrane walls and internal components.

- All living cells originate from other cells; cells reproduce themselves.
- For multicellular organisms, the individual cells are subordinate to the whole organism.

Some of Schleiden's and Schwann's original observations and ideas were incorrect due to the very limited power of the microscopes available to them. But over time and with additional research by others, their concept that cells are the basic units of life became an important step in understanding living organisms.

*See also* Hooke; Leeuwenhoek; Schwann; Strasburger; Virchow

**SCHMIDT'S THEORY OF THE EVOLUTION AND DISTRIBUTION OF QUASARS:** Astronomy: *Maarten Schmidt* (1929–), United States.

> *Quasars exhibit a greater red shift than regular stars. Therefore, they are younger, more distant, and more abundant stellar-like objects than stars.*

Maarten Schmidt expanded his research of our Milky Way galaxy to include the very dim and distant objects discovered in 1960 by Allan Sandage and Thomas Matthews. Sandage called these objects *quasars*, meaning "starlike." The first quasar, designated 3C 48, was identified by the radio signals that it emitted. Schmidt studied its light spectrum. Even though it had the luminosity of only a sixteenth-magnitude star (*see* Pogson), he found it exhibited the spectral lines of the element hydrogen. When viewing other quasars, their spectra became more confusing, until Schmidt realized that the hydrogen spectra lines shifted in wavelengths toward the red end of the spectrum. This Doppler red shift of light, which was greater than expected from a star, indicated that quasars are emitting great amounts of energy and light as they recede from Earth at fantastic velocities. The greater the red shift, the greater the speed at which they recede. Schmidt examined the hydrogen lines whose wavelengths had shifted to the red end of the spectrum, indicating that another quasar, named 3C 273 was receding as the universe continued to expand after the big bang. The extreme red shift could mean only that quasar 3C 273 was not only at least one billion light-years away, but its brightness was that of hundreds, or possibly thousands, of galaxies in a cluster. Schmidt's theory asserts that quasars are among some of the earliest types of matter formed when the universe was young, and because they continually recede from us (and each other), they become more abundant as the universe ages. Schmidt proceeded to map the quasars in the universe, leading him to conclude that the so-called steady-state universe cannot exist. He and others interpreted their red shift data as indicating that the number of quasars increases with distance and that no objects have been located at greater distances. The possible demise of the steady-state universe concept has generated theories about an infinite universe. More recently, the discovery of **black holes** has advanced the theory that these huge dark masses of matter, from which matter or light cannot escape, are the source of the tremendous energy of quasars, or possibly the source of new quasars or even new universes.

*See also* Doppler; Gold; Hawking; Hubble; Sandage; Schwarzschild

**SCHNEIDER'S THEORY OF BIOLOGICAL SYSTEMS AND CLIMATE CHANGE:** Meterorology: *Stephen Henry Schneider* (1945–), United States.

*Coupled with the fields of physical and biological scientific research, as well as social and political assessments, global climate studies can identify the factors, risks, consequences, and possible solutions associated with the phenomenon known as "global warming."*

Stephen Schneider is a climatologist who is currently on the faculty of Stanford University in California as a professor of environmental biology and global change. He is concerned with the interdisciplinary aspects that combine the sciences of physics and biology, along with social/public interests, related to global climate changes. In other words, the object is to combine the disparate factors and uncertainties and all the political assessments into a cohesive method to address the risks and benefits of global climate change, commonly referred to as global warming.

Schneider was the leader of a team at the National Center for Atmosphere Research at Boulder, Colorado, involved in the construction of a mathematical-based computer model that related a grid-like pattern of the world's climates at the surface of Earth. This imaginary grid consisted of about two thousand boxes of atmosphere, all connected together at about 30 kilometers in altitude (just think of the total atmosphere divided into two thousand partitions or boxes that surround Earth). Factors, such as temperatures and pressures in each box, were analyzed using the largest computer available in the late 1980s. This mathematical computer model could predict *weather* in broad cyclic categories and patterns of specific boxes, but not for the entire climate of the entire globe. It could determine *regional* future cooling in winter and warming patterns in summer. Benefactors who provided billions of dollars to fund this research wondered why the model only predicted what any grammar school student knew from the few years of experiencing temperature changes from summer to winter. Furthermore, this model could not predict accurately the climate change, if any, for a specific region. Since that time, computers have become much more technically advanced and the mathematical models more sophisticated to the extent that they can include more unknown variables in the models and provide improved output predictions. Unfortunately, there are still many factors that are not included in even the best computer climate models. For example, are all the factors that are involved in the continual evolution of our dynamic Earth incorporated in the model? Is the changing nature of the interior of Earth involved, and if so how? How do we account for several past global warming and cooling periods? And how and why did past global warming periods end? There are thousands of variables involved, but we do know that there are several gases (carbon dioxide, methane, nitrous oxide, sulfur dioxide, water, etc.) that nature and humans put into the atmosphere that affect *radiation forcing* which is defined as the change in the balance between radiation entering and escaping Earth's atmospheric system. A variety of research techniques are used to provide input for the study of global warming, such as the analysis of ice cores, deep ocean studies, and population studies. And yes, people, as do all animals and other living organisms contribute to the global climate change problems. The mathematical computer model capable of predicting long-term linear projections for future worldwide climates has yet to be constructed.

Schneider is a firm believer that the three main components of the climate change scenario (climate science, climate impacts, and climate policy) should be a major concern of all three parties in the triangle of "journalist-scientist-citizen." Before consensus is reached, all groups should insist that the most up-to-date scientific assessments of climate studies and global warming are incorporated into media reports that are disseminated to a concerned population. In other words, the natural and dynamic processes of Earth that have controlled cataclysmic climate changes long before human civilization must be factored into the current climate models that focus on manmade pollutants that may be hastening an otherwise inevitable shift.

**SCHRÖDINGER'S THEORY OF WAVE MECHANICS:** Physics: *Erwin Schrödinger* (1887–1961), Austria. Erwin Schrödinger shared the 1933 Nobel Prize for Physics with Paul Dirac.

*An electron's position in an atom can be mathematically described by a wave function.*

In 1925 the quantum theory was developed through efforts of Erwin Schrödinger, Niels Bohr, Werner Heisenberg, and others. Schrödinger was aware of Niels Bohr's application of the quantum theory to describe the nature of electrons orbiting the nuclei of atoms and Louis de Broglie's equation describing the wavelength nature of particles ($\lambda = \hbar/mv$, where $\lambda$ is the wavelength, $\hbar$ is Planck's constant, and $mv$ is the particle's momentum, i.e., mass times velocity). Schrödinger thought de Broglie's equation, which applied to only a single electron orbiting the hydrogen nuclei, was too simplistic to describe the state and nature of the electrons in the inner orbits of more complex atoms. Schrödinger laid the foundation of wave mechanics as an approach to quantum theory, resulting in his famous complex wave differential equation, a mathematical nonrelativistic explanation of quantum mechanics characterized by wave functions. Quantum theory is based on two postulates 1) energy is not continuous but exists in discrete bundles called "quanta" (e.g., the photon is an example of a discrete bundle of

Possibly the most famous cat in the history of physics was "Schrödinger's cat." He used this "cat" as a thought experiment to illustrate the incompleteness of quantum mechanics when physicists proceed in their thinking from microscopic subatomic systems to larger macroscopic systems. The cat thought experiment that grew out of a discussion between Albert Einstein and Erwin Schrödinger is a good example of the probabilistic outcomes in nature. The "experiment" depends on establishing a system where there is exactly a 50-50 chance of an occurrence of a particular quantum event, such as the decay of a radioactive nucleus of an atom. In essence, the experiment involved a cat that is enclosed in a small room or a box isolated from the outside environment. Enclosed with the cat are a Geiger counter, a small bit of radioactive material, and a bottle of poison gas. There is no given time, rate, or sequence at which the radioactive nuclei of atoms in the material will disintegrate. When will the Geiger counter that has been set to do so, in turn, release the poison gas that kills the cat and thus detect a single bit of energy radiating from a single nucleus? The question is: What is the *probability* that this sequence will occur in just one hour? Schrödinger stated that if the cat is truly isolated from external interference, then the state of the radioactive material, the Geiger counter, and the cat being either alive or dead are in a superposition of states. Each of these three states are 1) the radioactive material has either decayed or not decayed, 2) the poison gas has either been released or it has not, and 3) the cat has or has not been killed. The observer interfering with the experiment by making measurements can only determine the true states of these three positions. In other words, the observer becomes entangled with the experiment. Thus, subatomic particles can only exist in probabilistic states and the concepts related to quantum mechanics cannot be scaled up to large macro systems—such as cats.

light energy) and 2) subatomic particles have both wave (frequency) and particle-like (momentum) characteristics. His equation proved more useful in describing the quantum energy states of electrons in terms of wave functions than Bohr's quantum mechanical theory of particles orbiting around the nuclei of atoms. The wave function can be determined by the solution of a differential equation that has been named after Schrödinger. Schrödinger's theory for the wave nature of particles advanced the acceptance of the wave-particle duality of quantum mechanics.

*See also* Bohr; de Broglie; Dirac; Heisenberg; Nambu

**SCHWANN'S THEORY OF ANIMAL CELLS:** Biology: *Theodor Schwann* (1810–1882), Germany.

*The formation of cells is a universal principle for living organisms.*

In 1838 Mathias Schleiden proposed a cellular theory for plant tissues. At about the same time, Theodor Schwann made microscopic examinations of various animal tissues. He had already formulated the concept that animal tissues, particularly muscle tissues, were mechanistic rather than vitalistic (life based on some other agent than matter and energy). Schwann suggested the substances that compose animal tissues do not evolve directly from molecules but rather from cells, and the material contained in animal "cells" is similar to plant cells. He stated that just as plant cells are derived from other plant cells, so are animal cells derived from other animal cells. However, Schwann mistakenly assumed that the material inside cells did not have any structure of its own except for what he called a "primordial blastema." Schwann's theory stated that animal cells represented fundamental units of life. Although he did not believe in spontaneous generation, he at first claimed that cells arose from nonliving matter. This proved to be a paradox until it was determined that all cells originate from other cells. Schleiden and Schwann are both credited for developing the cell theory that states all organisms are composed of cells, which are the basic structural and functional units of life. *See* Schleiden for details of the cell theory.

*See also* Hooke; Virchow

**SCHWARZSCHILD'S "BLACK HOLE" THEORY:** Astronomy: *Karl Schwarzschild* (1873–1916), Germany.

*Once a star collapses below a specific radius, its gravity becomes so great that not even light can escape from the collapsed star's surface, thus resulting in a black hole.*

Karl Schwarzschild was an astronomer who, in addition to providing information about the curvature of space and orbital mechanics, studied the surface of the sun. His theoretical research indicated that when a star is reduced in size to what is now called the *Schwarzschild radius* (SR), its gravity becomes infinite. In other words, if a star with a specific mass is reduced in size to the critical Schwarzschild radius, its gravity becomes so great that anything entering its gravitational field will not escape. The edge of the **black hole**, which is referred to as its *horizon*, is the zone where the escape velocity from the hole exceeds the speed of light. The critical spherical surface region of a black hole where all mass and light are captured is called the *event horizon*. Schwarzschild used the sun to determine this critical radius for stars. The SR for the sun, when it does collapse, will be about 3 kilometers, and it will be incredibly dense. (Because the sun's current radius is about 700,000 km, it will be many billions of years before it "shrinks" to the SR of 3 km.) To determine the SR for other stars, divide the object's (star) mass by the mass of the sun and multiply by 3 km. This equation might be expressed as: $M_o \div M_s \times 3$ km $= M_o$'s SR ($M_o$ is the mass of the object for which SR is to be determined, $M_s$ is the mass of the sun). Therefore, the critical radius for a black hole is proportional to its mass. The current theory suggests that a black hole may be open at the bottom of its "funnel shape," where the mass that was captured by the black hole may reemerge as new stars or a new universe. Although the concept of black holes is demonstrated by mathematics, the idea that the "lost" mass will exit or escape to form another universe has not been proven.

*See also* Hawking; Penrose

**SCHWINGER'S THEORY FOR RENORMALIZATION:** Physics: *Julian Seymour Schwinger* (1918–1994), United States. Julian Schwinger shared the 1965 Nobel Prize for Physics with Richard Feynman and Sin-Itiro Tomonaga.

> *Electromagnetic theory and quantum mechanics can be combined into a science of "quantum electrodynamics."*

Julian Schwinger was somewhat of a child prodigy who attended the public schools of New York City and entered the City College of New York at the age of fourteen where he published his first scientific paper at age sixteen. After transferring to Columbia University, he received his BA in 1936 and his PhD in 1939. He then worked with Robert Oppenheimer at the University of California before moving to Purdue University in Indiana. During World War II he worked in the Radiation Laboratory at Massachusetts Institute of Technology (MIT) where he was sent to work on the atomic bomb at the University of Chicago. He so disliked working on the bomb project that he just drove back to MIT to work on radar. After the war, he accepted a professorship at Harvard University from 1945 to 1974. It was during his tenure at Harvard that he devised his first concepts of renormalization related to quantum electrodynamics while he continued his study of particle physics.

Several other scientists, including Paul Dirac, Werner Heisenberg, Wolfgang Pauli, and later Richard Feynman, Sin-Itiro Tomonaga, and Freeman Dyson, laid the groundwork for the theory of quantum electrodynamics (known as QED). They contributed to the understanding of the behavior of how atoms and atomic particles react in electromagnetic fields, whereas Schwinger's contribution was the combining of electromagnetic theory and the field of quantum mechanics into the new field of quantum electrodynamics in a way that was consistent with Einstein's theory of relativity. This is a good example of how modern scientists, and in particular physicists, build on past theories and the accomplishments of others.

*See also* Dirac; Dyson; Edison; Feynman; Heisenberg; Weinberg

**SEABORG'S HYPOTHESIS FOR TRANSURANIUM ELEMENTS:** Chemistry: *Glenn Theodore Seaborg* (1912–1999), United States. Glenn Seaborg shared the 1951 Nobel Prize for Chemistry with Edwin McMillan.

*The elements beyond uranium, atomic number 92, are similar in chemical and physical characteristics.*

In 1940 Glenn Seaborg and his colleagues discovered the first two elements beyond uranium (92). They are neptunium (93), a beta-decay element somewhat similar to uranium, and plutonium (94), a radioactive fissionable element used for nuclear reactors and bombs. The discovery of these new elements resulted in Seaborg's hypothesis that elements beyond uranium (92) formed a group of elements with similar characteristics that represented a new and unique series of elements. He compared this new series, named the *actinide transition series*, to the lanthanide series of rare earths, lanthanum (57) to lutetium (71), which also have very unique and similar characteristics. Seaborg used his hypothesis to predict the existence of many more "heavy" elements in his proposed *transuranic actinide series*. Still later in his career, he speculated there was a "superactinide" series of elements ranging in atomic number from about 119 to as high as 168 or even 184. All of these super-heavy elements, if discovered, will be radioactive, very short lived, and difficult to detect. In 1944, Seaborg and his colleagues discovered three new elements that are included in the **Periodic Table of the Chemical Elements**: plutonium (94), americium (95), and curium (96). Still later, Seaborg is credited with discovering berkelium (97), californium (98), mendelevium (101), nobelium (102), and lawrencium (103). The elements einsteinium (99) and fermium (100) were discovered after the detonation of the 1952 Hydrogen bomb and are artificially produced in nuclear reactors. Element 106, discovered in 1974, is currently named seaborgium (Sg), in honor of Glenn Seaborg. In 1999 Seaborg's Berkeley, California, laboratory announced the discovery of element 118, which has a half-life of about 0.00005 of a second. However, for years after its discovery, it has been mired in controversy amidst accusations of scientific misconduct.

*See also* Bohr; Fermi; Heisenberg; Lawrence

**SEEBECK'S THEORY OF THERMOELECTRICITY:** Physics: *Thomas Johann Seebeck* (1770–1831), Germany.

*If electricity can produce heat when flowing through a wire, then a reverse effect should be possible; that is, heating a circuit of metal conductors should produce electricity.*

Thomas Seebeck was familiar with Joule's law, which in essence states: that a conductor (wire) carrying an electric current generates heat at a rate proportional to the product of the resistance ($R$) of the conductor (to the flow of electric current) and the square of the amount of current ($I$ or A), (the current amperage). Using this information as a basis, in 1820 Seebeck joined the ends of two different types of metals to form a loop or circuit. When a temperature differential was maintained between the two different metal junctions, an electric force (voltage) proportional to the temperature differences between the two metal junctions was produced. This phenomenon is known as the *Seebeck effect*, where electricity is produced by temperature differences in the circuit. This device is now referred to as a *thermocouple*. If several thermocouples consisting of junctions between two dissimilar metals are connected in a series, a "thermopile" is formed, which can increase the voltage output equal to the number of junctions. It was later discovered that when the temperature of a single junction increases, the temperature decreases conversely at a second junction in the same circuit—the heat is transferred from one junction to the other. The rate of transfer is proportional to the current, and if the direction of the current is reversed, so is the heat (it is absorbed). In 1854 Lord Kelvin demonstrated that if there is a temperature difference between any two points on a conductor carrying a current, heat will be either generated or absorbed, depending on the nature of the material. It was later discovered that magnetism also affects this process. This principle has been applied to generate small amounts of electricity, as a "thermometer" to measure temperatures, and as a means of heating or cooling. Small heating and cooling devices in manned spacecrafts and portable refrigerators use the Seebeck effect.

*See also* Joule

**SEGRÈ'S HYPOTHESIS FOR THE ANTIPROTON:** Physics: *Emilio Gino Segrè* (1905–1989), United States. Emilio Segrè shared the 1959 Nobel Prize for Physics with Owen Chamberlain.

*If anti-electrons (positrons) exist and can be produced in particle accelerators, antiprotons should also exist.*

In 1932, Carl Anderson built on Paul Dirac's idea that antiparticles are similar to elementary particles except for their electrical charges. The existence of the positron (positive electron) established the existence of the antiparticle positron, which unlike the negative electrons, has a positive charge but is similar to the electron in all other characteristics. Emilio Segrè, along with Owen Chamberlain, hypothesized that if antiparticles, such as positrons, can be generated in particle accelerators, antiprotons will also be generated if the accelerator is powerful enough. In 1955 they used the Berkeley Bevatron accelerator to generate six billion electron volts (BeV) to bombard copper with high-energy protons, which produced only one antiproton for about forty-thousand or fifty-thousand other kinds of particles. They detected these few high-speed antiprotons by the unique radiation they emitted, which was later confirmed by exposing photographic plates to **antiproton** tracks. At the time antiparticles were discovered, it was also theorized these antiparticles annihilated regular particles when they met—for example: $e^- + e^+ \rightarrow$ energy. The question was: Why isn't all the matter in the universe that is composed of elementary particles obliterated into energy if their antiparticles annihilate them? The answer is: At the time the universe was "created," more regular particles (e.g., electrons with negative charges) were formed than antiparticles (e.g., positrons with positive charges); thus the negative electrons now dominate.

*See also* Anderson (Carl); Dirac

**SHAPLEY'S THEORY OF GLOBULAR CLUSTERS:** Astronomy: *Harlow Shapley* (1885–1972), United States.

*Masses of stars that are clustered together and known as "globular clusters" are found within the Milky Way galaxy.*

The first globular cluster was discovered by the German astronomer Abraham Ihle (1627–1699) in 1665 and was known as M22. However, his telescope was not powerful enough to separate the individual stars within the globular mass of stars. Later, using a more powerful telescope the French astronomer Charles Messier (1730–1817) observed the globular cluster known as M4 and was able to determine that the cluster was composed of many, many individual stars held together in a ball-shaped cluster by gravity. Note: The M before the number of a cluster refers to its entry in Charles Messier's star catalogue that was published in 1774. Other seventeenth- and eighteenth-century astronomers, including Edmond Halley, Gottfried Kirch (1639–1710), Philippe Loys de Chéseaux (1718–1751), Jean-Dominique Maraldi (1709–1788), as well as Abraham Ihle, were actually discoverers of the globular clusters. Stars in globular clusters are older and less dense than stars in open clusters that are found in the central disks of galaxies. The clusters are composed of many thousands of so-called low-metal stars (such as hydrogen and helium) and are similar to the stars found in the central core of spiral galaxies. They are surprisingly free of gas and dust found in other regions, possibly because the gas and dust were incorporated into stars eons ago. High-density stars, such as those found in globular clusters, do not have the conditions necessary to maintain planetary systems. The globular clusters within the Milky Way are usually found surrounding the core but in an asymmetrical distribution. Shapley estimated the size of the Milky Way galaxy by the amount of light from the central globular clusters. Although his estimations were not accurate due to dust in the galaxy, he did demonstrate that the Milky Way was much larger then previously believed.

*See also* Hubble

In 1920 a "great debate" took place between two famous American astronomers, Harlow Shapley and Heber D. Curtis (1872–1942) concerning the extent of the universe and the interpretation of galaxies. Harlow Shapley took two positions: 1) first, the sun was not at the center of the Milky Way galaxy and 2) that globular clusters and spiral nebulae are found within the Milky Way. Shapley's model of the galaxy moved the sun from the center to an outer location in a spiral arm. So he won on the first point of the debate, and his model for the location of the sun is still used today. On the second point Curtis believed that galaxies were not only smaller than proposed by Shapley, but also were located outside our universe, that is, Milky Way. It seems neither side won because it could not be determined at that time if these nebulae (nebulae are fuzzy luminous objects that later were resolved into groups of individual stars) were actually star systems. This debate was settled when Edwin Hubble discovered that spiral nebulae are galaxies located at great distances from our Milky Way galaxy. Hubble's basis for this was data that determined the distance of Cepheids which are stars with variable brightness. By 1947 a total of 151 globular clusters were discovered in the Milky Way. It is estimated that there are at least two hundred, many of which are hidden by dust and gases within the galaxy.

**SHARP'S THEORY FOR THE "SPLICING" OF DNA:** Biology: *Phillip Allen Sharp* (1944–), United States. Phillip Sharp shared the 1993 Nobel Prize for Physiology or Medicine with Richard Roberts.

*Messenger RNA found in eukaryotic cells hybridizes into four sections of DNA that loop from the hybrid regions of the DNA.*

As did other molecular biologists in the 1970s, Phillip Sharp believed **eukaryotic** cells (with nuclei) would act similar to prokaryotic cells (without nuclei), where the DNA would form triplets with RNA to provide the codes to form amino acids. After examining the results of his hybrid double strands of DNA/RNA in the adenovirus (the virus that causes common colds), he noted that small sections of the loops that formed from the hybrids broke off and became spliced with the messenger RNA. These then escaped from the cells to become templates for protein production. Sharp and his colleague Richard Roberts determined the "split genes" identified in the adenovirus were common to all eukaryotic cells (which include the cells in the human body). They concluded that over 90% of the DNA was "snipped" out of the strands and became "junk" DNA. This result promised to provide some answers for the problems of genetic splicing related to some hereditary diseases. If this splicing or segmentation of the DNA molecules is better understood and can be controlled, it may be possible to find a cure for some hereditary cancers.

*See also* Crick-Watson; Roberts; Sanger

**SHEPARD'S THEORY OF SUBMARINE CANYON FORMATION:** Geology: *Francis Parker Shepard* (1896–1985), United States.

*Several forces, including underwater erosion on steep slopes on continental coastlines, turbid underwater currents, underwater slumping/slope failure, and faulting, are responsible for the formation of underwater canyons.*

Francis Shepard collected data for fifty years that he used to indicate the possible causes for the giant underwater canyons that are found on the deep sides of continental slopes, often at depths greater than two kilometers below sea level. Many of these large submarine canyons have been carved out of sediment as well as hard crystalline rock. Some of the largest underwater canyons are larger than any land-formed canyon, including the Grand Canyon found in the western United States. There is some evidence that some of these are extensions of land-based rivers that, eons ago, carved out a deep channel. The largest underwater river-type canyon is formed by the extension of the Congo River. It is 500 miles long and 4,000 feet deep. Other examples of submarine river canyons are the Amazon Canyon; the Hudson River Canyon; the Ganges River canyon; the Indus River canyon; three found in California are the Monterey, the La Jolla, and Scripps canyons; and two in the Bering Sea are the Bering River and the huge Zhemchug River canyons. One concept is that margins of continents rose up and then sank. Another idea was that during the ice age sea levels dropped enough to dry up the Mediterranean Sea and exposed much of the major oceans' floors to geological and other type of erosion events. This concept assumes that the canyons were once above water and were formed by erosion of flowing river water.

In Francis Shepard's book *Submarine Canyons and Other Sea Valleys* he clarifies his theory and admits that exactly how submarine canyons formed is still a puzzle, but he offers several complex solutions to the puzzle. He proposes that gravity is basically the force that produces slow movement of coastal and ocean sediments and underwater landslides, as well as deep water turbid currents that erode the canyons over millions of years. Gravity is also responsible for slope failure resulting in mass wasting, slumping, and submarine landslides that occur on steep hills. He claims that these forms of erosion are mostly responsible for forming the submarine canyons. Many geologists do not accept this theory because past history indicates that the sea level has dropped only about 100 meters. Thus, the erosion of canyons many thousands of feet below sea level cannot be explained using Shepard's theory. It is now accepted that gravity related to the degree of down slope of canyon walls form channels that "dig up" and transport loose conglomeratic-type materials from the continental slopes over long time periods by gravity. This material consists of small fragments of larger rocks, sand, and silt that are transported by turbid underwater currents and waves over long distances and thus is the primary mechanism for the formation of these giant submarine canyons.

**SHOCKLEY'S THEORY OF SEMICONDUCTORS:** Physics: *William Bradford Shockley* (1910–1989), United States. William Shockley shared the 1956 Nobel Prize for Physics with John Bardeen and Walter Brattain.

> *In crystal form, the element germanium will carry an electric current less well than a metal but much more efficiently than an insulator, thus it can act as a semiconductor, which enables it to rectify and amplify electric currents.*

In 1948 William Shockley and his colleagues John Bardeen and Walter Brattain discovered that small impurities within the germanium crystal determine the degree of its conductivity or capacity to carry electricity. This type of material, which allows some electricity to pass through it, is called a *semiconductor*. Later they realized that other crystals, such as silicon, were even better and less expensive semiconductors. Because these devices are solid, they are also known as *solid-state semiconductors*. Shockley soon learned how to vary the small amount of impurities in the crystal's structure, enabling it to be used as a "switch," or as a **rectifier** or **amplifier**. These semiconductor elements when "doped" with small amounts of specific impurities that are composed of atoms with either four or five electrons in their outer shells (orbits) will act as electrical conductors. If arsenic with five electrons in its outer orbit is the impurity used, it will carry the current in the semiconductor, and thus the conductivity is named a n-type conductor (negative). When elements with fewer than four outer electrons, such as boron that has just three outer electrons, are introduced as an impurity, they act like "holes" where electrons are missing. Thus, this is referred to as p-type conductivity (positive, or lack of a negative). The current flows when one electron from a close atom is transferred to fill up this "hole." In doing so, it creates another "hole," which results in these successive positive

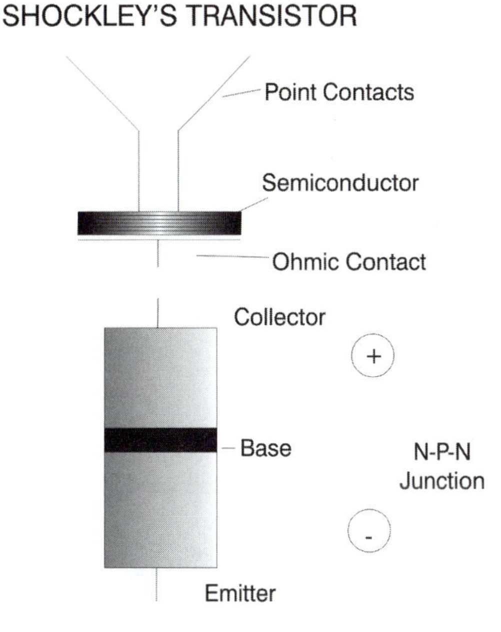


Figure S1. An artist's diagram of Shockley's transistor with an *n-p-n* junction. The *n*-type material carry electrons, while the *p*-type conductivity occurs when the "holes" in the material left by moving electrons are filled, thus allowing the junction to act as a rectifier and transistor to alter and amplify current.

holes being filled with electrons, forming a flow of electricity that can be regulated. Shockley made a "sandwich" of p-type material with n-type material to form a junction, which is known as an *n-p-n junction*, capable of amplifying electrical impulses (radio and TV). This n-p-n "sandwich" junction uses very little electricity as it transmits current across a resistor—thus the name *transistor* (*see* Figure S1).

Transistors replaced glass vacuum tubes in radio and television receivers, resulting in the tremendous miniaturization of electronic equipment, and formed the basis for the current electronics industry.

*See also* Bardeen

**SIDGWICK'S THEORY OF COORDINATE BONDS:** Chemistry: *Nevil Vincent Sidgwick* (1873–1952), England.

> *Two electrons from one atom can provide both "shared" electrons to form "coordinated" organic compounds.*

Nevil Sidgwick became interested in the concept of valence as the sharing of electrons in shells of atoms as proposed by Richard Abegg, Gilbert Lewis, and Irving Langmuir. The valence concept is built on Niels Bohr's quantized atom, where the electrons orbit in specific shells (orbits) based on their level of energy and was first proposed to explain how atoms combined to form molecules during inorganic chemical reactions. For example, each of two chlorine atoms shares an electron so that each can have eight electrons in its outer shell (orbit). This means each had seven electrons plus one shared with its close neighbor chlorine atom, thus forming the diatomic molecule of chlorine gas (*see* Figure S2).

Sidgwick's theory stated that similar "sharing" of electrons also occurred in the formation of both complex metal and organic compounds, but this sharing was different from the inorganic "covalent" electron bond, where each atom contributed one electron to the other atom. For this new type of bonding, one single atom could provide both electrons to combine with another atom, forming a new complex molecule. Therefore, he called these complex molecules *coordinated compounds* (*see* Figure S3).

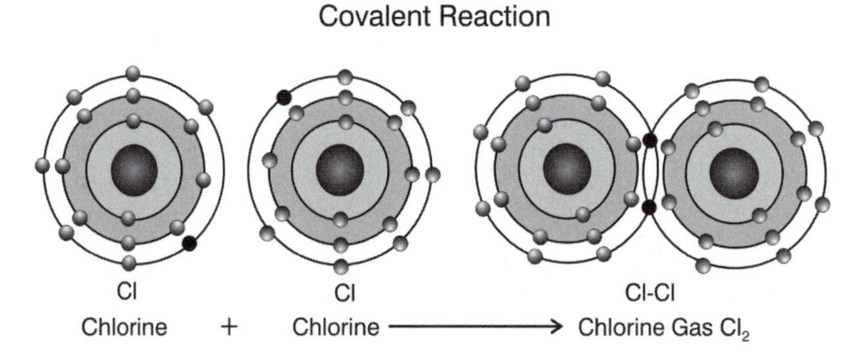


Figure S2. A depiction of a covalent reaction between two chlorine atoms (each with an outer orbit of 7 electrons) sharing two electrons to form a chlorine molecule ($Cl_2$).

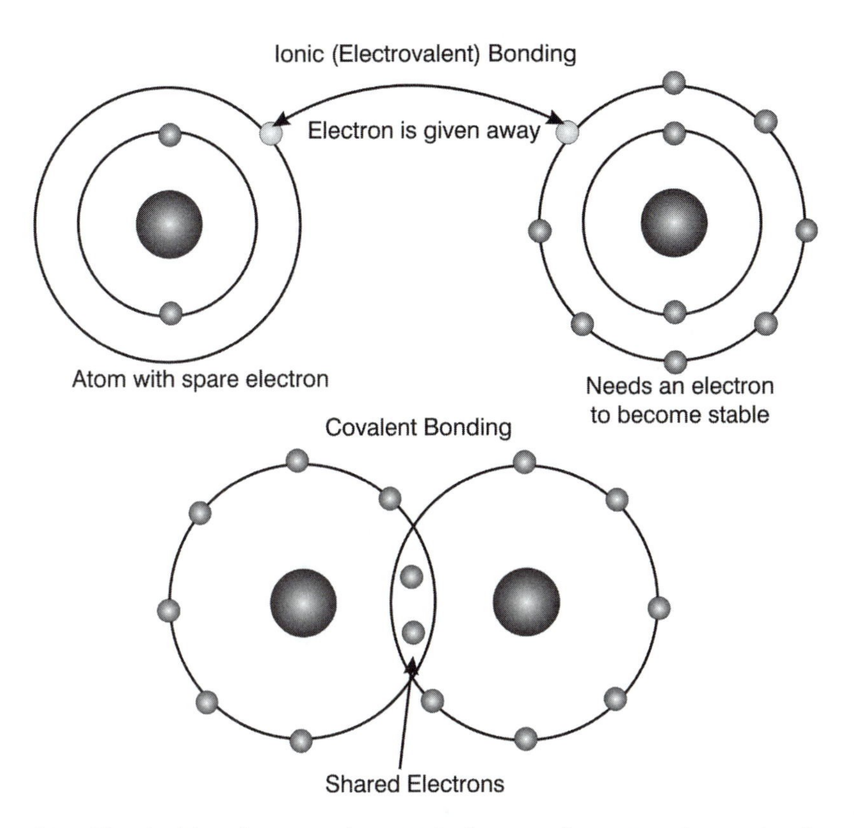


Figure S3. Chemical bonding uses electrostatic forces to form atoms into molecular compounds. Chemical reactions change, break, or re-form these bonds. There are two basic types of bonding: *Ionic* bonding between atoms occur when atoms with a dearth of negative charges in their outer orbits (valence) naturally attract electrons from other atoms to form ions and molecules. *Covalent* bonding occurs when atoms share electrons and each atom contributes one or more electrons to form the covalent bond. In both cases, the end results indicate the atoms have achieved an outer orbit electron configuration to the noble gases (Group 18, VIIA of the Periodic Table of the Chemical Elements).

The concept of coordinated bonds provided a better understanding of organic chemical reactions.

*See also* Abegg; Bohr; Langmuir; Lewis

**SIEMENS' THEORY FOR REGENERATING HEAT:** Physics (Engineering): *Carl Wilhelm Siemens* (1823–1883), Germany and England.

*Heat can be economized by a regeneration condenser process.*

Carl Wilhelm Siemens was one of fourteen children born to a farmer for the estate of the Crown in Germany. He and his older brother, Ernst Werner Siemens (1816–1892), an electrician and industrialist, became famous inventors. As a young man in 1843, Carl Wilhelm emigrated to England, where he lived until his death at the age of sixty. Carl Wilhelm did not believe in the existing theory that heat was a substance called "caloric" but rather accepted the new concept that it was a form of energy. Using this concept, he improved steam engines by economizing heat by condensing it and thus regenerating it. Along with superheated steam, he improved the efficiency of steam engines. Because of technical difficulties, his innovation was not financially successful. Even so, the Royal Society of Arts in Great Britain awarded him a gold medal for his development of the "regenerative condenser." He went on to develop the Siemens–Martin process that used his regenerative furnace, a much more efficient heating system than the open-hearth furnace in which pig iron was heated to the point that its impurities, such as carbon, were expelled. The Siemens regenerative furnace was able to recover enough heat to save about 75% of the fuel required for the process. It operated at a higher temperature by preheating the air used for combustion in a system that extracted heat from the exhaust gases passing through a specially designed brick chimney. By using this heat regeneration process, it became possible for the furnaces to melt steel, which was the first challenge to the Bessemer process. Today, the process of providing adequate heat for making steel is improved by using oxygen in an enclosed furnace or an electric arc-type furnace.

Carl Wilhelm Siemens and his older brother, Ernst, were interested in what at the time was the new field of telegraphy. Telegraphy referred to a new communications system that transmitted and received unmodulated electrical impulses via wire cable connected to transmission and reception stations. The end result: the telegram. Their first underground cable in Germany used **gutta-percha** for an insulation that was vulcanized. However, because the vulcanization process uses sulfur, it caused a reaction with their copper wire, thus destroying the insulation. They moved their business from Germany to England where they improved their techniques and received a contract to lay a telegraph cable from London to Calcutta, India. In the process Carl Wilhelm designed the first cable-laying ship in 1874, called the Faraday that laid 60,000 kilometers of cable throughout the world. The brothers' interest included improving electrical generators with a new type of self-activation dynamo that had numerous applications in the generation of industrial power, lighting, and so forth, for commerce as well as for individual homes. The electrical unit for conductance (the reciprocal of resistance, I or R) is called the *siemens*, in honor of Sir Carl Wilhelm.

*See also* Carnot; Joule

**SIMON'S THIRD LAW OF THERMODYNAMICS:** Physics: *Sir Francis (Franz) Simon* (1893–1956), Germany and England.

*The degrees of freedom for random paramagnetic molecules, which absorb heat from liquid helium, will become zero at the temperature of absolute zero.*

Walther Nernst claimed that for thermodynamic reasons, absolute zero (−273.16°C or −459.69°F) can never be reached because all materials at the absolute zero point would have no entropy, which he believed was impossible. Sir Francis Simon established the third law of thermodynamics (which simply states that it is impossible to cool any object to a temperature of absolute zero kelvin or −273.15°C.) by using a magnetic method of cooling as well as the use of liquid helium. Simon was able to reach the temperature of 0.0000016 K, within about 1/200,000 of one degree above absolute zero by surrounding liquid helium with a magnetic field that, when removed, causes the paramagnetic molecules to orient themselves in a random fashion while absorbing the small amount of remaining heat from the helium. Although Nernst received the 1920 Nobel Prize in chemistry for his concept that one could approach but never achieve absolute zero, it was Simon who established the third law of thermodynamics as the point where all molecular motion ceases—there is absolutely no heat (i.e., no kinetic energy, thus no molecular motion). This is the point where material substances have no degrees of freedom, which is absolute zero (0 K).

*See also* Nernst

Sir Francis (Franz) Simon was born into a wealthy German merchant's family and attended the universities in Munich, Göttingen, and Berlin, where in 1921 he received his PhD in physics. While at the University of Berlin, he worked with Walther Nernst who developed what was first known as the Nernst theorem or postulate, later known as the third law of thermodynamics. This law states that the entropy of a system at absolute zero degrees kelvin will not exist in a **ground state**. Entropy is the degree of "disorganization" or molecular motion (heat) within a system. In other words, absolute zero K of a system cannot be achieved.

During the 1930s anti-Semitism spread throughout fascist Germany causing Simon, who was Jewish, to emigrate to England. In 1933 he was invited to be an assistant professor at Oxford University. While there, his research interests were in using helium to achieve low temperatures, as well as in physical chemistry when he devised a method of separating the isotope uranium-235 from the element uranium-238. Later his work was transferred to the Manhattan Project in the United States. The gaseous diffusion process that he helped develop, that is, separating uranium isotopes, proved important in the development of the atomic bomb.

In addition to his accomplishments in developing a method of achieving low temperatures within a small fraction of a degree of zero degrees K, and his contributions to the gaseous diffusion process for separating U-235 from U-238, he is the only person to receive the Iron Cross from the German government before World War II, as well as a knighthood from the British government in 1954.

**SLIPHER'S THEORIES OF INTERSTELLAR GASES AND ANDROMEDA:** Astronomy: *Vesto Melvin Slipher* (1875–1969), United States.

***Slipher's interstellar gas theory:*** *There are enormous amounts of dust and gaseous material dispersed between and among the stars and galaxies.*

In the early 1900s Vesto Slipher was the first to make telescopic observations of the great clouds of interstellar material that reflects the stars' light and that is located between the stars. Up to that time, these clouds appeared to the unaided eye as "dust" or "gas." Slipher proposed that these observable but diffuse nebulae (clouds of gas and dust) become luminous due to light from nearby stars that is reflected off the dust and gases in space. He determined this radiation varies, thus altering the brightness of the night sky. Slipher also discovered the existence of the elements sodium and calcium dispersed in interstellar space (*see also* Rubin).

***Slipher's theory for the speed of the Andromeda Nebula:*** *The dark lines of the light spectrum of Andromeda indicate it is approaching us at a tremendous speed.*

Another one of Slipher's important achievements was his determination of the angular velocity of spiral nebulae as they rotate. He did this by measuring the displacement of their spectral lines by using the Doppler effect. By comparing this data, the velocity of moving stellar objects can be determined. One amazing result of using this discovery was that he was able to determine that about half of the spiral nebula that are observable are moving towards Earth, and the other half are moving away from Earth. His measurements of these great velocities helped prove that spiral nebulae were located outside our Milky Way galaxy and helped support the theory of an expanding universe.

For several hundred years, astronomers thought Andromeda was simply a large accumulation of gas concentrated at one location in the sky. In 1612, the German astronomer Simon Marius (1570–1624) was the first to view and describe this fuzzy luminous cloud that he called the Andromeda nebula (*nebula* means "cloud" in Latin). At one time it was also believed that Andromeda might be located in Earth's Milky Way galaxy. Slipher devised a technique using the Doppler effect to measure the radial velocity of spiral nebulae to determine the shift of their spectral lines. For instance, when an object moves away from us, its light's wavelength lengthens toward the red end of the spectrum—thus, the red shift. Conversely, if the object moves toward us, its light's wavelength is shortened to the blue end of the spectrum. In 1912 Slipher determined that the Andromeda nebula is not part of our galaxy but is rather a large galaxy moving toward Earth at a speed of more than 300 kilometers per second (about 200 miles per second). This theory, at first disputed by other astronomers, resulted in a better understanding of the nature of the universe. Slipher used the Doppler shift method to examine the spectra of several dozen extragalactic objects, predating Edwin Hubble's use of the red shift to measure the distance of far objects in the vast universe. Hubble estimated that Andromeda was seven hundred and fifty thousand light-years from Earth. Since then this estimate has been increased to over one million light-years, meaning the light from the Andromeda nebula now viewed by astronomers started its trip over one million years ago. Today, it is estimated there are over one hundred billion galaxies similar to Andromeda in the universe, each containing billions of individual stars.

*See also* Doppler; Hoyle; Hubble

**SMOOT'S THEORY OF A NONUNIFORM UNIVERSE:** Astronomy: *George Fitzgerald Smoot* (1945–), United States.

*There are "spots" in the universe that are slightly warmer than the average temperature of the universe. Therefore, the universe is not absolutely isotropic.*

Research from the early 1960s suggested that the universe must be isotropic (exactly the same in all locations). However, at the same time, the concept of an inflationary (ever-expanding) universe, proposed by Alan Guth, required the existence of areas throughout the universe that are less dense and/or with slightly different temperatures. In addition, Guth claimed that the observable inflationary universe could have originated from an infinitesimal "nothing," which later was known as a "singularity." In 1989 a satellite that carried instruments designed to measure differences in radiation at different locations in the universe, as well as the absolute brightness in the sky, provided reams of data that Smoot analyzed. His theory of island structures in the universe is at odds with the isotropic concept of space, but it does agree with recently discovered clusters of galaxies and even superclusters of galaxies. He determined some areas of space are slightly warmer (by 1/30,000,000 of 1°C) than other regions of the universe. Thus, this difference in radiation indicates the universe is not isotropic. His data supported the now widely held theory of the big bang followed by an inflationary universe.

*See also* Guth

**SNELL'S LAW:** Physics and Mathematics: *Willebrord van Roijen Snell (aka Willebrord Royen Snellius)* (1580–1626), Netherlands.

> *The refraction of light is the ratio of the sines of the angles of incidence (i) and the angle of refraction (r), and is a constant equal to the refractive index of the medium through which the light travels* (*see* Figure S4).

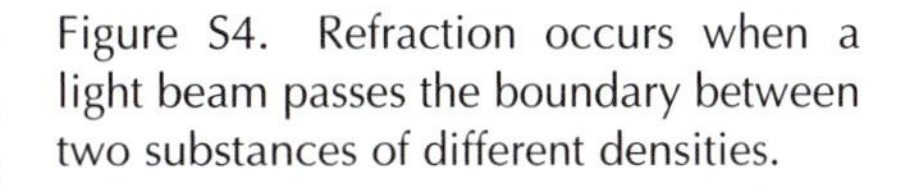
Figure S4. Refraction occurs when a light beam passes the boundary between two substances of different densities.

Refraction is basically the change in direction of a ray of light as it enters the boundary between two different mediums that have different refractive indexes. The refraction index of a substance is dependent on the ability of the substance to bend light.

In Figure S4 the incident light ray traveling through air enters the water at angle "i" to the perpendicular, while the light ray is refracted (bent) as it enters the water at the angle "r" to the perpendicular. The difference in the degree the light ray travels from the air into the medium (water) is based on what is known as the refractive index. This index can be measured and varies from 1 if the light ray travels between two mediums that have the same refractive index. (Note: *The scale for the Index of Refractions of light for some common substance ranges from 1.0000 for the refraction of light in a vacuum, to 1.0003 in air, 1.31 in ice, 1.33 in water, 1.46 in quartz glass, 2.11 in 50% sugar solution, and a diamond which has an index of refraction 2.42.*) Therefore, a light ray traveling from a less dense medium has a higher index of refraction when it travels through a substance of a greater density. This is why a diamond, which is denser than air, has a refractive index of 2.42 that is the greatest refractive index of all gemstones. In other words, the denser the medium compared to air, the greater the bending of light from the perpendicular. Snell's law is used to mathematically express this relationship, that is, $3.00 \times 10^8$ m/s (miles per second), which is the ratio of the speed of light in a vacuum to the speed of light in a denser medium. The phenomenon of refraction is responsible for many optical illusions when viewing objects under water—they seem bent and the object's apparent position is not its actual position.

> Willebrord Snell was born in the city of Leiden in the Netherlands. He originally attended the University of Leiden as a law student, but after presenting some lectures in mathematics at the university, he switched to his famous father's profession of mathematics. He contributed to the fields of optics, astronomy, and navigation, as well as mathematics and other areas of science. In 1621 he discovered the basic law of refraction, that is, the bending of light rays that occurs when a light ray changes its speed as it travels from one medium to another of a different density. Although he only lived to be forty-six years of age, he made many contributions to science and mathematics, as well as publishing five books. He improved the work of the ancient scientist Eratosthenes by using a method of measuring the size of Earth by triangulation, which became the basis for the modern science of *geodesy*. He also improved Archimedes' method of estimating pi by drawing a circle and adding polygons to the outside and inside of the circumference of the circle. Snell improved the accuracy of the techniques by using polygons with ninety-six sides that enabled him to correctly calculate pi to 7 places. His contribution to navigation involved the use of *loxodromes* (the path a sphere makes at a constant angle with Earth's meridians).

Optical illusions also occur in the atmosphere when there is a variation in layers of air through which an object is viewed. This optical illusion may account for some sighting of UFOs, "ghosts," and various unexplained objects.

*See also* Descartes; Fermat

**SODDY'S DISPLACEMENT LAW FOR RADIOACTIVE DECAY AND THEORY OF ISOTOPES:** Chemistry: *Frederick Soddy* (1877–1966), England. Frederick Soddy received the 1921 Nobel Prize for Chemistry.

***Soddy's radioactive displacement law:*** *A radioactive element that emits an alpha particle (a positive helium nucleus) is transformed into another element with a lower atomic weight, while a radioactive element that emits a negative beta particle (electron) will raise that element's atomic number.*

Frederick Soddy was aware of the particles and radiation that were emitted by radioactive elements and that as their atoms lost either positively or negatively charged particles, these radioactive elements were transmuted "changed" from the original. At this time, the existence of the neutron was unknown. The discovery of the neutron in 1932 by Sir James Chadwick explained the difference in atomic weights for isotopes (elements with the same atomic number but different atomic weights). But up to this time, some confusion persisted relative to what took place during the decay of radioactive elements. Soddy's displacement law of radioactive decay states that heavy radioactive elements, which emit **alpha particles**, will reduce that element's atomic weight by 4. Conversely, if the radioactive element emits an electron, it will have a higher atomic number. Soddy's displacement law provided the information needed for his original theory of isotopes.

***Soddy's theory of isotopes:*** *Some forms of the same elements have similar chemical characteristics (same number of protons) but exist with more than one atomic weight.*

Frederick Soddy believed that many elements were "homogeneous mixtures" of similar elements with similar atomic numbers but for some reason had slightly different atomic weights. He referred to these elements as isotopes, meaning the "same place" in Greek. At that time it was not known that the variances in weights of similar atoms were due to the different numbers of neutrons in atomic nuclei. Soddy demonstrated that two elements, uranium and thorium, are radioactive and will decay into isotopes of lead. Although these two radioactive elements decay in different sequences, their end products are isotopes of lead (lead with different atomic weights). By subjecting uranium and thorium to different chemical reactions, Soddy and his mentor, Sir Ernest Rutherford, arrived at what is known as the *radioactive series*, which explains in detail the decay sequence taken by these elements before becoming stable isotopes of lead. Later in his career, Soddy also believed there were limited sources of hydrocarbon energy on Earth, which culminated in his proposal to use the energy of radioactive elements as a solution for our energy problems.

*See also* Chadwick; Rutherford

**SORENSEN'S NEGATIVE LOGARITHMS REPRESENTING HYDROGEN ION CONCENTRATION:** Chemistry: *Soren Peter Lauritz Sorensen* (1868–1939), Denmark.

*The negative logarithm of the concentration of hydrogen ions in a solution can be used to measure the acidic or basic (alkaline) properties of solutions. This measure is called pH.*

Soren Sorensen's system was one of the first attempts to measure the extent to which a solution was either acidic or basic. To make his concept work, he used negative **logarithms** to determine the concentration of the hydrogen ions ($H^+$). If the solution has a greater concentration of $H^+$ ions than $OH^-$ ions, the solution is acidic. Conversely, if the solution has a higher concentration of $OH^-$ ions, it is basic (caustic or alkaline). The pH scale ranges from 0 to 14, with pH 7 being neutral (the $H^+$ ion and $OH^-$ ions are equal). Solutions decreasing from pH 6 to pH 0 indicate increasing $H^+$ ion concentration (greater acidity), while a reading ranging from pH 8 to pH 14 indicates solutions of increasing $OH^-$ ion concentration (greater alkalinity). Because the pH scale is logarithmic, each unit increase in pH represents a tenfold increase in the concentration of either the $H^+$ or $OH^-$ ions for each mole per liter (mol/L). For example, pH $0 = 1 \times 10^0$ $H^+$ concentration (mol/L), while pH $6 = 1 \times 10^{-6}$ concentration, and for *pH* 14 the $H^+$ concentration would only be $1 \times 10^{-14}$. Conversely, for pH 14, the concentration of the $OH^+$ ion would be $1 \times 10^0$, and for pH 0 the $OH^-$ ion would be only $1 \times 10^{-14}$ (mol/L). A simpler interpretation of the scale states that if a solution has a pH lower than 7, it is acidic; a pH higher than 7 indicates a basic; and a reading of 7 on the scale means the solution is neutral. This system measures the pH of all solutions in a number of ways, the simplest of which is the use of indicators, such as paper strips, that change color when wetted by the solution. The color change is matched with a color chart to determine the pH value of the solution. A more accurate and less subjective method is the use of a pH meter consisting of a glass electrode sensitive to the $H^+$ ions, where the reading can be compared to a reference electrode. The pH meter also permits continuous readings, which are not possible with the paper method. Determining the pH level of the acidity or alkalinity properties of solutions or other substances is important to many industrial processes dependent on the exact degree of pH required for some chemical reactions.

## SPALLANZANI'S THEORY REFUTING SPONTANEOUS GENERATION:

Biology: *Lazarro Spallanzani* (1729–1799), Italy.

*If solutions containing microorganisms are boiled over a long period of time and are not exposed to the air, all living organisms will be destroyed.*

The concept of spontaneous generation dates back to the earliest humans and their curiosity about how living things just seemed to appear and grow. In 1668 Francesco Redi, who studied insect reproduction, investigated William Harvey's concept that flies do not just spontaneously appear but are produced from eggs (*see* Redi for details on his classic experiment). In 1745 the British biologist and Roman Catholic priest John Needham (1713–1781) proposed that a "life force" was present in all inorganic matter, including air, which could cause life to occur spontaneously. Therefore, after boiling his chicken broth but leaving it exposed to air, microorganisms grew, "proving" his belief that life could be generated from this "life force." Needham later repeated this experiment, but he sealed off the glass spout of the flask containing the boiled broth. The organisms still grew. Again he claimed to have proved the viability of spontaneous generation. Lazzaro Spallanzani decided that

Needham had not boiled the broth long enough and had not removed the air from the jar before sealing it. Spallanzani improved the experiment by not only boiling the broth longer, but he drew out the air, creating a vacuum in the flask as he continued to boil the broth. He then sealed the flask. No microorganisms grew, which proved Spallanzani's theory that spontaneous generation could not occur without air.

*See also* Pasteur; Redi

**SPEDDING'S THEORIES:** Chemistry: *Frank Harold Spedding* (1902–1984), United States.

***Spedding's theory for separating the lanthanide elements:*** *The lanthanide series of elements (numbers 57 to 71) have very similar physical and chemical characteristics; therefore they can only be separated by using ion-exchange chromatography.*

Frank Spedding received his PhD degree from the University of California at Berkeley in 1929 at the beginning of the Great Depression and found that jobs for chemists did not exist. He spent a number of years on several low-paying fellowships including a National Research Fellowship that enabled him to continue his research at Berkeley. He was then hired as a chemistry instructor by the famous American chemistry professor, Gilbert Lewis, who encouraged Spedding to continue his work with the absorption spectra of solids. Spedding combined his ideas related to quantum mechanics to the spectra analysis of molecular structures of compounds that he knew demonstrated a sharp type of spectra in the atoms' and molecules' gaseous phases, whereas solids do not usually produce a sharp spectra. Spedding also knew about the problems of separating and identifying elements commonly called "rare earths" (which are not really rare, but are almost impossible to identify individually) that are found in the lanthanide series of elements 57 to 71. A chemistry professor at the University of Illinois, B. Smith Hopkins (1873–1952), isolated less than an ounce of a few rare earths, which were extremely difficult to come by in a purified form using laboratories procedures. Spedding convinced Hopkins to "lend" him a tenth of a gram from his samples with the agreement that he would do nothing to damage the samples and return them when he was finished with his research. Because he knew that a mineral containing a rare earth, when cooled to about 80 kelvin, gave a sharp-line spectrum, Spedding concluded that this was evidence of the peculiar arrangement of electrons in atoms of the rare earths. Also, by using this arrangement, he was able to determine the symmetry of the atoms in crystals of the individual lanthanide elements. Thus, he was able to separate individual elements of the lanthanides series by using ion-exchange chromatography—a commonly used process during which hard water is percolated down a column of minerals that exchange the ions of elements that cause the water to be "hard" with synthetic resins. Spedding then used this process to separate lanthanide chloride molecules where the lanthanide ions were separated from the chloride ions. In 1933 he received the Langmuir Award that, at that time, was awarded to chemists younger than thirty-one years of age (*see also* G. N. Lewis).

***Spedding's thermite theory for purifying uranium:*** *By using an exothermic type of chemical reaction between uranium and the oxide of another metal, pure uranium can be produced.*

A thermite chemical reaction is an exothermic reaction in which aluminum metal is oxidized by the oxygen of another metal, usually ferrous iron II or ferric iron III. The process also can be a similar mixture of two chemicals, one containing the element oxygen. In a typical thermite reaction powdered aluminum and powdered iron oxide are mixed with a small amount of binder to prevent the two metals from separating. When ignited, an extremely high temperature is produced that cannot be extinguished by water. The reaction with iron II oxide and aluminum follows: $Fe_2O_3 + 2Al \rightarrow Al_2O_3 + 2Fe$. It takes a source that will produce a very high temperature to start the reaction of the mixture. Either a "ribbon" of magnesium metal or a mixture of glycerin and potassium permanganate, when ignited, will produce heat high enough to start an exothermic reaction. The mixture of about 25% aluminum and 75% iron will produce heat capable of welding large steel objects outside the foundry site, such as the undercarriage of train after a wreck, or producing seamless railroad rails after they are laid. Thermite bombs also have been used as weapons of war. Because the fires they start are not easily extinguished (sand must be used instead of water), great damage to cities has resulted.

In the early days of World War II various ideas concerning the building of an atomic bomb were seriously discussed. One of the obstacles was that uranium-238 was not radioactive enough to provide a sustained chain reaction, but the isotope uranium-235 would be pure enough for a chain reaction. Researchers tried several methods of separating the two isotopes of uranium. Frank Spedding came up with the idea of using molten uranium metal in a thermite-like process. In 1942 the process was scaled up, and his team produced two tons of machined cylinders of U-235 that measured 2 inches in diameter and 2 inches long. Part of this production was sent to Stagg Field in Chicago where Spedding was director of the Chemistry Division of the Chicago Manhattan Project. Thus, the first successful atomic pile used the uranium he produced using a thermochemistry-type reaction.

**SPENCER-JONES' CONCEPT FOR MEASURING SOLAR PARALLAX:** Astronomy: *Sir Harold Spencer-Jones* (1890–1960), England.

> *An accurate astronomic unit (AU) can be calculated by using the position of a minor planet then comparing it to Earth's distance from the sun.*

Cognizant of the fact that the minor planet Eros was to approach Earth at a distance of only 16 million miles on a specific date in 1931, Harold Spencer-Jones received the cooperation of many astronomers worldwide who all at the same time photographed Eros' position. Using these data, he established the solar **parallax** by comparing the radius of Earth from the center of the sun. His figure, 8.7904 seconds of arc, was later corrected to 8.7941 seconds of arc. This provided a more accurate figure for the AU used to measure large distances for objects in the solar system. One AU is approximately 92,956,000 miles, which is the mean distance of Earth from the sun's center. The AU is much too small a unit for measuring great distances in space. The parsec, which equals 3.258 light-years and is the distance a star would be from Earth if it had a parallax of 1 second of arc, and the distance light travels in a vacuum over one year (a light-year) are the current units used for measuring distant objects in the universe. Spencer-Jones also used a very accurate quartz timepiece to determine the period of rotation of Earth on its axis. He concluded that the rotation of Earth is extremely regular but deviates very slightly each year.

*See also* Cassini

**STAHL'S PHLOGISTON THEORY:** Chemistry: *Georg Ernst Stahl* (1660–1734), Germany.

> *When substances burn, phlogiston is released. The more complete the burning, the more phlogiston the substance contains and releases.*

Since ancient times, the concept of objects' burning and rusting puzzled philosopher/scientists. In the seventeenth century Johann Becher (1635–1682), the seventeenth-century German alchemist, advanced the concept of phlogiston, which made sense to many people even though phlogiston had no color or taste. The word is derived from the Greek word *phlogistos* that means flammable. The concept was that when an object burned, heat and light were released, while at the same time it became lighter and produced ash. Therefore, the combustible substance "lost" something called phlogiston. About a hundred years later, the German chemist George Stahl improved the phlogiston concept to the point where it became the first rational theory of combustion. For example, when charcoal burned, almost no ash remained, meaning that charcoal contained a great deal of phlogiston, which was released during combustion. When metal was heated over charcoal, a coating or ash, which Stahl called **calx**, formed on the metal. Conversely, if the metal was consumed by a very hot fire, its phlogiston was freed, leaving behind the ash-like calx. He concluded that the metal must be composed of calx and phlogiston (metal → phlogiston + calx). Therefore, if the process was reversed and calx was heated over charcoal, it "absorbed" the liberated phlogiston from the charcoal to become metal again (phlogiston + calx → metal). He assumed charcoal was rich in phlogiston, and the release of phlogiston from burning substances seemed a rational explanation of combustion. For example, if a burning candle is placed in a closed jar, the air will soon become "saturated" with phlogiston because the candle is pure phlogiston, as evidenced by the absence of ash. This explanation is false. However, Stahl was correct when he claimed the rusting of iron was also a form of "combustion." The phlogiston theory was shattered in the late 1700s when Antoine Lavoisier developed the currently accepted theory of combustion.

*See also* Cavendish; Lavoisier

**STARK'S THEORIES:** Physics: *Johannes Stark* (1874–1957), Germany. Johannes Stark was awarded the 1919 Nobel Prize for Physics.

***Stark's theory of the Doppler effect on fast-moving particles:*** *The frequencies of radiation emitted by rapidly moving particles change as the speed of the particles change.*

The Doppler effect results from the change in frequencies for both sound and light. The red shift (Hubble effect) is used to measure the distance and motion of stars. This technique is based on the Doppler phenomenon, where the frequency of the wavelengths of light from a distant fast-moving source tends to "spread" and register as longer waves as it recedes from us and thus appears red. Stark's theory applies the Doppler effect to fast-moving particles such as electrons and photons. These fast-moving particles can be affected by both magnetic and electrical fields, which increase or decrease the frequencies of radiation they produce.

***The Stark effect:*** *A strong electric field can "split" spectral lines, which increases the number of lines.*

Johannes Stark was familiar with the Zeeman effect, which used strong magnetic fields to split the spectral lines of electromagnetic radiation with specific wavelengths. Because the electromagnetic spectrum consists of radiation of specific frequencies produced by the relationship between magnetism and electricity, Stark reasoned that if strong magnets can "split" a specific frequency of the spectrum, electricity could do the same. He used a strong electric field to produce a similar "splitting" or multiplying of lines of the frequencies of the electromagnetic spectrum. This was demonstrated to be a quantum effect (small changes), which Stark at first accepted but later rejected. The Stark effect was used to develop techniques to study electromagnetic radiation and subatomic particles.

*See also* Doppler; Hubble; Maxwell; Zeeman

**STEFAN'S THEORY OF BLACK BOX RADIATION:** Physics: *Jozef Stefan* (1835–1893), Austria.

> *The rate of emission of electromagnetic energy of a hot body is proportional to the radiating surface area.*

The black box theory is one of the radiation laws that apply to the properties of electromagnetic radiation when it interacts with matter in the universe. A "black box" is a hypothetical body that absorbs all the radiation that it receives. This theory can be expressed mathematically. The radiated *power* expressed as (P) is the rate of emission of the electromagnetic energy of a hot body and is proportional to the radiating surface area (A), and the fourth power of the thermodynamic temperature (T) as expressed mathematically as $j = \sigma T^{-4}$. Where *j* is the power density or heat loss, $\sigma$ is the Stefan constant, and $T^{-4}$ is the absolute temperature to the fourth power. Stefan's law is better known as the *Stefan–Boltzmann* law because Ludwig Boltzmann, who was one of Stefan's students, indicated that the law was only valid for blackbodies that either absorb or radiate all wavelengths of radiation (the theoretical hollow blackbody that would be a black box with a hole in it that would absorb or radiate all radiation entering it). The Stefan–Boltzmann law can be deduced from other theoretical principles. Even so, the constant proportionality ($\sigma$) in the formula was known as the Stefan constant, which, if the body is perfectly black $\sigma$ will equal 1. Thus, the value of the constant is equal to $5.6697400 \times 10^{-8}$ $Js^{-1}m^{-2}K^{-4}$.

Jozef Stefan was born into a lower-middle-class family who lived in the Austrian Empire. He excelled in his elementary classes and became the top student in physics in his gymnasium (high school). He graduated in mathematics and physics from the University of Vienna in 1857 and later became a professor in the same institution. He was associated with several other institutions including the Vienna Academy of Sciences. He published about eighty scientific articles and originated the law in physics known as the "physical power law" related to the radiation from a blackbody. His law of blackbody radiation applies to the radiation from stars that are almost in perfect equilibrium with their environments; thus, although not perfect, it satisfies the conditions as blackbody radiators. Thus, using his law, Stefan was able to determine the temperature of the sun's surface as 5430°C. The Stefan–Boltzmann law was also used to estimate the temperature of Earth as −6°C.

**STEINBERGER'S TWO-NEUTRINO THEORY:** Physics: *Jack Steinberger* (1921–), United States. Jack Steinberger shared the 1988 Nobel Prize for Physics with Leon Lederman and Melvin Schwartz.

*A beam of neutrinos will produce two different types of neutrinos.*

The law of conservation of energy states that in an isolated (closed) system, energy cannot be created or destroyed, although it may be changed from one form to another; but the sum of all forms of energy must remain constant. Physicists knew that when neutrons disintegrated and their "pieces" were measured, something appeared to be missing because the sum of the energy (or mass) of the "pieces" did not add up to the original for the neutrons. Physicists also knew that when a nucleus broke down, it emitted a beta particle (high-energy electron) plus a proton, but these two particles did not contain the total energy required by the law for the conservation of energy. In 1931 Wolfgang Pauli suggested that in addition to the electron (beta particle) and proton resulting from the disintegration of a neutron in the nucleus of an atom, another undetected particle must also be ejected. Because this undetected particle has no electrical charge and very little mass, it is difficult to detect. Pauli's theory stated that when the neutron of an atom breaks down, three, not two, particles are ejected (*see* Figure F2 under Fermi). Enrico Fermi mathematically described and named this theoretical third particle *neutrino*, meaning "little one" in Italian. Scientists tagged it as a "ghost particle" because it had yet to be detected except by the use of mathematics. Even so, mathematically it accounted for the missing energy when beta particles (electrons) are emitted from neutrons. Jack Steinberger developed a technique that produced and controlled a beam of neutrinos. This was somewhat of a surprise because neutrinos contain no electrical charge and practically no mass. He used this beam of neutrinos to demonstrate leptons, which are a group of particles that include electrons and neutrinos and come in pairs of opposites, thus confirming the accuracy of his theory. His theory expanded particle physics to explain the electron ($e^-$) and its opposite, the positron ($e^+$). In addition, two different types of neutrinos were discovered: the electron neutrino referred to as the *e* neutrino (v*e*) and the muon neutrino referred to as the μ neutrino (vμ). Steinberger's theory also helped to explain spin as a characteristic for these opposite particles. These two types of neutrinos each have a spin of one-half (opposite to each other) and a slight mass of 105.7 MeV. There are two types of muons or mu mesons: one with a positive charge ($\mu^+$), the other with a negative charge ($\mu^-$). Steinberger's theory advanced the concept of the duality of subatomic particles.

*See also* Fermi; Pauli

**STENO'S THEORY FOR FOSSIL FORMATION:** Geology: *Nicolaus Steno (aka Niles Stensen)* (1638–1686), Denmark.

*The age of fossils can be determined by the manner and time in which solid (formerly organic) bodies are imprinted on other solid (inorganic) bodies.*

Nicolaus Steno was one of the first to propose a theory for the origin as well as the nature of the formation of fossils and one of the first to propose that fossils were formed in sedimentary strata laid down in ancient seas. Two important principles are incorporated into Steno's theory of fossils. First, it is possible to identify the first to be solidified: the organic fossil material or the inorganic substance in which it was formed. Using this concept, he determined that *glossopteris*, which are seeds and remains of ancient fernlike plants, left their imprints first on the surrounding "mud" before the mud turned into rock. Because these fossils were a hard substance found within another hard substance, he concluded it was possible to establish the date of the fossil if the date of the rock was known. Glossopteris, also known as "tongue stones," were often confused with shark's teeth because of their shapes. At that time in history, it was assumed they fell from the heavens and were the actual objects, not fossils. Steno was criticized when he claimed that "sharks' teeth" were not actual teeth but rather fossils formed by minerals replacing the original substance. His second principle stated that if both the fossil and the rock material surrounding it were similar, they then could have been formed in the same way at approximately the same time. Steno's theory, an invaluable method for interpreting fossil records, is still used today and is an example of evidence supporting Darwin's theory of organic evolution. Steno is known as the father of paleontology.

*See also* Darwin

**STERN'S THEORY FOR THE MAGNETIC MOMENT OF THE PROTON:** Physics: *Otto Stern* (1888–1969), United States. Otto Stern was awarded the 1943 Nobel Prize in Physics.

> *Protons in atoms behave like small magnets and thus assume one of two orientations within a magnetic field.*

Space quantization theory states that atoms can align themselves in only a few directions when they are in a magnetic field. Otto Stern theorized that some atoms, such as those in silver, could align themselves in only two directions rather than in all directions, as proposed by Newtonian physics. Stern and his colleague, the German physicist Walter Gerlach (1889–1979), devised a unique experiment to test this theory. Their device was a magnet with the north pole as a flat surface and the south pole shaped as a knife edge placed close to but not touching the flat north pole surface (*see* Figure S5).

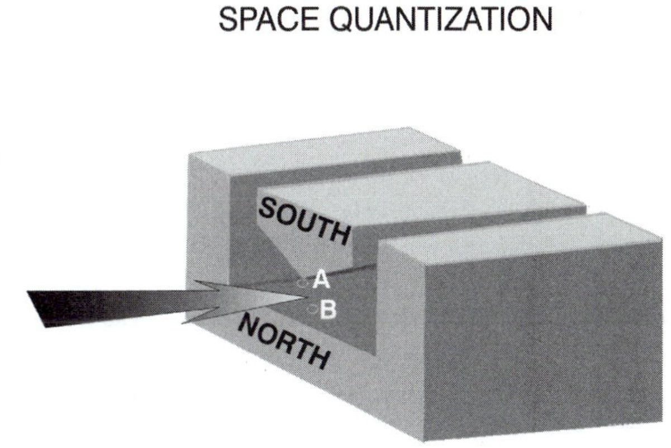


Figure S5. A much stronger magnetic field is produced at *A*, where the sharp edge of the south pole (*S*) of the magnet leaves a gap at *B* (the north pole *N*) which generates a nonhomogeneous magnetic field that produces opposite spin orientations in atoms and is known as space quantization.

This arrangement produced a nonhomogeneous magnetic field between the poles. They directed a beam of neutral silver atoms produced by heating silver metal in a vacuum, which acted like tiny atomic magnets (similar to micro compasses), through a slit between the center of the two poles of the magnet. The nonuniform magnetic field split the thin line of silver atoms being directed across the gap between the north and south Poles. This caused the narrow beam of silver atoms to split into two distinct paths, each representing opposite spin orientations of the atoms (spin-up and spin-down). If the thin line of silver atoms was divided into broad bands, it would have indicated more than two orientations. The restriction of the silver atoms to just two orientations is referred to as *space quantization*. Magnetic resonance imaging instruments used to examine atomic and molecular structures as well as for medical diagnoses are just two of the applications that resulted from Stern's theory.

*See also* Purcell; Rabi; Tyndall

**STOKES' LAWS OF HYDRODYNAMICS AND FLUORESCENCE:** Physics: *Sir George Gabriel Stokes* (1819–1903), England.

***Stokes' law of hydrodynamics:*** *A spherical body moving through a viscous fluid at a given speed produces a frictional drag.*

Sir George Stokes applied mathematics to solve many of the problems that concerned contemporary physicists. He developed a complicated equation to explain the hydrodynamics of fluids as a coefficient of viscosity: $6 \pi \eta rv$, where $\eta$ is the coefficient of viscosity, $r$ is the radius of the spherical body, and $v$ is the speed of the spherical body through the fluid. This computation for the coefficient of viscosity applies only for normal conditions; the law breaks down at extreme temperatures and pressures. Nevertheless, Stokes' law of hydrodynamics is applicable in many industries (e.g., sending oil through pipelines and mixing of liquids).

***Stokes' law of fluorescence:*** *The wavelength of fluorescent radiation is greater than the wavelength of the radiation causing the fluorescence.*

Michael Faraday created a vacuum inside a closed glass jar and then passed an electrical current through the vacuum. He assumed that because the air had been removed, nothing remained inside the jar to stop or slow the flow of current. However, he noticed a greenish glow formed on the inside of the glass as the current was turned on. Sir George Stokes named this phenomenon *fluorescence*, which now refers to any visible light produced by fast collisions of radiation (light, photons, electrons) with matter. It was impossible for Faraday to evacuate all the gas molecules from the glass vessel; therefore the concept of electricity as a fluid was not realized. Stokes proposed his law after another scientist noted that a fluorescent beam could be diverted by exposing it to a magnetic field. He concluded that the source of radiation (electricity) that caused the fluorescence was always of a lesser wavelength than was the actual wavelength of the fluorescent light itself. This law is applicable only under certain conditions of atmospheric pressure within an evacuated glass container and the concentration of gaseous molecules within that container. Stokes' law and this concept pioneered the development of the Geissler tube, computer monitors, and TV screens.

*See also* Faraday; Fraunhofer

**STONEY'S THEORY OF THE ELECTRON:** Physics: *George Johnstone Stoney* (1826–1911), Ireland.

> *Electricity is composed of small units of fundamental particles, as is matter, and these units can carry electric charges.*

George Stoney was aware of Svante Arrhenius' work on ionic dissociation as related to solutions of certain substances (e.g., salts that act as electrolytes and carry electric current by ionic dissociation). Arrhenius explained that the salt dissolved to form ions, which could then carry electrical charges (*see* Figure A7 under Arrhenius). Thus, a better understanding of the structure of the atom resulted. George Stoney's theory states there is an "absolute unit" of electricity with just one type of charge, and it is carried from atom to atom when an electric current flows through a conductor. In addition, these charges always exist in a ratio of whole numbers, never fractions of a unit. He based his theory on the calculated mass of the hydrogen ion given off during electrolysis. Stoney coined the term "electron" to describe this basic negative unit of electricity that could be carried by a single atom or even a group of atoms. Several years later the science community accepted his term electron (*see* Figure T1 under Thomson).

Using this knowledge, Stoney determined there were two different types of molecular motion. One relates to the relative motion of gas molecules to each other that does not result in a spectrum when exposed to radiation. The other is the internal, random motion of molecules in a substance that will produce spectral lines related to the type of substance involved.

*See also* Arrhenius; Helmholtz; Thomson

**STRASBURGER'S LAW OF CYTOLOGY:** Biology: *Eduard Adolf Strasburger* (1844–1912), Germany.

*New nuclei of cells arise from existing cell nuclei, and through a process of mitosis, they carry factors responsible for heredity.*

Eduard Strasburger based his research on plant reproduction. He was the first to describe the embryo sac in gymnosperms (conifer/pine trees) and to recognize that angiosperms (flowering plants) reproduce by double fertilization (when the two nuclei from a pollen grain fuse with two nuclei of the embryo sac). His law of cytology, in addition to cell division, described the division of nuclei to form new nuclei in plant cells. The law relates to the study of the growth, structure, reproduction, and chemical makeup of cells and basically states that new cells arise from existing cells. Strasburger's concept of mitosis occurring in the nuclei of plant cells was related to the division of these nuclei following the same principle as that of his law of cytology. From this concept of mitosis, he inferred there were factors of heredity, with which he was not completely familiar, that divided during the process of mitosis as the nuclei of plant cells divided. In mitosis, each chromosome is divided in half so that the two new (daughter) cell nuclei are exactly the same as the mother cell's nucleus. Strasburger also postulated that internal physical forces (i.e., hydraulics), rather than physiological factors (cell functions), are responsible for fluid (sap) being transported in the trunks of trees and stems of plants.

*See also* Schleiden; Schwann

**STRUVE'S THEORY OF INTERSTELLAR MATTER:** Astronomy: *Otto Struve* (1897–1963), United States.

*Interstellar matter appears more diffused than localized throughout the universe.*

Otto Struve was the grandson of the German astronomer Frederich Struve (1793–1864), who first used **parallax** to estimate the distance from Earth to the bright star Vega. The elder Struve's major contribution was the discovery and cataloging of over three thousand binary (double) stars. Otto Struve performed spectroscopic analyses of the binary stars recorded by his grandfather and also examined the structures and atmospheres of other stars and objects. His most important theory related to just what and how much "stuff" existed in the great distances of space separating the stars and galaxies. In 1937 he discovered that the vastness of space contained great amounts of ionized hydrogen. Other elements were also found, such as helium and calcium. More recently, some astronomers have estimated that over 90% of all the mass (matter) in space is "dark matter." Thus, it cannot really be seen because it neither gives off nor reflects light.

*See also* Arrhenius; Hoyle

**SUESS' THEORY OF CONTINENTAL DRIFT:** Geology: *Eduard Suess* (1831–1914), Austria.

*The southern continents of Earth (Africa, South America, Australia, India) were once combined as one large land mass.*

Eduard Suess spent years studying the similarities of geological and plant fossils on the continents of Africa, Australia, and South America and the subcontinent of India. He observed similar geological structures, including mountain ranges, regions of volcanoes and earthquakes, and coastlines for these landmasses, as well as examples of the ancient fossil ferns known as glossopteris that existed in the Carboniferous period of earth's development. Based on these clues, he derived his theory of a great supercontinent he named Gondwanaland, after the Gonds of ancient India. Alfred Wegener developed a theory for a similar supercontinent he named Pangaea (*see* Figure S6).

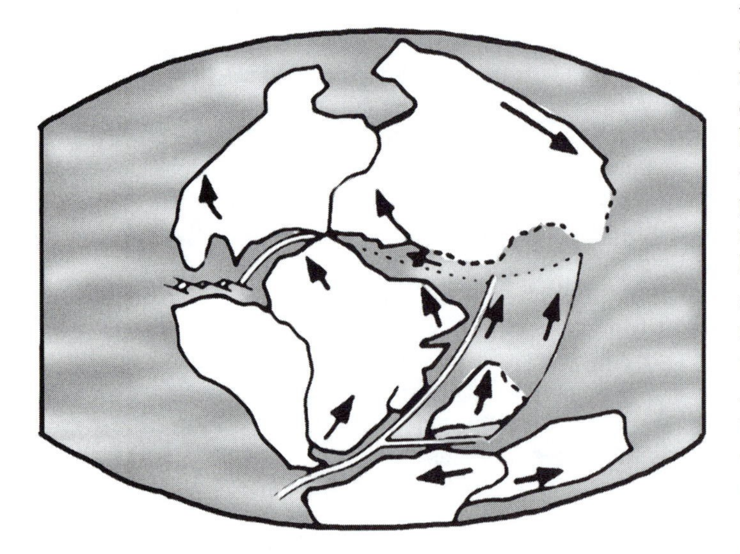


Figure S6. Eduard Suess' Gondwanaland, similar to Wegener's theory of the supercontinent Pangea, began to separate over 200 million years ago, resulting in the formation of the modern continents. Their concepts of continental drift led to the modern science of plate tectonics.

Suess' theory stated that these four southern land areas were once joined, and, since separation, still exhibit a pattern that indicates their common origin. More recently, computer models of the coast margins, mountains, and other formations of these continents validate his theory. Suess' theory is an important concept for understanding continental drift in the science of plate tectonics.

*See also* Wegener

**SWAMMERDAM'S THEORY OF PREFORMATION:** Biology: *Jan Swammerdam* (1637–1680), Netherlands.

*All parts of adult animals are formed at the beginning of the egg's development.*

The theory of preformation asserts that all the parts of animals are present in the female's eggs at the time of conception, or, as stated by some scientists in the seventeenth century, a tiny homunculus existed inside either the undeveloped female ovum (ovists) or the male sperm (spermatists) and was composed of all the parts of an adult human. Jan Swammerdam, a devoted microscopist, was the first to use the microscope in the study of zoology. He conducted excellent studies of insects and other smaller animals and was also the first to identify cells in frogs' blood. Swammerdam based his theory of preformation on the metamorphosis process of insects, which he dissected with excellent skill. He observed rudimentary parts (wings, legs, eyes, etc.) of adult insects inside the pupae and cocoons. He claimed a caterpillar did not metamorphose into a butterfly or moth but rather continued to grow into an adult from parts existing in the caterpillar stage. He claimed the same for a tadpole changing into an adult frog (i.e., all of the body parts were contained in the egg). The preformation theory was generally accepted into the early nineteenth century until the epigenesis theory developed. Proponents of the epigenesis theory believed the egg is undifferentiated, and development occurs throughout a series of steps after fertilization by a sperm. This is basically the theory that is accepted today. Karl Ernst von Baer's concept of the formation of a "germ layer" in the eggs of mammals, out of which all the embryonic organs develop, was the final death knell of the preformation theory. Moreover, Swammerdam's insect studies provided him with evidence that disproved the theory of spontaneous generation as an explanation for the origin of life.

*See also* Baer; Haeckel

**SZILARD'S THEORY OF NEUTRONS SUSTAINING A CHAIN REACTION:** Physics: *Leo Szilard* (1898–1964), United States.

*When the nuclei of certain heavy elements each absorb a single neutron, they will split into two nuclei of different elements, and in the process emit two additional neutrons. Thus, if such an element is concentrated in a critical mass, enough neutrons will be produced to create a sustaining nuclear chain reaction, releasing a great deal of energy.*

Leo Szilard followed up on the research on fissionable uranium conducted by Lise Meitner, Otto Frisch, and Otto Hahn. Once he knew it was possible to bombard uranium nuclei with slow neutrons to cause the nuclei to fission (split) and that this reaction would produce more neutrons than were absorbed by the nuclei, he was convinced a sustainable chain reaction was conceivable. In 1933 Szilard fled Germany to England, where he filed a patent for the neutron chain reaction, which he later assigned to Great Britain. In 1938, he arrived in the United States, where he attempted to convince the U.S. government and scientists to develop an atomic bomb because he was convinced the German government was doing the same. He and other scientists persuaded Albert Einstein to send the famous August 2, 1939, letter to President Franklin D. Roosevelt describing the potential for developing the atomic bomb and the urgency for doing so. Despite Szilard's continued work with the atomic bomb, including the idea of a breeder reactor that produces more radioactive material than it uses as fuel, he strongly opposed the use of atomic weapons. He is known as a "scientist of conscience."

*See also* Chadwick; Einstein; Fermi; Hahn; Meitner; Oppenheimer; Rutherford; Teller

# T

**TAMM'S THEORY OF THE CHERENKOV EFFECT:** Physics: *Igor Yevgenyevich Tamm* (1895–1971), Russia. In 1958 Igor Tamm shared the Nobel Prize in Physics with the Russian physicists Ilya Frank (1908–1990) and Pavel Cherenkov (1904–1990) for explaining the Cherenkov phenomenon of radiation.

> *When high-speed particles (electrons) pass through nonconducting transparent solids at speeds faster than light passes through the same solid, radiation is emitted.*

Igor Tamm's theory is based on the quantum theory of diffused light in solid bodies. Although these high-speed particles cannot travel faster than the speed of light in a vacuum (nothing can), they do pass through certain types of solids and liquids at speeds approaching that of light (186,000 miles per second in a vacuum). At the same time, light travels through the same substances at a slower rate of speed than do high-speed particles (electrons). For instance, light traveling through water or a crystal does so at a speed less than its (light's) speed in a vacuum, whereas the speed of a high-energy electron can surpass the speed of light in water or a crystal. (The speed of light in a vacuum is 299,793 kilometers per second; in water, light's speed is only 224,900 km/s, and when traveling through a diamond [with a high index of refractions], the speed of light is only 124,000km/s.) Tamm's theory explained Cherenkov radiation as being similar to a shock wave produced when an object moves faster than sound through air (e.g., a bullet or jet airplane going faster than sound; *see* Mach). For sound and Cherenkov radiation, the velocity of the object (particle) passing through the medium is greater than the shock wave created by the object's motion. This explains why water surrounding the core of nuclear reactors glows an eerie bluish/green color and why there are "showers" of cosmic radiation on Earth. Detectors designed to count Cherenkov radiation measure the strength of high-speed particles and can also determine their velocities, which almost reach the speed of light. Tamm's explanation of the

Cherenkov effect enabled physicists to understand better the operation of nuclear reactors, as well as the nature of cosmic radiation.

## TARTAGLIA'S MATHEMATICAL SOLUTION TO CUBIC EQUATIONS:

Mathematics: *Niccolo (Niccoló) Fontana, known as Tartaglia* (c.1499–1557), Italy.

> *The insertion of a cosa ("a thing"), which represents an unknown quantity, into an equation can assist in solving certain cubic-type equations.*

Born in Brescia in either 1499 or 1500, Niccolo was six years old when his father, a mail deliverer, was murdered during his rounds, leaving the family in poverty. When Niccolo was about twelve years old, he was almost killed during the French invasion of Brescia (the Republic of Venice) that killed forty-six thousand residents. During this slaughter, which reportedly only lasted seven days, a French soldier who attacked Niccolo with a saber sliced into Niccolo's jaw and palate. Too poor to afford a doctor, his mother nursed him back to health. He was left with a serious speech impediment and stutter, and thus accepted the name "Tartaglia," which means "stammerer," the appellation by which he was known for the rest of his life. Tartaglia became a self-taught mathematics teacher who had an extraordinary ability in advanced mathematics. During his years as a teacher in Venice he entered many debates among mathematicians. The first person to solve cubic equations algebraically was a Bolognese mathematician known as Scipione del Ferro (1465–1526). While del Ferro kept this information a secret during his lifetime, on his deathbed he told his assistant, Antoniomaria Fior (dates unknown) his solution. At that time only one solution was known because negative numbers were not used in those days. This type of equation is now known as $x^3 + mx = n$. Tartaglia discovered how to solve more than this single-type cubic equation using squares and cubes as related to numbers, that is, $x^3 + mx^2 = n$. A contest was arranged between Fior (who considered himself a great mathematician) and Tartaglia in which each would put up a sum of money to submit to the other thirty different problems involving cubic equations. Tartaglia submitted a variety of questions, all different and based on his use of squares. On the other hand, Fior had only been given by del Ferro the solution for one type of equation that did not use squares. Tartaglia was inspired and solved his thirty equations in about two hours, while Fior was unable to complete his set of questions.

This seemed to solve the question of who was the best mathematician, but the solution worked out by Tartaglia created another controversy. It seems that Tartaglia shared his solution in confidence with another mathematician by the name of Gerolamo Cardano, who promised to keep Tartaglia's secret. However, when Cardano learned (mistakenly) that it might have been del Ferro, and not Tartaglia, who first solved the cubic equations, he felt no longer bound by his secret agreement and published his methods. Cardano noticed that Tartaglia had used the square root of a negative number in his solutions. This led to a contest between Tartaglia and Cardano. Cardano did not show up for the contest. Instead, he sent his assistant Lodovico Ferrari (1522–1565) who won this second contest. This resulted in the end of Tartaglia's career as a professor of mathematics, as well as his source of income. An interesting consequence followed as a result. When Cardano published his results, it led to the establishment of the policy that the first person to publish the results of an experiment, discovery, or invention, and not necessarily the first person to actually conduct the experiment or who had made the discovery, is the one given credit. This policy is still in effect today. During the remainder of his life, Tartaglia harbored a great resentment toward Cardano. Today, Tartaglia, del Ferro, and Ferrari are all given credit for the work that resulted in the solutions of cubic and quadric equations.

Tartaglia made other contributions to mathematics, arithmetic, and geometry, including his study of Pascal's triangle and his study of tetrahedrons. His other accomplishment was his translation of Euclid's *Elements* into modern Italian. He also made mathematical improvements in the military sciences including ballistics. Tartaglia died in poverty in Venice where he had lived most of his life.

*See also* Cardano

**TATUM'S THEORY OF GENE-CONTROLLING ENZYMES:** Biology: *Edward Lawrie Tatum* (1909–1995), United States. Edward Tatum shared the 1958 Nobel Prize for Physiology or Medicine with George Beadle and Joshua Lederberg.

*Specific genes are responsible for the production of specific enzymes that control particular biochemical reactions in living organisms.*

Edward Tatum began his experiments by inducing mutations in the genes of the fruit fly. He extended this concept by exposing particular types of bread mold to X-rays to induce mutant genes in the mold. He discovered that when these mutant molds were grown in different types of media, they were affected differently according to varying types of nutrients in the growing medium. He then crossbred these distinct mutant mold genes, noticing that their diet peculiarities were inherited according to the standard Mendelian percentages (*see* Mendel). Tatum and a colleague theorized that particular genes are responsible for specific enzymes in living organisms. Enzymes act as organic catalytic proteins found in living cells that control and regulate the chemical process of life. From this they concluded that all chemical processes taking place in plant and animal cells are controlled and regulated by genes.

*See also* Clark; Delbruck; De Vries; Lederberg; Mendel

**TAYLOR'S THEORY OF GRAVITATIONAL WAVES:** Astronomy: *Joseph Hooton Taylor, Jr.* (1941–), United States. Joseph Taylor shared the 1993 Nobel Prize in Physics with Russell Hulse.

*When a binary pulsar is influenced by a nearby massive object, the pulsar changes its orbital period, thereby producing gravitational waves.*

Einstein used his theory of general relativity to predict that when a massive body rapidly accelerates, it will radiate gravitational waves. Such "waves" produced by accelerating stellar bodies are much too weak to be detected on Earth, and thus are still theoretical. Taylor and his student, Russell Hulse (1950–), observed a binary pulsar, which is a pair of massive bodies whose orbits intersect and whose gravities affect each other's velocities and thus their orbital periods. Taylor and Hulse's pulsar was located about sixteen hundred light-years distant from Earth in space, so the theoretical radiation of its gravity waves was much too weak to reach Earth. They continued to watch this observable pulsar and its companion, a dark neutron star, and assumed their near approach to each other would cause a slight change in the pulsar's acceleration and its orbit. This change in acceleration should be detectable over a period of time as a very minor alteration. After several years of analyzing his observational data, Joseph Taylor detected a very slight decrease in the pulsar's orbital period. He claimed the data supported Einstein's theory for the existence of gravitational waves. Even so, no direct radiation gravitational waves from pulsars or any other deep space objects have been measured on Earth.

**TELLER'S THEORY FOR THE HYDROGEN BOMB:** Physics: *Edward Teller* (1908–2003), United States.

*The production of X-rays from a fission bomb will produce the pressure and temperature required for a nuclear fusion reaction.*

As the atomic (fission) bomb was being developed, Edward Teller, a nuclear physicist, was already contemplating the design for a hydrogen (fusion) nuclear bomb. The distinctions between the two types of nuclear weapons are important. "Atom bomb" is really a misnomer because atomic reactions involve the outer electrons of atoms and molecules during ordinary chemical reactions. Thus, they are chemical in nature—not nuclear. It is the nuclei of atoms that are involved for both types of nuclear weapons—the so-called atomic and hydrogen bombs. What is commonly referred to as the "atom" bomb involves the fission (splitting) of nuclei of heavy, unstable radioactive elements (e.g., uranium-235 or plutonium-239), which releases enormous energy and radiation, whereas the "hydrogen" bomb is the fusion or combining of nuclei of lighter elements to form nuclei of heavier elements, which also releases great quantities of energy but less radiation.

After World War II, U.S. President Harry S. Truman, concerned that the Russians also had developed and exploded their first "atomic" bomb, encouraged the development of the "hydrogen" bomb for national security. Previously Teller and other scientists had studied various designs for such weapons. Teller proposed three different designs, two of which proved impractical. The third seemed promising until a theoretical mathematician, Stanislaw Ulam, pointed out that Teller's design was not only impractical but much too expensive. Together they developed a further model that overcame the physical and economic problems of the other designs. One problem was that fusion, unlike fission, could not occur under normal conditions of temperature and pressure. Great force was required to slam the positively charged nuclei of hydrogen (protons) together to overcome their natural repulsion. The fusion reaction, which is also referred to as **thermonuclear**, requires tremendous heat and pressure to complete the reaction (e.g., the sun's conversion of hydrogen nuclei into helium nuclei).

Ulam proposed construction of the fission (atomic) bomb around the H-bomb to provide the force necessary to fuse "heavy" hydrogen atoms (with two protons in their nuclei) together to form nuclei of helium. Teller improved Ulam's concept by devising a "mirror" to focus and concentrate the X-rays produced by the A-bomb surrounding the H-bomb to produce the force necessary to start the fusion reaction. The following nuclear fusion reaction occurs: ${}_1H^2 + {}_1H^2 \rightarrow {}_2He^4 + \rightarrow$ Energy. There are two types

There is a little-known story involving academic physicists, including Edward Teller and a number of other top tenured physicists who were called upon by the U.S. Department of Defense for advice on aspects of the physics of war. They met in secret to discuss not only the possibility of an H-bomb, but other physics-related problems beyond the capabilities of the Defense Department. Organized in 1960, this group called themselves "the Jasons," which is an acronym for July-August-September-October-November. "Jason" is also related to the Greek myth of Jason and the Argonauts. Top physicists in major research universities used the summer months to conduct most of their research work. The first groups did meet during the summer, and later for short periods during the academic year. However, they did not want it known that they were doing research defense work for the U.S. government, particularly because it was *applied* research—not *basic* research about the nature of matter and force that drives the universe. The applied research and technology for war not only makes use of knowledge gained by basic research, but involves personal moral decisions as well. Most of the thirty to sixty scientists who agreed to be part of the Jasons received top government clearance.

During World War II there were a number of top physicists who were patriotic and worked for or advised the federal government on war-related science policies, particularly on the Manhattan atom bomb project. After the war they returned to their universities and their pursuit of basic science. By the time of the Vietnam War and into the 1960s top scientists again became consultants to the Defense Department, but in a much more secret capacity, primarily because of the Cold War political mentality. A number of the former Manhattan Project scientists, as well as younger people, met in the summer of 1960, as well as subsequent summers, to help solve highly classified problems for the U.S. government in several war-related areas, including intelligence gathering. Although most of the scientists had security clearance, they were given additional freedom to highly classified information, leading to breakthroughs over the next forty-plus years in the areas of high-tech use of electronics on the battlefield that included advanced radar, as well as a means of underwater communications with our ships and submarines worldwide. They also worked on the concept of the "Star Wars" warning system, as well as on the now-timely issues of global climate change, electronic barriers that can be used on battlefields, as well as on the Mexican border, and numerous outer space–related operations. Animals, including dogs, honeybees, and others were trained to detect (smell out) hidden bombs and landmines. More recently, departments within various government agencies have taken over much of the work done by Jason scientists in the past. The number of Jasons who are Nobel Prize winners is impressive. They include: Steven Weinberg, Murray Gell-Mann, Hans Albrecht Bethe, Luis W. Alvarez, Eugene Wigner, Charles Hard Townes, and Val Fitch, among others.

of heavy hydrogen, deuterium D–2 and tritium T–3, both of which are used in fusion reactions because they contain extra neutrons in their nuclei (*see* Figure O1 under Oliphant). The atomic weight of two hydrogen-2 atoms is 4.0282, whereas the atomic mass of a helium-4 atom is 4.0028, representing the loss of 0.0254 mass units when hydrogen nuclei fuse to form helium nuclei. This may seem like a minute loss of mass to convert to energy ($E = mc^2$), but when trillions of nuclei are involved in a fusion reaction, this "extra" or leftover mass is converted to about ten times the energy released by a typical atomic (fission) bomb. The first successful fusion hydrogen bomb was detonated by the United States in 1951. Ulam and Teller both claimed it was their own concept to use an atomic bomb to "trigger" the H-bomb. Although many scientists gave Ulam the credit, by this time Teller had become a strong political advocate for developing thermonuclear weapons as a deterrent factor and became known as the father of the hydrogen bomb.

Teller and other advocates of national nuclear polices attacked those who opposed the U.S. policy on nuclear weapons. One of the major developers of the fission (atomic) bomb, Robert Oppenheimer, was horrified at the prospect of using the much more destructive hydrogen bomb in warfare and had his career cut short by opposing Teller's position on nuclear weapons. Edward Teller proceeded to encourage scientists to develop thermonuclear weapons and the strategic defense initiative as a means to protect the United States from long-range missile attacks.

*See also* Curies; Fermi; Hahn; Meitner; Oppenheimer; Pauli; Rutherford; Szilard; Ulam

**TEMIN'S THEORY FOR TRANSCRIBING RNA INFORMATION INTO DNA:** Biology: *Howard Martin Temin* (1934–1994), United States. Howard Temin shared the 1975 Nobel Prize for Physiology or Medicine with David Baltimore and Renalto Dulbecco.

*In addition to genetic information "flowing" from DNA to RNA, a special enzyme makes it possible for DNA to receive information from RNA, thus allowing DNA to provide crucial information needed for cell growth.*

While conducting research with cancer cells in chickens, Howard Temin discovered a new enzyme. He named it *reverse transcriptase* because it could reverse the flow of genetic information that at one time was believed to proceed only in one direction, from **DNA** to **RNA**. This DNA-to-RNA sequence was referred to as the "central dogma" because this sequence was required for DNA to replicate itself. This concept was generally accepted by all molecular biologists. Temin's new enzyme "transcribed" the RNA into DNA, which improved DNA's effectiveness in controlling the processes of cell metabolism. At about the same time, David Baltimore independently made the same discovery.

*See also* Baltimore; Dulbecco; Gallo

**TESLA'S CONCEPT OF HIGH-VOLTAGE ALTERNATING CURRENT:** Physics: *Nikola Tesla* (1856–1943), United States.

*High-voltage alternating current can be transported more efficiently over long distances through wires than can direct current.*

In the early 1880s Thomas Edison developed the direct current generators and distribution system used by his Edison Electric Light Company in New York City. It revolutionized the use of electricity but had one major drawback: it had to be generated near the site where it was to be used. This made it useful for lighting compact cities, but since direct current (**DC**) lost much of its energy when sent over wires for some distances, it was an impractical system. Another problem with direct current was that DC dynamos (generators) and motors required a commutator with wire "brushes" to provide electrical contact with the armature and terminals. This arrangement required constant maintenance because the brushes needed frequent replacement. Nikola Tesla's concept solved these problems by using a new system of dynamos (generators) and transformers that could produce current that alternates (**AC**) direction many times per

> With the scientific and engineering worlds, and the courts, extending to him (and not Edison) a clear title to the honor of being the great pioneer discoverer and inventor of the principles and machines that created the modern electrical system, Tesla stands without rival as the genius who gave the world the electrical power age that made our mass-production industrial system possible. The name Tesla should therefore, in all right and justices, be the most famous name in engineering world today. (From *Prodigal Genius: The Life of Nikola Tesla,* by John O'Neill. Cosmos Classics. 2006, p. 117).

second and could be "boosted" to high voltages by transformers, enabling it to be sent over longer distances. An interesting result was that in the 1890s William T. Love (dates unknown) began digging a canal on land that he owned to circumvent Niagara Falls. He planned to use this diverted water to generate direct (DC) electricity for industries that could be located near his canal and thus at a greater distance than industries restricted to the falls area due to the limits of direct current. Nikola Tesla's AC system interrupted the project, and the ditch was later filled in with waste material and became known as the infamous Love Canal. Today alternating current in the United States has a 60-cycles-per-second (Hz) rate of changing direction, at 120 volts with relatively high ampere current, while much of the rest of the world uses 50-cycle (Hz), 240 volts with low ampere current.

Nikola Tesla is best known for his insightful technical knowledge of electricity, which he applied in developing many inventions, some of which were years ahead of their time. Among them are the Tesla coil/transformer used in radios and TV sets, the induction (brushless) motor used in computer hard disk drives (as well as almost every other application where motors are required), telephone repeaters for long-distance phone lines and the transatlantic cable, and wireless communication devices now used in cellular phone systems. A partial list of Tesla's almost eight hundred patents include 7 patents related to direct current generators and motors; 39 patents related to electric transmission of power, dynamos, motors, and other systems; 28 patents for high frequency devices for control of electrical systems; 76 patents for wireless systems such as radio, wireless telegraph, and tuning devices; 26 patents for steam turbines, pumps, oscillators, and speedometers. In addition to his many inventions, he should be given credit for additional theories. For instance, he not only theorized but also demonstrated that Earth itself is a source of useful power. He experimented with using Earth as a resonator that could build up frequencies that might be able to communicate wirelessly worldwide as well as destroy surface buildings. His many inventions with light, wireless communications, and electrical systems were forerunners of our present-day technologies.

*See also* Ampère; Edison; Faraday; Oersted; Ohm; Volta

**THALES' THEORY THAT WATER IS THE BASIS FOR ALL THINGS:** Philosophy: *Thales of Miletus* (c.625–547 BCE), Greece.

*All material things are derived from water.*

Thales of Miletus is considered the first Greek philosopher/natural scientist who was educated in the Milesian School for natural philosophers. He also founded a school of natural philosophy in Ionia, a city on the Aegean Sea in Asia Minor. There is a distinction between the Milesian School and the Ionian School. The Ionian included the philosophies of both the Milesians and other Ionian philosophers, for example, Heraclitus of Ephesus (c.535–475 BCE) who believed in ethereal fire from which all things originated and returned in a never-ending process. Thales is known more by stories and folktales than he is by tangible evidence for his thoughts. His accomplishments led to the beginnings of geometry, astronomy, and natural science in general. He is given credit for establishing proof of the claim that a circle, when divided into two equal halves, is so divided by its diameter (this is one theorem neglected by Euclid three hundred years later). He is also credited with predicting a solar eclipse in the year 585 BCE. He believed the question of composition as well as the origin of all material things was connected as the same question, and the answer is water. He chose water because of its importance to the growth and nurture of all living things, and its importance in the lives of the ancient Greeks.

*See also* Euclid

**THEOPHRASTUS' CONCEPTS FOR PLANT CLASSIFICATION:** Biology: *Theophrastus* (c.372–287 BCE), Greece.

*Plants can be distinguished and classified according to their structures and physiology.*

Theophrastus, a student of Aristotle, became the head of the Lyceum in Greece upon the death of Aristotle. One of Aristotle's great achievements was his classification of animals as known at that time. Theophrastus used some of Aristotle's techniques in his own pursuits. He was a keen observer who wrote excellent descriptions of plants. In his two main books, one dealing with plant structures and the second with their functions (physiology), he described over five hundred plant species. His writings influenced botanists over several centuries. Theophrastus was the first to establish a relationship between the structure of flowers and the resulting fruits of plants, but his main theory distinguished between *monocotyledon* and *dicotyledon* seeds. After examining grass and wheat seeds and noticing they had only one seed "coat," he classified them as monocotyledons, while bean seeds, having two seed "coats," were classed as dicotyledons. Cotyledons are the first shoots of the new plant arising from the germinating seed; thus he considered them as "coats" or "covers." He described the differences between flowering plants (angiosperms) and cone-bearing plants (gymnosperms). He coined many new terms and names for plants and their parts and thus is considered by many biologists as the father of botany.

*See also* Aristotle; Linnaeus

**THEORELL'S THEORY OF ENZYME ACTION:** Biochemistry: *Axel Hugo Theodor Theorell* (1903–1982), Sweden. Hugo Theorell was awarded the 1955 Nobel Prize in Physiology or Medicine.

*The oxidation of the ADH (alcohol dehydrogenases) enzyme is responsible for breaking down alcohol in the kidneys.*

Axel Hugo Theodor Theorell was born in Linköping, Sweden. After finishing secondary school in Linköping, he studied medicine at the Karolinska Institute in

Stockholm where he received his bachelor of medicine degree in 1924 and his medical degree in 1930. At Karolinska his areas of interest were the lipids related to blood plasma. He also worked at the Pasteur Institute in Paris for several months in the mid-1920s where he studied bacteriology. As a result of illness, he abandoned the practice of medicine and accepted a teaching and research job in biochemistry and later became head of the biochemistry department of the Nobel Medical Institute in Stockholm. Later he moved to Theodor Svedberg's Institute of Physical Chemistry in Uppsala, Sweden, where Theorell discovered the properties of crystalline myoglobin. He also discovered a "yellowish" enzyme that he called "lactoflavin" but was later renamed "riboflavin" which is related to vitamin $B_2$. Later, by using an electrophoresis method, he purified the enzyme and separated the protein part of the pigment from the carrier. After determining that its structure is a lactoflavin phosphoric ester, he named it "flavin mononucleotide" (FMN). In 1941 he and his colleagues developed the dehydrogenizing oxidizing enzyme (ADH), a protein that is found in liver and yeast. They discovered that the enzyme is also responsible for the oxidation of alcohol. This led to a practical test for ethyl alcohol in the liver. After thirty-three years as director of the Nobel Medical Institute, he retired in 1970, while still continuing to be involved in enzyme research.

**THOMSON'S ELECTRON THEORY:** Physics: *Sir Joseph John Thomson* (1856–1940), England.

> *Rays from a cathode tube can be deflected and measured by using magnetic and electric fields. Thus, cathode rays must be particles smaller than an atom and which carry an electrical charge.*

J. J. Thomson experimented with radiation produced by a cathode ray tube. In 1876 the German physicist Eugen Goldstein tested a Geissler tube, a vessel that enclosed a vacuum and used an internal electrode to study the "flow" of electricity. Previously, several other scientists had observed a faint fluorescence within the tube when an electric current was sent through the internal electrode. Goldstein named the tube he used a *cathode-ray tube* because the glow originated from the negative cathode inside the tube. Contrary to Benjamin Franklin's theory that electricity "flowed" from positive to negative, Goldstein noted that electricity flowed from the negative to positive, while others noted this stream of glowing current could be diverted by a magnetic field. Thomson hypothesized that this beam must be composed of charged particles for it to interact with magnetic fields. In addition, he theorized that these charged particles should also react (be bent) in electric fields as well as magnetic fields. He installed an apparatus that measured the deflection of the cathode's rays by both electric and magnetic fields, which enabled him to determine that the ratio of the rays' electric charge to their mass was high. From earlier experiments on the unitary nature of an electric charge, he assumed that the charge he detected on the cathode rays was of the same unit. Therefore, because the ratio was high, this meant the electric charge (e) was much greater as compared to the mass (m) of these new particles, which must be much less based on the ratio e/m. He also assumed these low mass particles from the rays were parts of atoms, not the whole atom, and that they were about one thousand times less than the mass of the hydrogen atom (one proton). It was later determined that this mass equals 1/1837 of that of a proton.

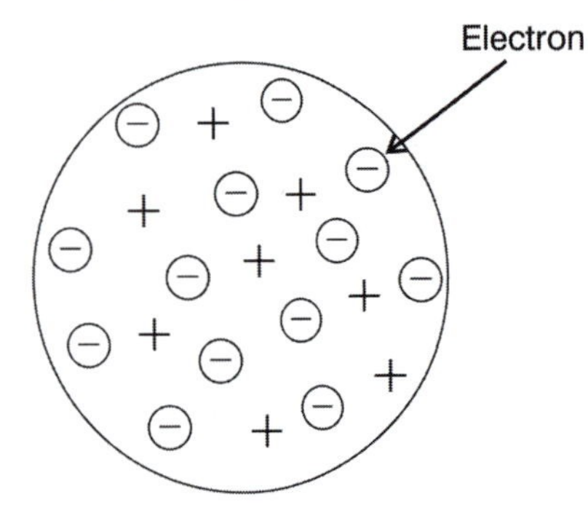


Figure T1. Thomson's early model of the atom was sometimes called the "raisins in pudding" model as he thought the negative electrons were randomly dispersed throughout the atom.

At first Thomson referred to these new particles as *corpuscles*. Later, he named them *electrons* because these new particles carried an electric charge and were detected as originating from the negative electric electrode of the cathode ray tube. Electrons are the fundamental unit of electricity. It is a basic unit. No smaller electrical charge has since been discovered. Thomson designed a model of the atom using electrons embedded in the atom whose sum charge matched the positive protons' sum charge, thus making it a neutral atom. He also investigated the role of the electrons in producing chemical reactions. Sometimes his model is referred to as "berries in muffins," "raisins in pudding," or "the fuzzy atom" because he presumed the electrons were more or less evenly distributed within a pool of positive charged particles throughout the structure of the atom (see Figure T1).

It was later discovered that electrons exist more as electrically charged particles in orbits, shells, or energy levels positioned at a relatively greater distance from the comparatively small, massive, positive, centralized nucleus. This concept of an electrically neutral atom with one or more of its outer electrons joining with those of other atoms is the basis of chemistry and the formation of molecules and compounds. Among other usages, the interaction of magnetic and electric fields on a stream of electrons is used to control the signals that produce the pictures on television receivers and computer monitors.

*See also* Bohr; Crookes; Faraday; Röentgen; Rutherford; Townsend

**TING'S THEORY FOR A NEW PHOTON-LIKE PARTICLE:** Physics: *Samuel Chao Chung Ting* (1936–), United States. Samuel Ting shared the 1976 Nobel Prize for Physics with Burton Richter.

*Bombarding a beryllium target with a stream of positive protons can produce a new subatomic particle with a longer lifetime than expected according to its mass.*

Samuel Ting's parents met as graduate students at the University of Michigan at Ann Arbor, but they returned to China to resume their respective academic careers. While they were on a return visit to Ann Arbor, his mother gave birth prematurely to Samuel, hence his U.S. citizenship. However, at the age of two months, he and his parents returned to a China that was at war. He received his early education as home schooled by his parents and maternal grandmother. Therefore, he did not receive any formal schooling until the age of twelve. Upon returning to the United States in the 1950s, he entered the University of Michigan where he earned a PhD degree in physics in 1962. He worked in various physics laboratories (including CERN in Switzerland and the Deutsches Elektronen-Synchrotron [DESY] in Hamburg, Germany) and has been a professor at Massachusetts Institute of Technology (MIT) since 1969. His research work was more experimental than theoretical, even though his main achievement was a theory based on experimental data (it usually proceeds in the other direction—theories lead to experimental data). He spent most of his time on the physics of quantum electrodynamics and the processes that subatomic particles, such as the photons, undergo as they decay into electrons and muon pairs. His research at the Brookhaven National Laboratory at Long Island, New York, led to the discovery of a new particle that had a lifetime longer than was expected by its heavy mass. About the same time, the American physicist Burton Richter (1931–) at the Stanford Linear Accelerator Center (SLAC) in California independently discovered the same particle. Because Ting called his particle "J" and Richter named his "psi" (Ψ), they combined the two names to call the new particle the JΨ or the J/psi particle. These discoveries were soon confirmed by other high-energy physics laboratories and also have led to the discovery of many more "heavy" subatomic particles.

**TISELIUS' HYPOTHESIS FOR PROTEIN ANALYSIS:** Chemistry: *Arne Wilhelm Kaurin Tiselius* (1902–1971), Sweden. Arne Tiselius was awarded the 1948 Nobel Prize in Chemistry.

*Protein molecules carry an electrical charge; thus, it should be possible to separate them by the use of electric fields.*

Arne Tiselius was familiar with electrophoresis, a procedure used to analyze chemical substances by the use of a weak electrical current. Most particles of matter contain a very small electrical charge on their surfaces. If these substances, as solutions, are applied to a special paper strip where one end of the strip is connected to a direct current source with a small negative charge and the other end of the paper is attached to the positive pole, the current will attract or repel the components of the sample substances at different rates. Depending on the size and individual characteristics of the component chemicals in the substance, these individual atoms and molecules will spread out on the strip of paper in very specific patterns, which can then be identified. Tiselius theorized that if this system could be improved to separate the proteins of blood, which also carry a small electrical charge, it might be possible to identify specific components of blood. He developed an improved electrophoresis system consisting of a U tube in which the proteins could be tracked as they separated. The tube could also be "disjoined" to extract specific components of the proteins for analysis. He further designed a lens system for refracting light through the different fractions, enabling a quantitative measurement of the particular protein fraction. Using this system, called the *Tiselius tube*, he identified four major components of blood serum proteins: albumins and the alpha, beta, and gamma globulin proteins. The best known is gamma globulin, whose chemical structure was first detected by American biologist Gerald Maurice Edelman (1929–); it stimulates antibodies in the immune system to protect against several diseases, including hepatitis and AIDS-related infections. Gamma globulin does not protect against the HIV virus.

**TODD'S THEORY FOR THE STRUCTURE AND SYNTHESIS OF NUCLEOTIDES, NUCLEOSIDES, AND NUCLEOTIDE CO-ENZYMES:** Biochemistry: *Alexander Robertus Todd (Baron of Trumpington)* (1907–1997), Scotland. Alexander Todd was awarded the Nobel Prize for Chemistry in 1957.

*Nucleotides, nucleosides, and nucleotide co-enzymes are found in chromosomes (the units of heredity) and cell plasma. They are constructed of three different substances: phosphoric acid, a sugar substance, and a nitrogen heterocyclic base that are all combined in a large macromolecule.*

Todd was born outside of Glasgow, Scotland, in 1907 and graduated with a bachelor's degree from the University of Glasgow in 1928. He received his PhD in natural philosophy from the Johann Wolfgang Goethe University of Frankfurt-on-Maine, Germany, in 1931. He also earned another PhD in chemistry from Oxford University in England in 1933. His research on nucleosides led to his investigation of the compounds that formed the units of DNA and RNA. By 1949 he had synthesized adenosine triphosphate (ATP) and flavin adenine dinucleotide (FAD). Some of his major research was related to determining the structure of vitamins to synthesize them. He was the first to determine the structure of vitamin $B_{12}$ and was able to synthesize vitamin $B_1$ and vitamin E. He also studied the alkaloids contained in marijuana as well as the alkaloids found in plants and insect pigments. Todd was knighted in 1954 and made a lifetime peer in 1962. He was elected president of the Royal Society in 1975. Todd has received numerous other honors, including the Nobel Prize in Chemistry in 1957 for his work on nucleotides and nucleotide co-enzymes.

**TOMONAGA'S THEORY OF RELATIVISTIC QUANTUM ELECTRODYNAMICS:** Physics: *Sin-Itiro Tomonaga* (1906–1979), Japan. Tomonaga shared the 1965 Nobel Prize in Physics with Richard Feynman and Julian Schwinger.

*Quantum mechanics can be applied to subatomic particles by the exchange of another "virtual" particle between two particles, thus developing a quantum field theory consistent with the theory of special relativity.*

Sin-Itiro Tomonaga's father was a philosophy professor at Kyoto Imperial University in Japan. After attending the top high school in Tokyo, Sin-Itiro entered Kyoto University, receiving his BA degree in 1929. He spent the rest of his career at Kyoto University, becoming first a professor of physics in 1941 and later in 1956 the president of the university. His research interests were similar to other theoretical physicists in the United States in the area of relativistic quantum field theory of electromagnetism (QED) that was first being investigated as early as 1929. The period of years during World War II prevented the exchange of research ideas between United States and Japanese scientists.

One of the early concepts of light was that it traveled in straight lines, always taking the shortest distance from the source of the light to its reception point. A classical argument was: Because light from a point source travels out in all directions, how does it know where it is going, even if the light's starting and ending positions are known? This is when QED theory was introduced as a possible answer. When light starts from its point of origin, it does not know its end point, but it always takes the shortest path, which is also the quickest path to its end point. QED theory explains this phenomenon by the interaction between light (photons) and charged particles, or just between two charged particles. QED describes the interactions of particles and antiparticles with each other by their exchange of photons, which use a complex set of formulae that have been visualized by the use of Feynman diagrams that assign the best path to all possible paths for the light to take (*see* Feynman). Using QED explains or predicts the probability of what will happen with a high degree of accuracy, making it a very highly accurate and useful physical theory. For instance, according to QED, light can go faster or slower than 186,000 miles per second "c" but on the average will travel at the speed of "c." QED cannot predict exactly what will happen during an experiment, but it can predict the most probable outcome.

*See also* Dehmelt; Dyson; Feynman; Planck

## TONEGAWA'S THEORY OF ANTIBODIES AND THE IMMUNE SYSTEM:

Biology: *Susumu Tonegawa* (1939–), Japan/United States. Susumu Tonegawa received the 1987 Nobel Prize for Physiology or Medicine.

> *The B-lymphocyte cells found in the immune system are able to produce billions of different antibody genes, thus providing the organism with protections from a multitude of pathogens.*

The research into the mechanism related to the generation of the diversity of the immune system's production of antibody genes was divided into two opposing camps of theorists. One group was known as the "somaticists" (also known as the *paucigene* group) and the other "germliners" (known as the *multigene* group). As their research progressed, it seemed they both were partially correct. The debate became known as the "generation of diversity" (or GOD). The first group's theory (somaticists) stated that the immune response of the body was dependent on the action of specific cells or antibodies that circulated in the body, whereas the second group's theory (germliners) asserted that there was a specific mechanism (currently unknown) that was responsible for the body's immunological reactions. Louis Pasteur's early work demonstrated how to induce acquired immunity by the use of attenuated pathogens. His theory was that toxins related to specific microorganisms caused infectious diseases and that the immune response was a natural reaction to counter the pathogen. More recent research found that this was not always the case with certain microbacteria, tropical diseases, and parasites where the immune system could not protect the living body. It

> Susumu Tonegawa was born in Nagoya, Japan, in 1939. He attended Kyoto University where he received his bachelor's degree in 1963. He received his doctorate from the University of California, San Diego in 1978. After finishing his PhD, he was employed by the Salk Institute, followed by a period at the Basel Institute for Immunology in Switzerland where he did his major research. In 1981 he became a professor at Massachusetts Institute of Technology (MIT) where he founded and became director of the Picower Institute for Learning and Memory. In the year 2006 he objected to the hiring of a female tenured faculty member for a different neuroscience institute at MIT, namely the McGovern Institute for Brain Research. He ostensibly informed the potential candidate and junior faculty member that he and she would become competitors at MIT. As a result, eleven tenured female MIT colleagues wrote a letter to the university's president, Susan Hockfield, requesting that an investigation be undertaken to review Dr. Tonegawa's alleged unethical conduct. Dr. Hockfield acceded to their request, and an internal MIT investigatory committee was formed, which later found no evidence of gender bias in Dr. Tonegawa's behavior. Nevertheless, he chose to resign his position as director of the Picower Institute at the end of 2006.

was not until the mid-twentieth century when DNA and RNA were discovered and later understood that research confirmed their involvement in the formation of specific antibodies that were developed by genetic mechanisms. Starting in 1976 Tonegawa conducted a series of important experiments that indicated that genetic material rearranges itself to form a multitude of different antibodies. The B-lymphocytes are capable of manufacturing billions of different antibodies that attack specific diseases even though, in humans, this type of cell only carries about one hundred thousand genes in the body's chromosomes.

**TORRICELLI'S VACUUM AND THEOREM:** Physics: *Evangelista Torricelli* (1608–1647), Italy.

***Torricelli's vacuum:*** *By filling a long glass tube, closed at one end, with mercury and inverting it into a dish of mercury, all but 760 mm of mercury will drain out of the tube, leaving a vacuum in the space above the column of mercury.*

Galileo, who had earlier demonstrated that air had weight, employed Evangelista Torricelli to work for him. Also sometime earlier, Jan Baptista van Helmont claimed air was not an element but rather a mixture of gases. These two concepts were incorporated into the answer as to why a pump could not raise water higher than 30 feet, which was a serious problem when draining mines that were flooded. Torricelli realized it was not the vacuum that "pulled" the water up, but rather the weight of the air (pressure) outside the pump that "pushed" the water up the pipe when low pressure was created inside the pump. In 1643 Torricelli calculated that since mercury weighs about 13.5 times that of water, air should lift quicksilver (mercury) only about 1/13.5 times as high as water. He tested this concept by filling a glass tube, closed at one end with mercury and then inverting the tube into a dish filled with more mercury. He removed the cork from end of the tube immersed in the dish of mercury, allowing the mercury in the vertical tube to settle as gravity pulled it down into the pool of mercury. Torricelli theorized that all the mercury in the tube would not exit the tube and end up in the dish of mercury because the weight of the mercury in the tube was the same weight as air outside the tube. The mercury in the glass tube maintained a height of about 760 millimeters (see Figure T2).

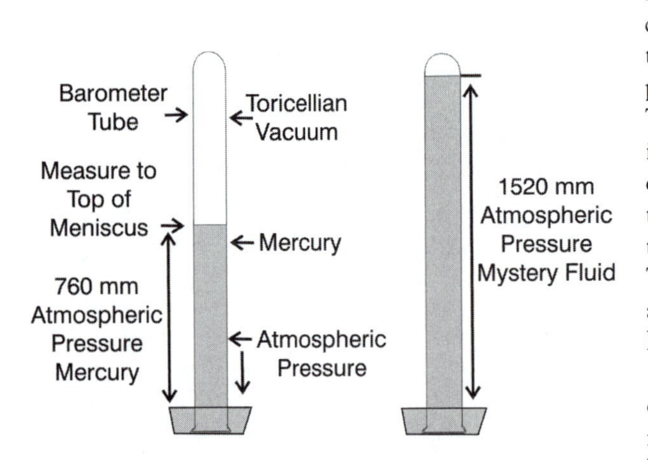


Figure T2. Torricelli's experiment demonstrated that the weight of air exerted pressure on the mercury in the dish, and it disproved the idea that the vacuum at the top of his closed tube "pulled" the mercury up the supporting column of mercury. This was the forerunner of the modern barometer.

The vacuum created over the mercury in the closed end of the tube was named the *Torricellian vacuum* or *torr*, in Torricelli's honor. One torr equals 1 millimeter of mercury (760 torr = 760 mm Hg = 1 atm or atmospheric pressure) (see Figure T3).

After viewing the column of mercury for a few days, Torricelli noticed the

PRESSURE UNITS AND VALUES
FOR STANDARD ATMOSPHERIC PRESSURES

| Pressure Units | Abbreviation | Standard |
|---|---|---|
| Atmosphere | atm | 1.00 |
| Millimeters of Mercury | mm Hg | 760 |
| Inches of Mercury | in Hg | 29.92 |
| Pounds per Square Inch | psi | 14.70 |
| Newtons per Square Meter = Pascal | $N/m^2$ = Pa | 101,300 |
| Millibar | Mb | 1,012.3 |

Figure T3. A table of the various units used to express standard atmospheric pressure.

height of the column varied slightly. He related this to changes in the weight of the air outside the tube of mercury caused by changes in the pressure (weight) of the air at the surface of Earth. This discovery was an unintended consequence of the search for a more efficient water pump, resulting in an instrument (the barometer) that could accurately measure air pressure.

*See also* Galileo.

***Torricelli's theorem:*** *The flow of a fluid through an opening in a standing pipe is proportional to the square root of the height of the liquid.*

Evangelista Torricelli's experience with the limitations of conventional water pumps and his concept of air pressure as the force that raises the water in "suction" pumps enabled him to develop his theorem. He observed that water escaping through holes at different heights in a standing pipe or container also escaped at different rates of flow. Being a mathematician, he believed there existed, and therefore calculated, a definite square root relationship between the height (depth) of the water and the rate of flow of water from the openings. His theorem is an important concept for industries that handle various types of fluids because the rate of flow as related to the height of the source of the fluid can be measured. This principle is used in elevated community water storage tanks, where pipes carry water under pressure created by the height of the water in the tanks.

*See also* Galileo

## TOWNES' THEORY FOR AMPLIFYING ELECTROMAGNETIC WAVES:

Physics: *Charles Hard Townes* (1915–), United States. Charles Townes shared the 1964 Nobel Prize for Physics with Nicolai G. Basov and Aleksandr M. Prokhorov.

> *Molecules that exist in discrete energy states and absorb discrete frequencies of electromagnetic energy will emit photons of the same frequency.*

Albert Einstein pointed out that if an electromagnetic photon of a specific frequency struck a molecule that was in a high-energy state, the molecule would proceed to a lower-energy state while emitting a photon of the same wavelength as the one striking the molecule. Charles Townes realized this would produce two electromagnetic photons of the same frequency, which then could strike other high-energy molecules to produce more photons, resulting in a type of "chain reaction" that would produce a multitude of photons of the same wavelength and frequency. The consequence would be a flood of monochromatic (one-color) single-wavelength photons of the spectrum, all proceeding as coherent radiation (in the same direction). Townes demonstrated this theory by sending small amounts of microwave photons of a given frequency into energized ammonia molecules. The energized molecules had previously been produced by intense broadband irradiation. The microwave photons caused these molecules to drop back to their original energy level, producing new microwave photons. The result greatly amplified the original weak microwave radiation, which resulted in a flood of coherent electromagnetic radiation called the microwave amplification by stimulated emission of radiation (maser). In 1958 Townes proposed this process could be applied to any wavelength within the electromagnetic spectrum. The concept was later improved by using just the section of the electromagnetic spectrum representing visible light, first called the "optical maser" and later named the light amplification stimulated emission of radiation (laser).

*See also* Turner

**TOWNSEND'S THEORY OF COLLISION IONIZATION:** Physics: *Sir John Sealy Edward Townsend* (1868–1957), Ireland.

*As an electric current passes through gas, some molecules become ionized; they then collide with and ionize other molecules, thus multiplying the original charge.*

Sir John Townsend followed up on J.J. Thomson's discovery of the electron, at which time Thomson estimated the electron to be about 1/1000 the weight of the hydrogen nucleus, later revised to 1/1837 the mass of the proton. Thomson also determined that the electron carried the basic negative charge. Townsend calculated the amount of strength of this negative charge by forming a charged cloud of water droplets and measuring the rate of fall of the charged particles as they passed a source of electricity.

Townsend's "collision" theory answered the question of how an electric current could pass through a gas that supposedly had a weak electric field. The explanation was that as the current's electrons passed through the gas, some of the gas molecules became ionized (each gas molecule carried a charge) in the electric field. This created collisions with other gas molecules, which then became ionized, and so on, until an "avalanche" or multiplication of electrons proceeded through the gas despite the weakness of the original electric field. This theory is important in the fields of electronics and communications, where electrons cascade through multiplier tubes used to measure the radiation tracks of subatomic particles.

*See also* Millikan; Thomson

**TURING'S THEORY FOR TESTING COMPUTER INTELLIGENCE:** Mathematics: *Alan Mathison Turing* (1912–1954), England.

*Because the brain is computable, it must be possible to program computers to acquire human intelligence and devise tests that will verify computer intelligence.*

In 1937 Alan Turing designed a computing device called the Turing machine that was connected to several devices including an input device, a long-tape operating program divided into sections, a printer, and a correcting device. The machine had five symbols programmed to control the machine, similar to the operating systems of modern desktop computers and which could be used to make mathematical calculations. After World War II Turing developed several types of computers. He named one the automatic computing engine (ACE) and the other Manchester automatic digital machine (MADAM). By 1950 he argued that a computer could be designed to imitate human intelligence. He then proceeded to design a computer test to prove this concept. The test was based on his idea called an "imitation game," later called the *Turing test*. The Turing test required an interrogator to ask one person and a computer (the interrogator could see neither the computer nor another person) a question that could be answered by typing out a textual answer. Turing claimed that if the interrogator was unable to judge which answer came from the human and which from the computer, then computer intelligence was proven. A similar test is used today to determine if an artificial intelligence (AI) computer program can really imitate human intelligence or "think" (so far, modern computers are only partially successful in imitating the human brain). Alan Turing made distinctions between computer intelligence (AI) and thinking, emotions, and other human attributes.

*See also* Babbage

> Aside from his contributions to the field of artificial intelligence and computer science, there is another, but less well known, aspect to Alan Turing's short career. During and following World War II, as well as during the period of the cold war in Europe, only a few of Turing's friends knew of his homosexuality. Even so, this fact excluded him from many cooperative sensitive research projects with the United States. At that time, the societies of Great Britain and the United States regarded homosexuals as security risks. In 1952 Turing had an affair with a young man who stole some of Turing's belongings. Turing reported the theft to the police who then arrested Turing. He was given the choice of going to jail for gross indecency or agreeing to probation for one year while undergoing hormone (estrogen) treatment, which was intended to neutralize his libido. At first, Turing tolerated the treatment, but as time went on, he became despondent and eventually took his own life by eating an apple that was laced with cyanide poison. To make the suicide look more like an accident (for his mother's sake), he left the apple by the side of his bed.

**TURNER'S THEORY FOR MEASURING OUTER ENERGY LEVELS OF MOLECULES:** Chemistry: *David Warren Turner* (1927–1990), England.

*The energies of outer electrons ejected from ionized gas atoms or molecules can be measured by deflecting these electrons with an electrostatic charge.*

David Turner devised a technique that used a narrow beam of monochromatic ultraviolet radiation (maser) to eject outer electrons from ionized atoms and molecules of gas. The energies of these ejected electrons can then be measured by their degree of deflection as they pass through an electrostatic field. This procedure is known as *molecular photoelectron spectroscopy*. Applying his theory, he assisted in developing a microscope that uses X-rays to "kick" out electrons from the sample, thus measuring characteristics of the sample based on the degree of deflection.

*See also* Townes

**TYNDALL'S THEORY FOR THE TRANSMISSION OF LIGHT THROUGH GASES:** Physics: *John Tyndall* (1820–1893), England.

> *Light passing through a clear solution of dissolved substances is not scattered, while light passing through cloudy water containing large molecules and clusters of molecules (colloids) will be scattered.*

John Tyndall, experimenting with the transmission of radiant heat through different types of gases and vapors, measured the absorption and spreading out of the radiation through these gases. In 1859 he studied the effects on light when it passed through various liquids and gases and noted the degree of scattering in the path of the light. This scattering was named the *Tyndall effect*, after his theory that particles in the path of the light cause the scattering that renders the light beam visible. Nephelometry, a field of physical chemistry that examines the scattering properties of small particles in air, is similar to the Tyndall effect. The scattering of the beam of light off minute particles suspended in air is more pronounced and effective when shorter-wavelength ultraviolet radiation is used. Tyndall used this effect to explain why the sky is blue overhead and why sunsets appear red. This occurs because sunlight passes through a greater number of dust particles, filtering out the ultraviolet light, allowing the longer wavelengths of light (orange to red) to be seen on Earth.

Tyndall is credited with first explaining the greenhouse climate effect as being a natural phenomenon. He also measured the air pollution in London using the scattering of infrared light. He was among the first to determine that the dust in the atmosphere contains microorganisms, as well as identifying the ozone molecule as a cluster of three oxygen atoms, not the normal two-atom molecule of oxygen. Tyndall is also credited with many inventions, including the fireman's respirator, an improved foghorn, and the gastroscope that enabled physicians to observe the inside of patients' stomachs without surgery. He made studies that led to improved knowledge of thermodynamics, solar energy, the transmission of light in space, and the structure of the Earth's atmosphere. He is thought of as the father of science education. He conducted many popular and exciting science demonstrations and was responsible for the teaching of the physical sciences in public schools and universities.

*See also* Ramsay

# U

**UHLENBECK'S THEORY OF ELECTRON SPIN:** Physics: *George Eugene Uhlenbeck* (1900–1988), Netherlands and United States.

> *A basic property of an electron is its spin around an axis that results in self-induced angular momentum associated with its magnetic dipole moment.*

In 1974 George Uhlenbeck, in cooperation with the Dutch-American physicist Samuel Goudsmit (1902–1978), observed the spectra anomalies known as the Zeeman effect (*see* Zeeman) in the spectral lines of X-rays. The quantum number ($s$) is always $1/2$. The spin of angular momentum of the electron is related to its spin around its axis and is not to be confused with the orbital angular momentum of an electron as it moves in its orbit around the nucleus. This phenomenon cannot be determined by using methods related to classical physics. Rather, the atomic beam method of spectroscopy gives greater precision in the measurement of the frequencies of spectral lines and thus provides greater sensitivity to the factors affecting the magnetic moment of the electron.

George Uhlenbeck's relatives, although of German origin, lived in the Dutch colonies. His father was born in Java in the Dutch East Indies (now called Indonesia) and served in the Dutch East Indian Army. George was born in Batavia, now known as Jakarta in Indonesia. When he was six years old, George's family moved to the Netherlands where he attended elementary, high school, and the University of Leiden. He received his PhD in physics in 1927, the same year in which he emigrated to the United States. He was appointed professor of theoretical physics at the University of Michigan, followed by a position at Columbia University in New York. During World War II he worked on a team out of Massachusetts Institute of Technology (MIT) who were developing radar. In 1960 he moved to Rockefeller Medical Research Center (now Rockefeller University) in New York where he remained until he retired in 1974.

His main interests were atomic structure and the kinetic theory of matter, and his aim was to understand the relationship between physics at the atomic level as well as at the macroscopic level. His students considered him an excellent teacher who was well organized and who made clear the elegant structure of statistical mechanics based on the past work of Maxwell, Boltzmann, Gibbs, and others. He received many awards during and after his career and was responsible for educating several generations of students in the modern esoteric field of statistical mechanics.

*See also* Boltzmann; Gibbs; Maxwell

**ULAM'S "MONTE CARLO" SYSTEM:** Mathematics: *Stanislaw Marcin Ulam* (1909–1984), United States.

> *It is possible to obtain a probabilistic solution to complex mathematical problems by using statistical sampling techniques.*

In the early 1940s, Stanislaw Ulam, a Polish Jewish mathematician who was born in Galicia (formerly Austria, now the Ukraine), was asked by the Los Alamos, New Mexico, nuclear development project administrators to develop a mathematical theory for nuclear reactions as applied to nuclear weapons. Before the days of analytical and digital computers, mathematically gifted people performed the tedious mathematical computing tasks and were called *computers*. Ulam's wife, Françoise, was a computer who assisted him in this task. Rather than tracking every uranium or plutonium atom in the atomic bomb models that were being devised by physicists (an impossible task), Ulam used statistical methods to simulate behaviors of individual nuclei in the reaction. He selected, at random, variables of the possible interactions of the nuclei, computed possible outcomes, and analyzed the results using probability statistics. Because his system was based on the *odds* as related to the probabilities of gambling odds, his method became known as *Monte Carlo statistics*, named after the famous gambling casino in Monaco. Ulam's Monte Carlo system of statistical probabilities is employed in many fields other than nuclear energy and has proven a valuable addition to our understanding of several biological processes, including life. In the late 1940s Ulam was involved with Edward Teller and others in the development of the hydrogen (thermonuclear) bomb. Ulam developed the mathematics that ensured success for the final design of the fusion H-bomb, which used a fission A-bomb as the "trigger" to provide the heat, the X-rays, and the pressure required to accomplish the thermonuclear reaction required for the H-bomb. Teller revised Ulam's concept to "focus" the X-rays produced by the A-bomb to trigger the fusion reaction.

*See also* Teller

**UREY'S GASEOUS DIFFUSION AND ORIGIN-OF-LIFE THEORIES:** Chemistry: *Harold Clayton Urey* (1893–1981), United States. Harold Urey was awarded the 1934 Nobel Prize for Chemistry.

***Urey's theory of gaseous diffusion:*** *Isotopes of an element in the gaseous state can be separated according to their different atomic weights.*

In 1932 Harold Urey was the first to discover and isolate deuterium (heavy hydrogen) from heavy water ($D_2O$). He knew liquid heavy hydrogen evaporated at a slower rate than did ordinary liquid hydrogen because regular hydrogen's nuclei are composed of just a single proton, whereas heavy hydrogen's nuclei contain one proton plus one neutron. Deuterium still has an atomic charge (atomic number) of +1, but an atomic weight of 2 (*see* Figure O1 under Oliphant). Using this concept, Urey distilled several liters of liquid hydrogen to about 1 cubic centimeter of deuterium whose existence was confirmed by spectroscopic analysis. Again, using the same principle, he separated the isotopes of uranium-235 (U-235) from uranium-238 (U-238) by converting regular U-238 into a gas and "filtering" it in such a manner that the lighter, unstable U-235 was collected. The isotope U-235, when reaching a critical mass, can be used in a self-sustaining chain reaction. In the early 1940s U-235 was used in the first A-bomb tested in White Sands, New Mexico. The same gaseous diffusion process, based on isotopes of radioactive elements having different atomic weights, was used to separate the isotopes of plutonium, used in the second A-bomb dropped in Japan in 1945. At the end of World War II, Urey's mass production of deuterium made possible the development of the hydrogen fusion bomb (*see also* Fermi; Teller; Ulam).

***Urey's theory for the origin of life:*** *If the right mix of organic molecules existed on primitive Earth, an energy input (lightning, geothermal, or ultraviolet) may have brought life from this "soup."*

Harold Urey and others believed life began on Earth some three to four billion years ago, which is approximately one or two billion years after Earth was formed. These figures are today's best estimates based on fossil and cosmological research. Disagreement still exists as to the source of the organic chemicals that first self-assembled to produce organic polymers and later cells. One possibility being revived is that bacteria and microorganisms arrived on Earth from comet and meteor ice and dust. Urey, along with his graduate student, Stanley Miller, set up an experiment to determine if several chemicals assumed to be on the **prebiotic** Earth could, under laboratory conditions, be converted into organic polymers, which might, under ideal conditions, self-organize into primitive life forms. Urey and Miller formed an atmosphere of methane ($CH_4$), hydrogen ($H_2$), ammonia ($NH_3$), and water ($H_2O$) in an enclosed glass flask that could be heated. The hot evaporated gases were collected in another flask and exposed to an electric spark between two tungsten electrodes. Following this, the gases were cooled and condensed back to a liquid. They actually produced over twenty-five amino acids, some purines, and other large organic molecules, but no evidence of life itself. An important part of the experiment was the formation of amino acids that combine with some ease to form complex proteins, which are essential for life. Urey's and Miller's work led to the idea that there are at least four stages for chemical evolution of life on Earth: 1) There is a nonbiological synthesis of simple organic molecules, 2) followed by the molecules forming more complex polymers (chains), 3) which form into pre- or probiological "clumps" or simple cells, and 4) some of the first organic substances are primitive RNA followed by DNA, which has the capability to pass on the chemical and living nature of cells from one generation to the next.

*See also* Crick; Darwin; Miller; Pasteur; Redi; Watson (James)

# V

**VAN ALLEN RADIATION BELTS:** Physics: *James Alfred Van Allen* (1914–2006), United States.

> *Earth's magnetic field should react with and trap high-speed charged particles originating from space into a concentrated zone above Earth's atmosphere.*

James Van Allen first used high-altitude balloons to study cosmic rays, which are high-energy particles (mostly protons) from space that constantly penetrate Earth. Three scientists—S. Fred Singer (1924–) from the University of Maryland, Paul Kellogg (dates unknown) from the University of Minnesota, and Sergei N. Vernov (1910–1982) of the Soviet Union—first proposed the idea that radiation from space surrounded Earth. These energetic particles from the lower altitudes were confirmed by sending photographic film into space in weather-sounding rockets. In 1958 the United States sent its first satellite, *Explorer I*, which weighed thirty-one pounds, into orbit by using a captured German V-2 rocket. Its purpose was to detect high-energy particles in near space. It found many particles at altitudes between 200 and 300 miles but, surprisingly, none were recorded above that region. James Van Allen theorized that this low-radiation "belt" was due to Earth's magnetism, and the reason particles were not detected at higher altitudes was that the radiation counters in the rockets were "jammed" by overwhelming masses of radiation and particles. In 1958, *Explorer IV* was launched containing a radiation counter surrounded by a lead shield that filtered out much of this radiation, thus providing a more accurate count. Additionally, the shielded instruments recorded an increasing amount of high-energy radiation above 300 miles. After World War II the Soviet Union and the United States exploded small atomic bombs in space to track the neutrons and energetic particles that were released. The United States also exploded three small nuclear bombs 300 miles above the South Atlantic to produce energetic particles in the upper atmosphere, which could then be detected and studied. Some radiation and particles from these nuclear explosions in space persisted for several weeks to several years and were strong enough to disable a number of satellites.

By 1967, this practice of detonating nuclear bombs in space was banned worldwide. Van Allen theorized that high-altitude radiation composed of charged particles was concentrated in areas or "belts" trapped by the magnetosphere, which rotates with Earth's magnetic axis. He based his theory on the concept that Earth's magnetic field extends far out into space, and these particles followed the magnetic field as evidenced by their alignment with the poles of this field. This region of concentrated radiation or "belt" was first called the *Van Allen belt*. However, when it was discovered there was more than one radiation belt, the name was changed to the *magnetosphere*. The *inner radiation belt*, the one detected by the Geiger counters used by Van Allen, is a rather compact area of particles in the general magnetosphere region over the equator. This belt is the result of cosmic radiation. Later, an outer radiation belt was discovered, which is composed of "plasma" of energetic charged particles trapped in Earth's magnetosphere at a higher altitude and responsible for magnetic storms on Earth. This magnetosphere phenomenon was later confirmed for other planets as well.

**VAN DE GRAAFF'S CONCEPT OF PRODUCING HIGH VOLTAGE:** Physics: *Robert Jemison Van de Graaff* (1901–1967), United States.

*High voltages can be produced and sustained by electrostatic generators.*

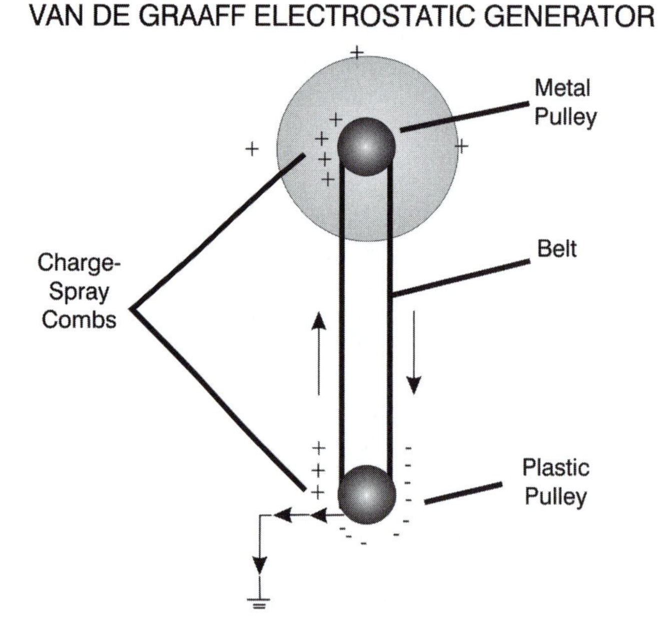


Figure V1. The Van de Graaff electrostatic generator produces an electrical charge from the ground and transfers it by the belt to the metal sphere, where it can be used as an electron source for a variety of research applications.

Robert Van de Graaff was cognizant of the need to produce very high voltages to accelerate subatomic particles, making them useful as "bullets" to bombard the nuclei of elements to produce isotopes and nuclear changes. Regular **AC** and **DC** generators and dynamos were incapable of producing voltages in the million-volt ranges that nuclear physicists needed to create new, heavier elements as well as basic subnuclear particles. Older type "wheel-and-brush" static electricity machines were developed soon after it was learned that static electricity could be stored in a **Leyden jar** by rubbing glass or rosin with silk or wool. These machines created a spark across an inch or two in dry air. In 1931 Van de Graaff devised an improved model that used a hollow sphere with a rotating insulated rubberized belt to transfer the charge to the surface of the metal sphere (*see* Figure V1).

Although his first models generated up to 100,000 volts with just a fraction of amperes (current), they were inadequate for use as particle accelerators. Van de Graaff's later models were enclosed in a tank with the insulated belt extracting negative ions from the ground (Earth), where they were stripped of their electrons and thus became positive ions. As the charges return to the grounding terminal, the negative ions are accelerated to achieve up to 10 million volts or 10 MeV of energy (but with very low amperes). Recently, even higher voltages have been achieved. This high voltage can accelerate charged particles (electrons and positrons) to very high energies, which can then be used as "bullets" to bombard target nuclei. Two of the first applications of this very high energy were the exploration of the nature of uranium nuclei fission and the production of high-energy X-rays for medical and industrial use. Van de Graaff generators, when used in combination with other types of particle accelerators, generate the tremendous speeds (energies) of charged particles needed to knock out the many particles from atomic nuclei. Van de Graaff generators also produce high-energy X-rays to detect flaws in machinery and small cracks in airplane structures, and to inspect the interiors of explosive weapons.

*See also* Tesla

**VAN DER MEER'S THEORY OF PARTICLES TO CONFIRM THE "WEAK FORCE":** Physics: *Simon Van der Meer* (1925–), Netherlands. Simon Van der Meer shared the 1984 Nobel Prize for Physics with Carlo Rubbia.

*The unification of the electromagnetic and weak forces requires the existence of three heavy particles—one negative, one positive, and one neutral.*

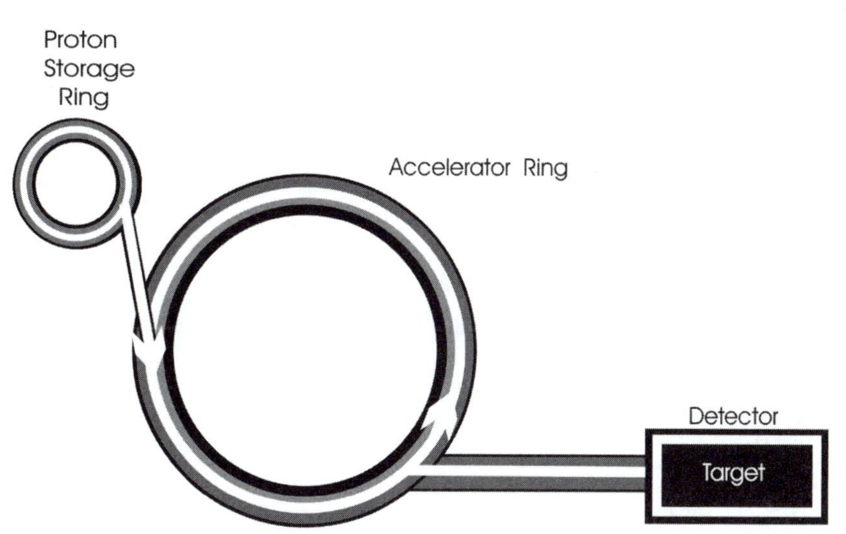


Figure V2. Particle accelerators are used to analyze and understand the nature of matter and energy. Subatomic particles are accelerated around the ring by electromagnetic forces until they approach the speed of light, which is required to smash into targets that break up into smaller particles, as well as energy that can be recorded and analyzed.

CERN is the main research organization for nuclear and particle research in Europe. It occupies the grounds of former German dairy farms outside Geneva, Switzerland, and consists of several dozen buildings on the border of Switzerland and France. Van der Meer, who confirmed the "weak force," and Carlo Rubbia used the present particle colliders at CERN for their research through the 1990s. Physicists realized that even with as much as they already know about submicroscopic particles of matter and energy, something more fundamental to matter in the universe remains missing. CERN is almost finished constructing the most ambitious research apparatus to conduct experiments to explore particle physics. It is called the Large Hadron Collider referred to as L.H.C., which, as of today, is the only attempt to test theories of "new physics," including the controversial "string theory." String theory claims that the ultimate particle/energy bits of matter consist of "strings" of various shapes too small to see that may exist in multiple universes. Also, they might exist beyond the four dimensions we already know and experience, that is, height, width, depth, and time. The L.H.C. is a circular tunnel 17 miles in circumference and 300 feet deep. An elevator is needed to descend to the research level. Once it is operational in late 2008, the first expected experiment will involve a group of scientists analyzing a four-million-megabyte-per-hour flow of data that will attempt to address many unanswered questions in the field of particle physics.

Van der Meer predicted that theoretical particles related to weak interactions were about eighty times as massive as protons, thus requiring a tremendous energy source to produce and detect them. The major obstacle to detecting these particles and confirming the theory of the weak interactions was their theoretical mass. In 1989, Van der Meer and his colleague, Carlo Rubbia, succeeded in generating energies sufficient to produce these massive particles. Arranging for a super-synchrotron (circular particle accelerator) to provide two beams of particles to collide head-on, they accelerated a beam of protons in one direction and a beam of antiprotons in the other (*see* Figure V2).

The result was a collision of particles, each at great energies, detected in the synchrotron's target. This collider concept greatly increased the energy to over 150,000 GeV, which was adequate to produce and detect the three heavy theoretical particles. They were named $W^+$, $W^-$, and the neutral $Z^0$ **bosons**, which coincide with and confirmed the weak interaction now proven by Van der Meer but first suggested by Enrico Fermi. The weak interaction is similar to electromagnetic interactions of particles except the weak interaction involves neutrinos, making it a much weaker force than electromagnetic forces. At the same time weak interactions are much stronger than gravitational interactions on particles. The weak interaction (and related particles) is one of the basic forces of nature, as is the "strong force," which binds neutrons and protons together in atomic nuclei.

*See also* Fermi; Rubbia

**VAN DER WAALS' EQUATION FOR GAS MOLECULES:** Physics: *Johannes Diderik Van der Waals* (1837–1923), Netherlands. Johannes Van der Waals was awarded the 1910 Nobel Prize in physics.

*Electrostatic forces are responsible for the attraction between gaseous molecules, thus affecting the corresponding relationships between the temperature, pressure, and volume of gases.*

The ideal gas law combines the Boyle–Charles gas laws and is limited to gases at "normal" pressures and temperatures. This law is not effective for "real" gases under other than "normal" conditions—very high or low temperatures or pressures. The ideal gas law equation relates the three properties of temperature, pressure, and volume for a chemical gas, pV = nRT, under normal conditions (*see* Ideal Gas Law). Johannes Van der Waals related the ideal gas law to the kinetic-molecular theory, which could account more accurately for the behavior of real gases and liquids by considering the attractive forces between molecules as well as their actual (but limited) volumes under other-than-normal conditions. The ideal gas law might be considered the equation of the "first state," whereas Van der Waals' equation is an equation of the "second state," which more accurately relates the behavior of gas molecules to kinetic energy under a variety of corresponding states of temperature, pressure, and volume. Van der Waals' equation is $(p + na/V^2)(V - nb) = nRT$. In this equation the *n's* are amounts, *a* and *b* are constants, $na/V^2$ accommodates the attractive forces between molecules of gases that may be more than zero, and $V - nb$ states the volume of a real gas is never zero, which restricts the gas's molecular motions in its actual volume. The weak electrostatic attractive force between the atoms and molecules of all substances is called the *Van der Waals force*. His equation enabled other scientists to solve the problems of how to liquefy gases found in the atmosphere.

*See also* Boyle; Charles; Gay-Lussac

**VAN'T HOFF'S THEORY OF THREE-DIMENSIONAL ORGANIC COMPOUNDS:** Chemistry: *Jacobus Henricus Van't Hoff* (1852–1911), Netherlands. Jacobus Van't Hoff was awarded the first Nobel Prize in Chemistry in 1901.

> *The three-dimensional symmetrical structure of organic carbon compounds accounts for their optical activity.*

Until 1874 molecular structures were depicted as two-dimensional. Also in 1874, Friedrich Kekule proposed his famous structure for the carbon atom as having four electrons oriented to the corners of a square, which explained left- and right-sided isomers of some elements. While contemplating this structure for the carbon atoms, Kekule dreamed of a snake eating its tail (*see* Figure K1 under Kekule). This gave him the insight for forming a ring of carbon atoms to form the benzene molecule. The benzene ring is a molecule that has six carbon atoms, each of which shares its electrons with its neighbors. The problem with his depiction was that the ring was two dimensional, which did not explain how certain molecules (isomers) polarize light in solution. In the same year, Jacobus Van't Hoff recast the organic carbon atom into a tetrahedral three-dimensional structure with the four bonds of the carbon atom pointed toward the vertices of the tetrahedron rather than to the corners of a two-dimensional square (*see* Figure V3).

Van't Hoff's model placed the atom as suspended in the central area of the three-dimensional figure. This was not only a unique insight but explained how some organic isomers are structured and react in solutions. Certain isomers do polarize light in solution; others do not. The difference is in the two- or three-dimensional structures of the molecules. Van't Hoff's theory of asymmetrical three-dimensional optically active carbon (organic) compounds provided the basis for modern stereochemistry (the study of how atoms are arranged [structured] within molecules and how this affects chemical

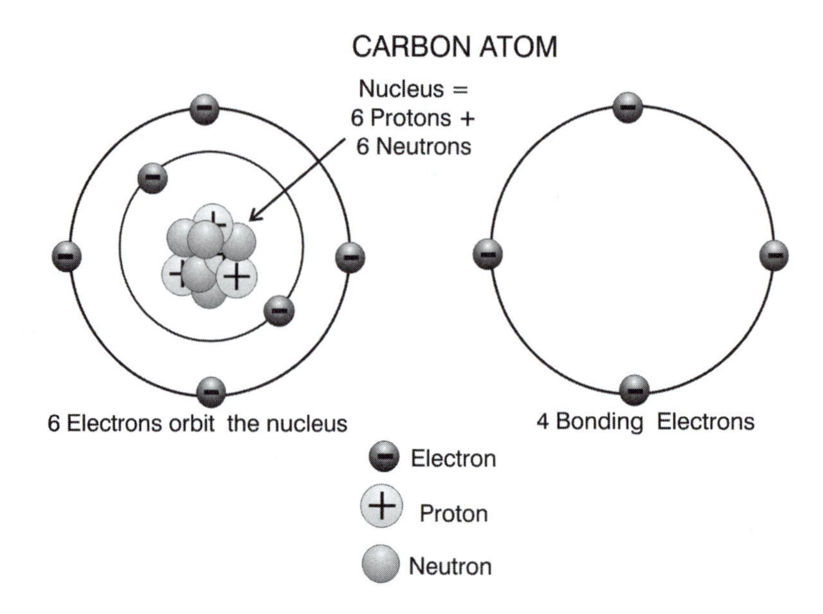


Figure V3. Van't Hoff determined that there are 4 electrons in the outer shell of the carbon atom. The carbon atom's structure with four valence electrons may be thought of as a tetrahedron with four vertices representing the bonding electrons. The tetrahedron has six edges and six line segments to join each pair of vertices. It is a representation of a three-dimensional triangle with the center being the carbon atom. This structure gives the carbon atom its unusual versatility and importance in forming organic and inorganic compounds

reactions). Van't Hoff's theory resulted in his law of chemical dynamics, which is important in the study of osmotic pressure in solutions.

*See also* Baeyer; Fischer; Kekule; Pasteur; Van der Waals

**VAN VLECK'S THEORY OF PARAMAGNETISM:** Physics: *John Hasbrouck Van Vleck* (1899–1980), United States. John Van Vleck shared the 1977 Nobel Prize for Physics with Nevill Mott and Philip Anderson.

> *Paramagnetic substances are independently susceptible to magnetic induction according to the temperatures involved.*

John Van Vleck extended the concept of quantum mechanics associated with particles to include not only quantum aspects of waves but also magnetism. There are two basic types of paramagnetism—the first involves electrons and the second involves nuclei of atoms. Atoms of elements that have an odd number of electrons, according to quantum mechanics, cannot have a spin of zero. This results in atoms with a magnetic moment that can be affected by a magnetic field. Examples are atomic and molecular radicals (with a charge). Paramagnetic materials are magnetized parallel to the magnetic field to which they are exposed. In general, they do not become as highly magnetized as do ferromagnetic materials, and they behave differently at very high and very low temperatures. The best examples of paramagnetic materials are the atoms and compounds of the rare earths located within the transition elements of the

**Periodic Table of the Chemical Elements.** Other examples are free organic radicals, nitric oxide, some low-conducting metals, and molecular oxygen. The effect of temperatures on the quantum nature of paramagnetic materials is known as *Van Vleck paramagnetism*. Paramagnetic materials are used in combination with liquid helium to remove additional heat in attempts to reach absolute zero. First they are magnetized. Then, when the magnetic field is removed, the molecules become randomly disorganized and remove more heat from the helium. A temperature of less than 0.5 kelvin has been achieved.

**VESALIUS' THEORIES OF ANATOMY AND PHYSIOLOGY:** Biology (Anatomy): *Andreas Vesalius, aka, Andre van Wesel* (1514–1564), Belgium/Flanders.

*More accurate anatomical features of human organs can be observed by actually performing surgery on human cadavers rather than using animal bodies.*

Andreas Vesalius is rightfully known as the father of modern human anatomy. This is a title he well-deserved because he spent most of his life not only studying the human body, but also making astoundingly accurate drawings of various parts of the anatomy. Many of these drawings are still in use and are highly prized today.

His father Andries van Wesel was the illegitimate son of the Holy Roman Emperor Maximillian's royal physician whose name was Everard van Wesel (dates unknown). In time, Andries became the apothecary to Emperor Maximillian I of Habsburg (1459–1519) while young Andre was being educated to follow in his father's footsteps. Vesalius entered the University of Leuven in Belgium in 1528, majoring in the arts, but soon decided to pursue a career in medicine at the University of Paris. While at Paris, he studied the theories of several well-known physicians, including Galen and the French anatomist Jacobus Sylvius (1478–1555). Due to the war between the Holy Roman Empire and France, he was forced to return to Leuven to

Vesalius was one of the first to use cadavers of executed criminals to perfect his knowledge of anatomy. Because refrigeration was not available in those days, he would work day and night to complete the job before the corpse putrefied. Because Leonardo da Vinci's anatomical drawings had not yet been published, Vesalius was not able to benefit from them. Even so, Vesalius' drawings were superb as well as accurate. This caused him a great deal of trouble because they challenged the anatomical misconceptions of Galen. One reason for the discrepancy was that Galen used animals (mainly pigs) for his dissections and Vesalius used humans. Vesalius' anatomical drawings and his accurate descriptions that appeared in his publications were the beginning of the end of the fifteen-hundred-year long "tyranny of Galen" that lasted well into the Renaissance period. Even so, Vesalius could not entirely break away from Galen's hold as the last word in human anatomy. The excuse that some physicians used for the discrepancies between the Galen and the Vesalius publications and drawings of the human body was that the body went through a series of dramatic changes in structure during this short period of history. The argument became more serious after Vesalius pointed out more errors by Galen. For instance, Galen believed that there were microscopic holes in the tissue (septum) that separated the heart's left and right ventricles that allowed the passage of blood from one side to the other. And even though he knew better, Vesalius accepted Galen's position. In Vesalius' seven-volume book *De Humani Corporis Fabrica* (usually referred to as *De Fabrica*) published in 1543, the author made many more corrections to Galen's teachings. Many physicians of the day criticized Vesalius' books to the point that he burned his remaining books as well as his unpublished material. It was not until after his death that Vesalius' work was recognized and accepted.

complete his studies. After an argument with one of his professors, he accepted the chair of Surgery and Anatomy at the University of Padua in Italy. For this time in history, his methods of teaching anatomy were most unusual. He performed his own surgery with the students gathered around him so they could better see what was being discussed. Generally, older instructors had assistants do the surgery while the professor read from out-of-date textbooks. In addition, they accepted the writings of Galen and others as correct with no effort to verify these statements with the facts before their eyes. Vesalius used dissection as a major hands-on teaching tool. He believed that direct observation was the only reliable source of information. Vesalius maintained a file of his superb drawings as large illustrated anatomical teaching tools for his students. When he found that they were being copied, he decided to publish them in 1538. These drawings were used in a debate as to the best method of "letting" blood to be used as a treatment for many illnesses. The ancient method advocated by Galen was to draw the blood from a site near the illness (which Vesalius supported); the other method, which was preferred by ancient Muslim physicians, accepted that the blood could be drawn from a more distant site and be just as effective. Vesalius' diagrams won the argument. He also rectified several other incorrect assumptions, for example, that the heart has four chambers—not two, that the liver has two lobes, that blood does not pass through the septum of the heart, and that blood vessels do not originate in the liver, but rather in the heart.

*See also* Galen

**VIRCHOW'S CELL PATHOLOGY THEORY:** Biology: *Rudolf Carl Virchow* (1821–1902), Germany.

*Cells are the "seat" of disease, thus diseased cells arise from other diseased cells.*

Rudolf Virchow accepted the concept first stated by William Harvey and the Scottish anatomist John Goodsir (1814–1867) that *cells are derived from preexisting cells.* Virchow's contribution to the field of pathology was his belief that disease is a pathological state of cells based on observations that abnormal cells found in particular diseases arose from normal healthy cells. He believed living cells could originate only from living matter. Virchow noted this was not a rapid process. Rather, once disease infected just one or only a few cells, these cells infected other healthy cells over a period of time. Some years later, the germ theory of disease, which he rejected, made Virchow's theory less important because this new theory provided a more rational explanation for disease. Even so, Virchow is regarded as the father of pathology.

*See also* Harvey; Lister; Pasteur; Schleiden; Schwann

**VOLTA'S CONCEPT OF AN ELECTRIC CURRENT:** Physics: *Count Alessandro Giuseppe Antonio Anastasio Volta* (1745–1827), Italy.

*A flowing electric current is not dependent on animal tissue and can be produced with chemicals.*

Count Alessandro Volta was acquainted with Benjamin Franklin's single-fluid theory of electricity and the French chemist Charles du Fay's (1689–1739) two-fluid electricity. He also knew of Luigi Galvani's belief that moist animal tissue was required to produce a continuous flow of electricity. Volta decided these theories had weaknesses and that a combination of the correct chemicals and materials could produce an electric current. In particular, he believed that dissimilar metals, not the animal tissue of Galvani's frog experiment, generated Galvani's electricity. In 1800, Volta separated alternating sheets of zinc and silver with sheets of cardboard soaked in concentrated saltwater, which acted as the electrolyte. This was called a *voltaic pile* and resulted in a revolution in the source and use of small amounts of electricity. His "pile" enabled others to develop mechanical clappers on electric bells, the telegraph, modern dry and wet cells, and batteries of cells, which are a combination of a group of cells connected in series or parallel. The unit of electric potential (force or pressure of the current) is named after Count Volta (volts = amps × ohms).

*See also* Faraday; Franklin (Benjamin); Galvani; Henry; Ohm

**VON LAUE'S THEORY FOR THE DIFFRACTION OF X-RAYS IN CRYSTALS:** Physics: *Max Theodor Felix Von Laue* (1879–1960), Germany. Max Von Laue was awarded the 1914 Nobel Prize in Physics.

> *If the wavelength of X-rays is similar to the space between atoms in crystals, X-rays passing through a crystal composed of atoms in a lattice arrangement should produce a diffraction pattern.*

It was known for some time that X-rays consist of electromagnetic waves similar to light and that electromagnetic wavelengths of X-rays are much shorter than waves of light. Max Von Laue was aware of this fact, even though it was not yet established that X-rays were one form of radiation exhibited in the electromagnetic spectrum. He also knew of the research indicating that the atoms in crystals may be arranged in very regular patterns, similar to a lattice structure, where they were lined up in rows. His theory stated that if the small crest-to-crest distance of the short wavelengths of X-rays were the same as the small distances between the atoms that make up crystalline substances, then diffraction of the X-ray beams should occur. He proceeded to "shoot" X-rays through crystals and record the diffracted beams on photographic plates. His first attempts produced rather blurred but nonetheless expected diffraction patterns, thus proving his theory. Von Laue's theory established that though X-rays were part of the electromagnetic spectrum, they are of a much shorter wavelength than visible light. Just as important, his work demonstrated that atoms in organic crystals and some organic substances are arranged in a symmetrical and regular order. His X-ray techniques aided researchers in decoding the structure of DNA, and his work with crystals ushered in the field of solid-state physics, leading to the development of modern electronics, including semiconductor microchips and computers.

*See also* Crick; Franklin (Rosalind); Maxwell; Röentgen

**VON NEUMANN'S THEORY OF AUTOMATA:** Mathematics: *John von Neumann* (1903–1957), United States.

***Von Neumann's theory of "artificial automata" (computers) might be expressed as:*** *(a set of inputs) to → (a set of internal states) yields → (a set of outputs).*

After emigrating to the United States from Germany in 1930, John von Neumann attempted to solve complicated mathematical problems related to the development of the atomic bomb, hydrodynamics of submarines, missiles, weather predictions, and military strategy. Much of this work required deciphering complicated nonlinear equations, which proved to be difficult and time-consuming. Von Neumann developed a systematic mathematical theory in logic that he called "automata," which he reasoned would help develop a better understanding of natural systems and what he called "artificial automata" (computers). The automata theory relates to three states of a system that involve three sets: namely, 1) the input, 2) the current internal state of the system, and 3) the output from the system. These operations can be thought of as three sets of information and two functions. In essence, this is von Neumann's design for a computer, which is also the basic logic for current computers: the program (input), the operating system (internal state), and the data produced (output). In 1952 von Neumann developed the Mathematical Analyzer, Numerical Integrator, and Computer (MANIAC), the first modern computer using an internally stored program (operating system). It was a huge machine that filled a room and required extensive cooling to keep the vacuum tubes from overheating. It provided the basic logic (automata) and design for modern computers.

*See also* Turing

**WADDINGTON'S THEORY OF GENETIC ASSIMILATION:** Biology: *Conrad Hal Waddington* (1905–1975), Scotland.

*By means of natural selection, acquired characteristics can be inherited genetically and through the process of evolution.*

During Darwin's lifetime, the science of genetics had yet to be developed; therefore the old Lamarckian belief that characteristics acquired after birth could be inherited was still considered viable. Once research and evidence that genes are the carriers of physical characteristics became known, Lamarckism became heresy. Waddington conducted an experiment that he claimed proved his theory of "genetic assimilation" (of acquired characteristics) by exposing the pupae stage of the fruit fly (*Drosophila*) to heat. He noted that a few exposed flies exhibited a different pattern of veins in their wings. Waddington separated and bred these different flies in an attempt to increase their numbers. After repeating selective breeding of flies for several generations, he observed that a large number of offspring manifested this same pattern; thus they seemed to be breeding true. Therefore, Waddington concluded that genetic assimilation of imposed characteristics resulted through the process of natural selection. Most scientists discredited his experiment and theory.

*See also* Darwin; Lamarck; Lysenko; Wallace; Zuckerandl

**WALDEYER-HARTZ NEURON THEORY:** Biology: *Heinrich Wilhelm Gottfried von Waldeyer-Hartz* (1836–1921), Germany.

*The nervous system is composed of individual cells whose fine extensions do not join cells adjacent to them but still communicate with a neighboring cell.*

Heinrich Waldeyer-Hartz studied animal tissue cells and their structures, defining the "colored bodies" in cells as *chromosomes*. He also studied nerve tissue and was the first to realize that nerves are not only composed of cells, just as in other animal tissue, but that the individual nerve cells are not in contact with each other; there is a gap where one nerve cell ends and the next begins. He named these individual nerve cells *neurons*.

*See also* Dale

**WALLACE'S THEORY OF EVOLUTION BY NATURAL SELECTION:** Biology: *Alfred Russel Wallace* (1823–1913), England.

*The tendency for species to produce variations as they drift from their original types is due to a separation of their ecologies.*

In the late 1800s Alfred Wallace, a contemporary of Charles Darwin, collaborated with Darwin on the development of the theory of organic evolution. Wallace proposed his concept, known as the *Wallace line*, where the separation of geographical landmasses results in the development of distinct species. His theory was based on the differences of animal species that he observed in Australia and Asia. Wallace claimed this "line" between species was created by the separation of the two landmasses, which, over a long period of time and many generations, also separated individual species that developed in very different directions due to natural selection created by disparate ecologies. His theory that varieties of a species tend to drift apart indefinitely from the original type is generally accepted today. The study of the geographic distribution of plants and animals is known as biogeography.

*See also* Darwin; Waddington

**WALLACH'S THEORY FOR THE MOLECULAR STRUCTURE OF ORGANIC COMPOUNDS:** Chemistry: *Otto Wallach* (1847–1931), Germany. Otto Wallach received the 1910 Nobel Prize in Chemistry.

*Pharmaceutical medications as well as the essences of many oils are composed of a variety of related forms of the hydrocarbon molecules.*

While studying pharmacology, Otto Wallach removed the essential oils from plants using steam distillation. Many of the resulting organic compounds were used as medicine as well as in the production of perfumes, creams, and flavorings. He theorized that many of these organic substances were chemically related but was unsure of how or why their molecular structures differed. Wallach, however, identified the great variety of a particular group of compounds, all of which possessed the same general formula but with different molecular weights (isomers). One of these, terpene ($C_{10}H_{16}$), is an unsaturated hydrocarbon found in some plants and has a unit structure containing five carbon atoms ($C_5$). A group of organic compounds similar to terpene that has the same molecular weight but different structures is referred to as an *isoprene*. He expanded this terpene example of a hydrocarbon isoprene with the general formula $(C_5H_8)_n$ to include other hydrocarbon compounds. Wallach also discovered that these molecules could be polymerized to form other higher-molecular-weight molecules, resulting in other larger organic (hydrocarbon) molecules with formulas that are multiples of the basic terpene $C_{10}H_{16}$ formula. Some examples of these isoprenes are camphene, citrine, cinene, eucalyptine, and common terpentine. Wallach's work with the basic structure of various $(C_5H_8)_n$ isoprenes was instrumental in improving and expanding several industries, including pharmaceuticals and perfumes.

*See also* Kekule; Pauling

**WALTON'S CONCEPT FOR TRANSMUTING ATOMIC PARTICLES:** Physics: *Ernest Thomas Sinton Walton* (1903–1995), Ireland and England. Ernest Walton shared the 1951 Nobel Prize for Physics with John Cockcroft.

> *See* Cockcroft for details of the theory proposing that accelerated protons split lithium nuclei into alpha particles (e.g., lithium + proton → alpha + alpha + energy).

**WATSON–CRICK THEORY OF DNA:** Biology: *James Dewey Watson* (1928–), United States. James Watson shared the 1962 Nobel Prize in Physiology or Medicine with Francis Crick and Maurice Wilkins.

> *See* Crick–Watson for details describing their theoretical model of the DNA double-helix molecule.

**WATSON'S THEORY OF ELECTRICITY AS A FLUID:** Physics: *Sir William Watson* (1715–1787), England.

> *Electricity is an "electrical ether" or single fluid of various densities that is contained in different material bodies.*

William Watson improved the effectiveness of the **Leyden jar** that was independently invented by Dutch scientist Pieter van Musschenbrock (1692–1761) and Prussian scientist Ewald Georg von Kleist (1700–1748) by lining the interior of the glass jar with metal foil. This improved device enabled Watson to store a large charge of static electricity and study the resulting larger electrical discharges, ultimately leading to his belief that electricity is a single "fluid." Watson theorized that different materials contained differing densities of this "electrical **ether**." If the density of two objects was equal, there was no sparking discharge, but if the "fluid" densities were unequal, the one with the greater density would discharge to the object with lesser density, until they were again equal. Although this theory is incorrect, it might be considered a forerunner of the concepts of equilibrium and the conservation of energy, including the concept entropy.

*See also* Ampère; Faraday; Franklin (Benjamin)

**WATSON-WATT'S CONCEPT OF RADAR:** Physics: *Sir Robert Alexander Watson-Watt* (1892–1973), England.

> *The interference in radio reception caused by airplanes flying over transmitting stations can be used to detect approaching aircraft.*

Robert Watson-Watt knew that some radio engineers complained about radio signal interference caused by passing airplanes. In the late 1930s, he theorized that this phenomenon might be used to detect enemy aircraft. In addition to this radio "interference" by aircraft, he based his concept on the results of two other research projects: 1) the use of radio waves to determine the range in miles of different layers of the atmosphere and 2) the use of radio signals to determine the existence and distance of thunderstorms. There were two main problems with using this concept for a reliable aircraft-detecting device at a distance of more than a few miles: 1) the need for a very high-powered transmitter and 2) the fact that only a very small, weak signal was "bounced" back to the receiver. Therefore, the receiver had to be capable of amplifying the signal by many factors greater than what was required for normal radio receivers. Along with scientists and engineers from the United States, he continued to develop a workable system he called *radio detection and ranging* (radar). By late 1938 several radar units were placed on the east coast of England to aid in the detection of approaching German bombers. Since that time, radar units have become much smaller and more sensitive. Radar has found many uses, including handheld units to detect speeding vehicles on roads and highways.

*See also* Doppler

**WEBER'S THEORY OF GRAVITATIONAL WAVES:** Physics: *Joseph Weber* (1919–2000), United States.

*Gravity waves should have the same characteristics of energy and momentum as do electromagnetic waves, and thus be detectable.*

Joseph Weber accepted Einstein's theory of general relativity, which included the concept that any accelerating mass generates gravitational waves as well as electromagnetic waves. Photons of light (electromagnetic radiation) exhibit wave and momentum characteristics; thus electromagnetic radiation must have mass. Weber reasoned that gravity waves should also exhibit momentum and thus be detectable; however, gravitational waves have not been detected because gravity is one of the weakest forces in nature. It may not seem so weak when falling to the ground, even from a low height, but compared to other forces of nature, such as the binding force of nuclear particles or even the forces that forge molecules out of chemical atoms, gravity is not very strong. In the mid-1960s, Weber designed a special barrel-like "antenna" detector that was three feet in diameter, constructed from aluminum and weighing more than three tons. He placed a series of piezoelectric crystals in its interior to detect gravity waves. He figured that any force, no matter how small, would alter the shape of these crystals; if there was even the slightest pressure exerted by an oscillating gravity field, the crystals would convert this distortion to an electric current that could be detected and measured. This instrument was so sensitive that the piezoelectric crystals could detect any deformity in their shape as little as 1/100th the diameter of an atom. To ensure gravity waves were being detected and not some other phenomenon, Weber erected a second detector at some distance from the first so that each antenna detector could be oriented in various directions. After several months, he claimed to have received what are called "coincident readings," meaning that when both cylindrical detectors were oriented toward the center of our galaxy, the same readings were recorded. Weber's results were never duplicated by other scientists despite the thousands of dollars spent on improved detectors. In fact, no "coincident readings" were ever recorded, even when one gravity wave detector was placed on the East Coast and one on the West Coast of the United States. Scientists have not abandoned the theory that gravity waves exist but rather assume they are too weak to detect with current instruments. A more recent experiment "shoots" two laser beams of monochromatic light at each other from a distance of several miles. When the beams collide, they interfere with each other. If gravity waves exist, they may possibly alter the interference pattern and thus be detected. Theoretical physicists predict that sometime in the early twenty-first century, gravitational waves will be detected, a belief based on their confidence in the proven reliability of the theory of general relativity.

*See also* Einstein; Curies

**WEGENER'S THEORY OF CONTINENTAL DRIFT:** Geology: *Alfred Lothar Wegener* (1880–1930), Germany.

*All land at the surface of Earth was once connected with the configuration of a "supercontinent," which, over time, separated into large sections that drifted apart to form the present continents.*

Scientists have long speculated on the shape of the world's landmasses and why this shape changed over eons. Sir Francis Bacon was the first to notice the similarity of the coastlines of eastern South America and western Africa and to suggest that they were once joined. Building on Eduard Suess' theory that western and eastern landmasses were once joined to form the hypothetical continent Gondwanaland, in 1924 Alfred Wegener called his "supercontinent" *Pangaea*, which means, "all-land" or "earth" in Greek (*see* Figure S6 under Suess). Wegener based his theory on four important observations:

1. There is a more accurate "fit" of the edges of the underwater continental shelves of the current continents than there is on the above-water coast lines.
2. Current measurements indicate Greenland is moving westward from the European continent.
3. Earth's crust is composed of a lighter granite-type rock material, which floats on the heavier inner basalt material. Thus, the crust is composed of two layers, and the continents formed of granite "float" over the heavier basalt ocean floor.
4. Although there are significant differences in plant and animal species found on various continents, there are also great similarities of species found on the now-separated continents, indicating these continents were once connected.

At first, many scientists disagreed with Wegener's theory of continental drift. Today it is accepted in an updated version to conform to the new science of plate tectonics.

*See also* Ewing; Hess (Harry); Suess

**WEINBERG'S GRAND UNIFICATION THEORIES:** Physics: *Steven Weinberg* (1933–), United States. Steven Weinberg shared the 1979 Nobel Prize in Physics with Sheldon Glashow and Abdus Salam.

***Weinberg's theory of the unification of electromagnetic and weak forces:*** *The interchange of photons and the weak force with the W and Z bosons results in the electroweak force combining with the electromagnetic force.*

Steven Weinberg knew of the dilemma of symmetry relating to photons, which are practically weightless, whereas bosons, which have intrinsic angular momentum, are a bit heavier than positive protons. He explained this conundrum by recounting the outset of the big bang. Weinberg used the idea of spontaneous symmetry breaking (where the symmetry of particles and energy was disturbed—chaos) to illustrate what occurred during the cooling-off period that followed the tremendous temperatures created at the outset of the big bang. This resulted in many fundamental particles assuming very different characteristics, leading to the belief that the current four primary natural forces were combined as one major force at that time. The four natural forces are:

1. *Gravity*. Although gravity is the weakest of these four forces and exhibits only an attractive force and acts over infinite distances, it is the predominant force over the entire universe.
2. The *weak nuclear force* causes the beta (electron) decay of a neutron into a neutrino and electron (neutron $\rightarrow$ proton + beta + neutrino). It is one of the fundamental interactions of elementary particles (*see* Figure F2 under Fermi). Essentially, it involves leptons and acts over extremely small distances, ranging between $10^{-9}$ and $10^{-10}$ cm.
3. The *electromagnetic force* acts on particles with electric charges and holds electrons to their orbits around the nuclei of atoms. It exhibits both attractive and repulsion forces and acts over infinite distances. The electromagnetic interactions are limited to atomic and molecular particles.
4. The *strong nuclear force* is the strongest of these four natural forces. It binds protons and neutrons (and quarks) together by gluons in the nuclei of atoms. It mostly involves hadrons and acts over small distances, ranging from about $10^{-6}$ to $10^{-9}$ cm.

***Weinberg–Salam greater unified theory:*** *The four fundamental forces of nature interact to make up all the forces found in the universe and thus may be integrated into a basic unified force.*

Weinberg became involved with issues related to cosmology, the origin of the universe, and the big bang. Weinberg, his colleague Abdus Salam, and other scientists continue to search for "superstrings" that may link these four basic forces. Their superstring theory states that all the known small particles of matter are not really the basic fundamental particles. Rather, an extremely small (not yet detected) vibrating string is the basic particle or energy unit, possibly only $10^{-35}$ cm, which is smaller than anything yet to be conceived. Instead of three or four dimensions, the "strings" supposedly may have six dimensions, each of which is curled up into each string. The advantage of the string theory is its ability to explain the unification of all four of the natural forces, the big bang theory, and black holes. Weinberg currently believes we are on the verge of uncovering the final theory, which Albert Einstein referred to as the unified field theory. It has also been referred to as the grand unification theory (GUT), the theory of everything (TOE), and the "answer."

*See also* Einstein; Glashow; Hawking; Salam; Witten

**WEISMANN'S GERM PLASM THEORY:** Biology: *Friedrich Leopold August Weismann* (1834–1914), Germany.

> *Germ plasm (today known as genetic resources or DNA) that is found in the ovum and sperm, which, in turn, are part of the chromosomes of living cells of organisms, are responsible for the continuity of characteristics from parent to offspring.*

Weismann stated that there were two types of cells in multicellular organisms: germ cells and somatic cells. The distinction that Weismann made between the two types of cells was that the "germ plasm" or protoplasm found in germ cells is passed unchanged from one generation to the next. In other words, germ plasm (not to be confused with "plasma") is responsible for inheritance of characteristics from one generation to the next. He considered somatic cells merely as a vehicle (means) to transport the germ plasm that is supposedly "immortal" because it passes inherited characteristics from generation to generation of a species. He was one of the first to propose that the germ plasm in the germ cells is protected from any type of modification by environmental (external) effects on the cells. This is now called the "Weismann barrier" that became a strong point in Darwin's theory that acquired characteristics cannot be inherited. The Weismann barrier led to a renewed interest in Gregor Mendel's work with inheritance that had been more or less ignored by scientists for many years. It also disproved Lamarckian theory of inheritance, which incorrectly stated that acquired characteristics can be passed on through germ plasm and thus inherited (as an example, Lamarck believed that if a woman dyed her hair red before conception, the resulting child would have red hair). The Soviet biologist Trofim Denisovich Lysenko, who was the minister of agriculture under Stalin, attempted to implement some of Lamarck's ideas of how to use environmental factors to improve genetics of the seeds of agriculture crops with catastrophic results. These measures resulted in the repeated failure of several generations of food crops and starvation in Russia. Weismann conducted an experiment designed to disprove Lamarckism by cutting off the tails of twenty-one generations of mice. He found that the offspring of the twenty-second generation still had tails, thus proving that injury (an acquired characteristic) is not heritable.

Although Weismann did not completely understand genetics, he did make some contributions to evolutionary biology. For example, he discovered what became known as "crossing over during meiotic division of gametes," and he understood the process of genetic variability as the basis of natural selection. He was the first to suggest that sexual reproduction was a means of providing new variations required for natural selection to work. More recently it has been found that the organism may be affected by some modification of germ plasm, such as changes in the immune system. Even so, the Weismann barrier is fundamental to Darwin's theory. Weismann wrote several important books in his lifetime. *Studies in the Theory of Descent* in 1882 contained a preface by Charles Darwin. In 1886 he wrote *The Germ Plasm:A Theory of Heredity*.

*See also* Darwin; Lamarck; Mendel

**WEIZSÄCKER'S THEORIES OF STAR AND PLANET FORMATION:** Physics: *Baron Carl Friedrich von Weizsäcker* (1912–2007), Germany.

***Weizsäcker's theory of star formation:*** *A nuclear chain reaction involving a "carbon-cycle" occurs inside a condensed mass of gas, resulting in the formation of a star that produces heat and light.*

In 1929 George Gamow was the first to propose that the source of a star's "core" energy is a nuclear reactor where hydrogen nuclei are converted into helium nuclei by the process of nuclear fusion, resulting in the release of vast amounts of energy. Several years later, when more was known about nuclear reactions, Hans Bethe provided the details of how hydrogen fusion could occur in the sun's core without exploding the star. Baron Carl Weizsäcker advanced a similar theory with the addition of what is known as the "carbon cycle" or the "carbon-nitrogen cycle" (not to be confused with the biosphere carbon cycle). Weizsäcker believed that in massive stars, a carbon molecule attracts four protons (hydrogen nuclei), and through a series of theoretical nuclear reactions, it produces one carbon nucleus and one helium nucleus while emitting two positrons and tremendous heat and light energy ($C + 4\ ^{+}H \rightarrow C + {}^{++}He + 2\ ^{+}p$). Because stars are composed mainly of hydrogen, this process, which takes place at their centers, can continue until all the hydrogen is converted to helium. However, a time span of billions of years must pass before "death" occurs for most stars. In addition to great amounts of heat and light produced by stars, the triple-alpha process, a nuclear reaction in stars, fuses three helium atoms to form carbon, which makes carbon-based life on Earth possible.

***Weizsäcker's nebula/planetary hypothesis:*** *As a nebula of swirling gases and small particles condenses, turbulence is created that will form the planets in their orbits.*

There is a long history of ideas, concepts, hypotheses, and theories to explain the origin of the solar system and its planets. One of the more popular ideas was the "passing star" theory, which explained how matter was pulled off two stars as they passed close to each other to form planets orbiting around one or both stars. In 1944 Weizsäcker applied mathematics related to the science of **magnetohydrodynamics** to explain how a mass of thin gas moving in a giant magnetic field in space could, through angular momentum, "push" the energy of the moving gas outward, thus providing angular momentum for the planets to remain in their orbits. One problem was that planets exhibit more angular momentum than Weizsäcker predicted, and angular momentum is always conserved; it cannot be created or destroyed, just transferred. Weizsäcker's nebula theory, with modifications that use the sun's magnetic field to increase the angular momentum of the planets (provided by Fred Hoyle), is currently the best explanation for the formation of our solar system.

*See also* Bethe; Gamow; Hoyle; Laplace

**WERNER'S COORDINATION THEORY OF CHEMISTRY:** Chemistry: *Alfred Werner* (1866–1919), Switzerland. Alfred Werner was awarded the Nobel Prize in Chemistry in 1913.

> *Metals have a primary and secondary valence—the primary valence involves the binding of ions (charged atoms), while the secondary valences involve atoms and molecules to form "coordinated compounds" of metals.*

Alfred Werner received his PhD in chemistry from the University of Zurich in 1890. Several years later in 1895 he returned to the university as professor of chemistry where his work revolutionized inorganic chemistry by distinguishing between a primary and secondary valence for metal atoms. He showed that the primary valence of a metal (mostly a transition metal) was involved with the binding of ions (atoms with a charge), whereas the secondary valence of a metal was applied to both the atoms and molecules. This characteristic made it possible for some metals, by the use of secondary valences, to join with themselves to form what Werner called "coordination compounds." A coordination compound is a metal surrounded by molecules or ions that are called "ligands" or complex agents. When a metal ion has an empty valence orbit, it can form an acid. Werner's theory of the number of atoms or groups of atoms, involved with a central metal atom led to the concept of a "coordination number" of 4 or 6 up to maximum number of 8, which then led to Abegg's rule known as the "rule of eight." The Werner coordination theory for transition metals was the beginning of modern inorganic chemistry in the early twentieth century.

*See also* Abegg

**WERNER'S NEPTUNIAN THEORY (NEPTUNISM):** Geology: *Abraham Gottlob Werner* (1750–1817), Germany.

> *As the great flood that covered the ancient Earth subsided, the dissolved minerals were chemically precipitated out to form the different types of rocks, minerals, and surface features including the mountains.*

Although a discredited scientific theory, Werner named his concept for the formation of rocks after the ancient Roman name (Neptunus) for the Greek god of the sea, Poseidon. The theory was also based on biblical scriptures and Werner's observations of the indigenous minerals and ores of the mining regions of his birth. As a young boy he became an assistant to his father who was a supervisor of an ironworks operation in northern Germany. Abraham entered the Freiberg Mining Academy in 1769 and later transferred to the University of Leipzig from which he graduated in 1775. He retuned to teach at the mining academy where he developed his theory. Neptunian theory explained how an Earth-covering sea receded in several steps by varying rates of chemical precipitation, thus forming different types of crystalline rocks. The first layer of the crust consists of very old igneous rocks such as granite, gneiss, and slates that contain no fossils. The precipitates from the oceans formed these old rocks before dry land appeared. Next was the transitional strata consisting of shale that contained fossils of fish. This layer was followed by the secondary strata of various types of limestone and sandstones that made up the secondary rocks. The alluvial or tertiary strata was next and consisted of gravels, sand, and clays that were formed as the oceans receded from the continents. Finally, once the water subsided, the exposed dry land contained lava produced by volcanoes, as well as other deposits. For a time, Werner's theory was accepted and even displaced other older theories for the formation of rocks such as the Plutonist's theory proposed by the Scottish geologist James Hutton (1726–1797) who suggested that igneous rocks were formed by molten matter. Werner was also a mineralogist who published the first textbook on minerals based on his classification of minerals. Although he eventually realized such a classification should be based on chemical characteristics, his book emphasized the need for correct classification based on the external characteristics and physical properties of minerals. All three factors (chemical, external characteristics, and physical factors) should be considered for any accurate classification. Also, he did not recognize the importance of the various types of crystallization in rocks as a means of identifying different types of minerals and ores. His theory for the origin of mineral and ore deposits followed his general theory of geology. He stated that precipitates filled the fissures that developed on the worldwide oceans' seafloors forming veins of minerals. This idea was opposite of the Plutonists who claimed that molten matter from the center of Earth filled these cracks with vapors to form deposits of minerals. Werner, who suffered from poor health his entire life, retired in Dresden, never married, and died in 1817.

**WHEELER'S "GEON" THEORY:** Physics: *John Archibald Wheeler* (1911–2008), United States.

*Geometrodynamics (geon) is an electromagnetic field maintained by its own gravitational attraction.*

John Wheeler searched for a theory to unify two seemingly unrelated fields: gravity and electromagnetism. This involved a method to demonstrate the concept of "action at a distance." Since the days of Aristotle, it was believed that something had to push or pull an object continually to make it move or cease moving. Neither did the ancients believe an object could be moved by a force not in direct contact with it. This was implicit in Newton's third law of motion (for every action, there is an equal and opposite reaction). Wheeler and his colleague, Richard Feynman, offered a solution that proposed a retarded effect on an object rather than an instantaneous effect. Their solution, somewhat like one of Einstein's "thought experiments," does not require any laboratory or equipment. Wheeler and Feynman suggested that two objects (1 and 2) be set up exactly one light-minute apart (1/525,600 of a light-year). Then any light (or any electromagnetic wave) sent from object 1 will take exactly 1 light-minute to reach object 2. Thus, it could be said that there was a delay from the signal to the reception of 1 light-minute, or because the action was received after it was sent, it was "retarded." In addition, there was no direct or instantaneous contact between the forces exerted by object 1 with the retarded action by object 2. Gravity and electromagnetism exhibit some properties of "action at a distance," which is one reason Wheeler attempted to unify them into a single theory. However, one problem was Newton's third law (if there is a "forward" effect from object 1 to object 2, there should also be an effect acting "backward" from object 2 to object 1). This problem could be eliminated only if "retarded" effects were considered. Geon unification theory has never been proved. Wheeler made contributions to the area of nuclear fission and from 1940–1950 worked at the Los Alamos Laboratory exploring the possibility of using heavy hydrogen to make a hydrogen (fusion) bomb.

*See also* Feynman

**WHIPPLE'S "DIRTY SNOWBALL" THEORY OF COMETS:** Astronomy: *Fred Lawrence Whipple* (1906–2004), United States.

*Comets are composed of ice, dust, gravel, some gases, and possibly a small rocky core. They are similar to a dirty snowball.*

In 1949, astronomer Fred Whipple based his comet theory on the spectroanalysis of their light and their evolution as they made return trips on elliptical paths through the solar system. He theorized that comets are basically formed of ice and contain a mixture of sand-like dust, gravel, and possibly some gases, such as carbon dioxide, methane, and ammonia. Some comets may have a rocky core. Whipple explained that when a comet approached the sun, even millions of miles distant, the comet's ice vaporized, expelling the dust and gas to form a hazy tail, which always pointed away from the sun as it continues on its orbit. This is a major feature of Whipple's theory. Comets have three basic parts: the head, which is the brightest, varies in size from 0.5 to about 5 or

7 miles wide; a halo, which may be 50,000 to 75,000 miles wide, that glows around the head; and the tail, which is a much fainter glow and may extend 50 to 75 million miles in front of the head. Sunlight and solar wind create radiation pressure on the comet, which forces the gaseous ice/dust of a comet's tail always to point away from the sun. Thus, the "tail" of the comet always precedes the head of the comet because of the solar pressure on the comet's tail, which is less dense than the comet's head. In 1986 a U.S. spacecraft investigating and gathering data on Halley's comet confirmed Whipple's theory of a comet's structure, with one exception: rather than being a "dirty snowball" of dust, it is now believed to be more like an "icy dust-ball" because the ice is condensed on the outside of the dust particles, and after each pass around the sun, more and more of the ice is lost, meaning that the comet becomes less and less brilliant as it ages.

*See also* Halley; Oort

**WHITEHEAD'S "ACTION-AT-A-DISTANCE" THEORY OF RELATIVITY:** Mathematics and Physics: *Alfred North Whitehead* (1861–1947), England and United States.

> *Action-at-a-distance is the interaction of two objects in space that are separated from each other but still interact with no mediator or connection.*

The ancients, including Aristotle, believed that for an object to move either on Earth or in the heavens, something had to either push or pull it; or if the push or pull was removed, the object ceased moving and no interaction was present. When more was learned about gravity and electromagnetism, these theories were used to partly explain this phenomenon. Einstein referred to this as "spooky action at a distance" and claimed it was evidence of quantum theory, general relativity, and gravity. Whitehead disagreed with Einstein's theory of relativity and developed his own "action-at-a-distance" theory that was based on philosophical principles. It was and never has been well accepted because it lacked any evidence.

Alfred North Whitehead was one of the most famous and last of the nineteenth century's philosopher/scientists who approached science more by using philosophical reasoning than by gaining evidence through research. He adopted a belief in "atomic occasions," which were different and succeeded one to another endlessly as a way to explain time and nature as well as his belief in a supreme being. He related the ultimate uniformity as in the nature of God. His father, also named Alfred Whitehead, was an Anglican preacher who home-schooled his son until the age of fourteen. The younger Whitehead was considered "sickly" by his parents, but after he entered public school, he excelled in sports and seemed to be a healthy child. In 1884 he graduated with a PhD from Cambridge University in England where he later became a teacher. One of his students, Bertrand Russell (1872–1970), became a fellow philosopher of science and a mathematician who objected to the materialistic and deterministic direction of nineteenth-century science that developed scientific theories where patterns were derived from the perceptions and measurements of the world rather than the basic properties of reality. This philosophical viewpoint was expressed in Whitehead's first book *Treatise on Universal Algebra* published in 1898. In 1910 he published his most important book, *Principia Mathematica* coauthored with Bertrand Russell. Whitehead did not contribute to the second edition of this book published by Russell in 1925. The original volume is considered one of the most influential works in the field of logic, on a par with Aristotle's *Organon*.

**WIEN'S DISPLACEMENT LAW:** Physics: *Wilhelm Carl Werner Otto Fritz Franz Wien* (1864–1928), Germany. Wilhelm Wien received the 1911 Nobel Prize in Physics.

*As the temperature rises for electromagnetic radiation, the total amount of radiation increases, while the wavelength of the radiation decreases.*

Wilhelm Wien knew that the amount of electromagnetic radiation increases as temperatures rise (a glowing red-hot stovetop element feels hotter than one that is not glowing and appears black). He also knew that very long and very short wavelengths are less abundant in nature than those near the center of the electromagnetic scale. After measuring various wavelengths, he determined these central "peak" wavelengths vary inversely with the **absolute temperature.** This is known as *Wien's displacement law*: the temperature of the radiation determines wavelength and amount of thermal radiation. Heating a hollow metal ball with a hole in it, called a "blackbody," and then measuring the wavelength and amount of radiation emitted demonstrates this law. As the temperature of the blackbody increases to the red-hot stage, longer wavelength radiation is emitted. When the temperature becomes even greater, white-hot shorter wavelength radiation is detected. The "amount" of radiation peaks at about the same wavelength range as that of visible light on the electromagnetic radiation scale. The law can be expressed as: $\lambda \mathrm{T} = \text{constant}$, where $\lambda$ is the wavelength, $T$ is the temperature, and the constant is equal to 0.29 cm k. Wien used this law to indicate the distribution of energy in the spectrum as being a function of temperature. The law is applicable for shorter wavelengths but breaks down for longer wavelengths. The blackbody radiation distribution law (the beginning of quantum theory), developed by Max Planck, is equivalent to Wien's displacement law when the frequency is very large. Planck's law is correct at any frequency, whereas Wien's is correct only for high frequencies.

*See also* Bohr; Boltzmann; Einstein; Helmholtz; Planck; Schrödinger

**WIGNER'S CONCEPT OF PARITY/SYMMETRY IN NUCLEAR REACTIONS:** Physics: *Eugene Paul Wigner* (1902–1995), United States. Eugene Wigner shared the 1963 Nobel Prize for Physics with Maria Goeppert-Mayer and J. Hans Jensen.

*Parity is conserved in nuclear reactions because nature cannot differentiate between left and right orientations or between time periods.*

Eugene Wigner, a theoretical physicist, contributed to the understanding of nuclear physics by applying quantum theory to fundamental symmetry principles. He stated that parity is conserved in nuclear reactions. (Any two integers have parity if they are both even or both odd, and fundamental physical interactions do not distinguish between right or left or clockwise or counterclockwise, thus ensuring symmetry, which is a major physical concept.) Wigner's theory stated that for all matter, energy, and time in the universe, nature makes no distinction between the physical orientation in space of particles, or of more or less time. This relates to nuclei and subnuclear particles' having mirror images, as they are involved in all types of chemical and nuclear reactions. In other words, it does not matter if the molecules or nuclei of matter are oriented as mirror images of each other. The results will be identical in the same time period. Or if a particle is ejected from a nucleus, no distinction is made as to whether it leaves from the right or left. This theory was accepted until 1958, when weak nuclear reactions were discovered. An example of a weak nuclear interaction is the decay of a neutron into a proton plus a beta particle (electron) and a neutrino; parity is not conserved. Even so, this exception does not eliminate the concepts of a parity or symmetry. Wigner's concepts of parity and symmetry are related to the premise that the greater the "cross section" of a nucleus, the more likely it is that the nucleus can absorb a neutron. This idea contributed to the successful production of a sustained chain reaction in the first nuclear pile located under Alonzo Stagg field stadium at the University of Chicago in 1942.

*See also* Boltzmann; Fermi; Schrödinger; Weinberg; Wu; Yang

**WILKINSON'S CONCEPT OF "SANDWICH COMPOUNDS":** Chemistry: *Sir Geoffrey Wilkinson* (1921–1996), England. Geoffrey Wilkinson shared the 1973 Nobel Prize in Chemistry with Ernst Fischer.

*Homogeneous catalysts can be formed by adding hydrogen to the double bonds of alkenes.*

Geoffrey Wilkinson, primarily an inorganic chemist, explored the attachment of hydrogen to metals to form complex compounds (hydrides) composed of molecules sandwiched together with hydrogen bonds that could be used as catalysts, later known as *Wilkinson's catalysts*. Using these catalysts, he developed systems that could alter the nature of organic compounds by adding hydrogen to the double bonds of some hydrocarbon type molecules. By bonding hydrogen to compounds known as *alkenes*, which have unsaturated molecules, he converted them into branched-chained hydrogen-saturated, paraffin-type compounds. Known as *addition hydrogenation*, this process converts unsaturated vegetable liquid oils (e.g., corn oil) to solid fats (e.g., margarine) by adding hydrogen to the double bonds of the oil molecules. This hydrogenization process may also rupture these organic bonds, resulting in hydrocracking, hydroforming, or catforming, which splits off sections of hydrocarbon molecules by using low heat and a platinum catalyst. This process converts crude petroleum into more useable branched-chained fractions (e.g., gasoline, ethane, propene). It is also known by a more generic name, *hydrogenolysis*, which converts bituminous coal into a variety of useful hydrocarbon products, including coal tar dyes, medicines, cosmetics, lubricants, and other petroleum-like products (e.g., "coal-oil" or kerosene).

**WILLIAMSON'S THEORY OF REVERSIBLE CHEMICAL REACTIONS:** Chemistry: *Alexander William Williamson* (1824–1904), England.

*A chemical reaction will reach dynamic equilibrium when, under correct conditions of concentration, temperature, and pressure, it becomes reversible.*

Alexander Williamson demonstrated that it was possible to produce a number of different organic compounds by replacing one or more hydrogen atoms in organic compounds, thus forming organic radicals. He based his idea on the work of the French chemists Charles Gerhardt (1816–1856) and Auguste Laurent by replacing one or more hydrogen atoms in inorganic compounds that form typical radicals. From this, he developed chemical formulas for a number of compounds, such as alcohols and ether. While experimenting with these new substances, he discovered that some chemical reactions are reversible. A mixture of two compounds will react to form two very different compounds. However, using the correct amounts of the initial substances along with the correct temperature, concentration, and pressure, the reaction will reverse itself, and the new compounds will revert to the original substances. In other words, once the first two compounds form the second two, the second ones will revert to the original two compounds ($A + B \leftrightarrow C + D$). Under these conditions the entire system is considered to be in dynamic equilibrium. This process is known as *Willliamson's synthesis* and is used in the making of ethers. This concept is vital to the chemical industry concerned with the conditions necessary to ensure a chemical reaction is not in equilibrium so that it will proceed in the desired direction, resulting in the preferred product.

*See also* Laurent

**WILSON'S HYPOTHESIS OF CLOUD CONDENSATION:** Physics: *Charles Thomson Rees Wilson* (1869–1959), England. Charles Wilson received the 1927 Nobel Prize in Physics.

> *If dust-free, supersaturated moist air is rapidly expanded, the moisture condenses on both fine nuclei and ions (particles).*

While experimenting with supersaturated water vapor in a laboratory vessel, Charles Wilson rapidly expanded the volume of this moist air, which formed a cloudlike formation in the chamber. Because the air was dust free, he assumed some type of "nuclei" were present, which provided a base for the moisture to condense into water droplets. He theorized that the recently discovered radiation called X-rays might also cause condensation tracks to form in the moist air in his chamber. Subsequently, he discovered that supersaturated air became conductive when X-rays passed through this moist air, and much more condensation was produced than could be caused by just expanding the air's volume. Wilson then developed his famous *Wilson cloud chamber*, based on the work of J.J. Thomson and Ernest Rutherford, which detects radioactive radiation of all kinds as well as very small, almost weightless subatomic particles as they form ionized curved paths through supersaturated air in the chamber. In the early part of the twentieth century the Wilson cloud chamber became a valuable research tool for the study of subatomic particles. These ionized paths in the chamber made by radiation form water droplets and can be photographed and studied to determine the characteristics of the radiation or nature of the subatomic particle.

*See also* Compton; Millikan; Rutherford; Thomson

**WILSON'S "OUT-OF-AFRICA" THEORY:** Biology: *Allan Charles Wilson* (1934–1991), New Zealand.

*The ratio of mitochondrial DNA differences between humans and great apes indicates a divergence of lineages five million years ago, which achieved a complete separation between species two hundred thousand years ago.*

Allan Wilson studied the DNA found in the mitochondria of cells that, unlike regular DNA, is found outside the cell nucleus. Mitochondria exist in the **organelles**, which are structures located in the cytoplasm that produce the energy required for cell growth and life. This extranuclear DNA, referred to as $_{mt}$DNA, is carried only in the mother's cells. It is also believed that genetic variations arise from mutation of the $_{mt}$DNA and accumulate through the maternal side at a rather steady rate, which provides a means to calculate statistically, through maternal $_{mt}$DNA, the age of ancestors. In other words, $_{mt}$DNA becomes a *molecular clock.* Wilson therefore theorized that all human mitochondrial $_{mt}$DNA must have originated with a very old, common, female ancestor. He collected a sample of mitochondria cells from individuals of all races from all parts of the world and discovered there are just two basic genetic branches, both of which originated in Africa. His theory that the maternal ancestor for all humans lived on the African continent became known as the "out-of-Africa" theory and was later dubbed the "Eve hypothesis" by journalists.

Wilson's next research dealt with the age of this "common" female ancestor. He found the ratio of $_{mt}$DNA between chimpanzees and humans was 1:25. Also, because the beginning of the separation of the *Homo* species from the great apes was about five million years, he calculated that 1/25 of this five million years was equal to two hundred thousand years. (More recently it has been established that chimpanzees and humans share over 98% of the same DNA.) He theorized this was the time a complete separation from our nonhuman ancestors resulted in a human species. Some scientists claim humans diverged from apes and became a separate species in more than one geographical region. Some paleontologists claimed that the divergence of humans from apes occurred over five million years ago; other scientists claim this divergence occurred no more than one million years ago. In 1980 Wilson wrote that the years of divergence was more than two hundred thousand years ago, whereas others believe it occurred fewer than two hundred thousand years ago. A competing theory, called "multiregionalism," proposes that ancient humans originated in several different regions of the world and over time migrated, interbred, and produced hybrids that became some of the now-extinct species of the *Homo* group (e.g., Neanderthal man). Most scientists now accept that the extinct species of *Homo* who walked erect on two legs developed about 100,000 to 200,000 years ago in Africa, 60,000 years ago in Australia, 40,000 years ago in Europe, and 35,000 years ago in Northeast Asia and appeared in the northwestern part of the North American continent about 15,000 to 30,000 years ago. The more recent species of man, *Homo sapiens-sapiens* (intelligent man), appeared in Europe or Eurasia about ten thousand to fifteen thousand years ago. More fossil evidence will need to be found and analyzed before the argument concerning the origin of humans as a separate species can be settled.

## WILSON'S THEORY OF DYNAMIC EQUILIBRIUM OF ISLAND POPULATIONS: Biology: *Edward Osborne Wilson* (1929–), United States.

*Geographically isolated species will establish a dynamic equilibrium of their populations.*

E. O. Wilson is an entomologist and sociobiologist who studied ants and other social insects. Wilson and his colleague, the ecologist Robert MacArthur (1930–1972), theorized that, in time, related species would develop distinct differences to resist interbreeding, and a "dynamic equilibrium" of their populations would naturally be established. They based their concept on what they called "character displacement," which takes place when isolated species are once again brought back into close geographic proximity to each other. To prove their theory, they eliminated all insects on a small island off the south Florida coast and waited to see how it would be repopulated as compared to the original number of species. After several months, they returned to find that the same number of species in the same ratios had repopulated the island, as before, thus proving their theory that for isolated geographic areas, a dynamic equilibrium among species populations will develop (they assumed the insects' eggs were not completely destroyed or adults arrived from other nearby land areas). Wilson also contends that individual animals and groups (insects and humans) use their genetically driven cultural attributes, which are the result of natural selection, to control their population and make sacrifices for the group.

*See also* Darwin; Wallace

**WITTEN'S SUPERSTRING THEORY:** Physics: *Edward Witten* (1951–), United States.

> *Events at the nuclear level unify general relativity by combining gravity, quantum mechanics, and space in ten dimensions.*

There are two major theories of physics: 1) the very small (quantum theory and the uncertainty principle as related to atoms, molecules, subatomic particles, and radiation) and 2) the very large (Einstein's theory of general relativity, gravity, the cosmos, black holes, etc.). Both are related to the Standard Model of quantum mechanics (*see* Schrödinger). Edward Witten was convinced that the string theory could resolve the problems encountered when combining these two great theories that deal with the small and large. Usually a minute elementary particle is defined as a "point." Witten redefined fundamental particles as a vibrating string or looped string that has different states of oscillation with harmonics similar to a vibrating violin string, making them somewhat "fuzzy" point sources. Therefore, a single string can have several harmonics and can consist of a large grouping of different types of elementary particles. This results in a spectrum of particles that then can be "quantized" and related to the "graviton," referred to as the *quantization of gravitational waves*, which in itself makes gravity a priori of the string theory. All types of minute particles and subatomic particles (e.g., electrons, protons, muons, neutrinos, and quarks) fit into the string theory. Thus, Witten claims to have combined the quantum mechanics aspect of the electromagnetic/small with the quantum of relativity/gravity of the large. Based on pure mathematics, Witten proposed how a space consisting of two, four, six, or ten dimensions can explain superstrings and how particles can interact within such a geometric formation, and that six of these dimensions are "folded" into the four known dimensions (height, width, depth, and time). The original string theory was based on the notion that just after the big bang, as the universe cooled, cracks and fissures formed in space that contained great masses, energy sources, and gravitational fields. Recently, Steven

Hawking claimed that no evidence exists to support the existence of strings; thus the great unification of Einstein's general relativity and gravity with electromagnetism remains elusive. However, string theory remains a somewhat contentious subject in the science community. There are hundreds of young theoretical physicists who continue to work on the string theory because they believe it will lead to a grand unification theory or a theory of everything, while there are a few others who claim that there is no physical observable evidence to justify the acceptance of a theory based on tiny strings and multiple dimensional universes.

*See also* Einstein; Hawking; Schrödinger; Weinberg

**WOHLER'S THEORY FOR NONLIVING SUBSTANCES TRANSFORMING INTO LIVING SUBSTANCES:** Chemistry: *Friedrich Wohler* (1800–1882), Germany.

*By applying heat, it is possible to convert nonliving (inorganic) molecules to living (organic) molecules.*

Friedrich Wohler's experiments challenged "vitalism," the prevailing theory dealing with organic chemistry of the 1800s that stated it was not necessary to explain the compounds that make up living organisms because there was a "spirit" connected to life. This spirit was the God-given "vital essence" in all living things, including organic molecular compounds. Vitalism was accepted as the reason humans cannot and should not transform nonliving chemicals into living substances. Wohler proved otherwise, which not only resulted in the beginning of the end of vitalism, but provided understanding of new concepts for inorganic and organic chemistry. In 1928 he used heat to decompose ammonium isocyanate, an inorganic chemical, into urea, an organic chemical found in urine ($NH_4NCO$ + heat $\rightarrow$ $NH_2CONH_2$) (these are two very different compounds but with the same molecular formula, thus they are isomers). Vitalism is a persistent theory and still has adherents. In the field of chemistry, Wohler's work pioneered modern organic chemistry (carbon chemistry), particularly as related to human physiology of respiration, digestion, and metabolism.

**WOLFRAM'S THEORY OF COMPLEX SYSTEMS:** Physics: *Stephen Wolfram* (1959–), England.

*Complex systems are driven by one-dimensional cellular automata that follow specific rules.*

Stephen Wolfram was interested in a theoretical model for parallel computing that would increase computational power. This idea is based on the ability to understand entities consisting of a group of "cells" (not to be confused with living cells) that are controlled by a series of rules leading to complexity and chaos. Complex systems are based on the concept of *cell automata* first proposed by John von Neumann. There are several rules for one-dimensional automata cells:

- All cells in a "set" may or may not be filled in.
- Patterns may alternate from "filled in" to "not filled in" and continue to change, but each set must be one way or the other.

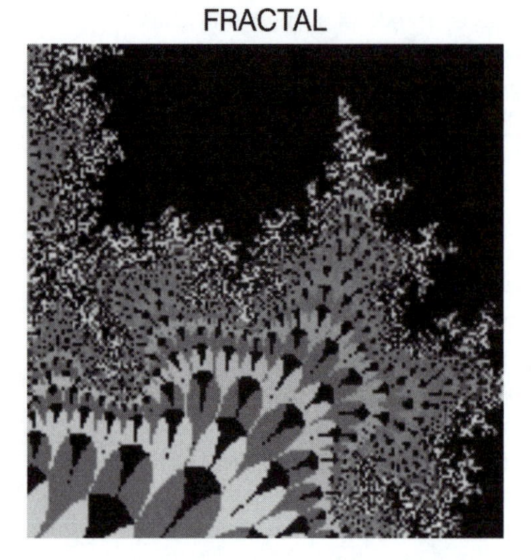


Figure W1. Fractals are "self-similarities" or "self-replacing" patterns that become increasingly complex and chaotic as they progress. They are similar to Penrose tiles connected in repeated diminishing patterns of geometric shapes as related to chaotic behavior.

- These patterns may "grow" and continue to form the same patterns in ever increasing complexity, as in self-replicating fractal patterns.
- These patterns will continue to become increasingly complex and chaotic.

This theory of complexity and self-organization explains many natural systems, including how simple organic molecules combined to form increasingly complex patterns until they could become self-replicating and thus living. The theory also describes formal language theory related to the evolution of grammar and original languages into the modern languages of the world.

*See also* Penrose; von Neumann

**WOLF'S THEORY OF THE DARK REGIONS OF THE MILKY WAY:** Astronomy: *Maximilian Franz Joseph Cornelius Wolf* (1863–1932), Germany.

*The dark areas of the Milky Way galaxy are regions where dense "clouds" obscure some of the stars.*

Maximilian Wolf designed the *Wolf diagram* used to measure not only the absorption of light but also the distance from Earth to what was called "dark nebula." He attached a camera to the eyepiece of a telescope, enabling him to expose photographic plates and thus record observations over long periods of time. By using time exposure, more light from distant and dim objects was gathered, allowing him to view images on the photographic plates that could not be seen otherwise. Using these methods, Wolf theorized that the dark areas in the Milky Way are gas clouds. More recently it was proposed that over 90% of all matter in the universe is composed of **dark matter**, gas, or even **neutrinos**, none of which can be seen by visible light and which outweighs all the trillions of stars.

**WOODWARD'S THEORY OF ORGANIC MOLECULAR SYNTHESIS:** Chemistry: *Robert Burns Woodward* (1917–1979), United States. Robert Woodward received the 1965 Nobel Prize in Chemistry.

*Molecules of organic substances maintain an orbital symmetry enabling them, by geometric orientation, to rotate 180° degrees on their axes and thus become a "negative" mirror image of the organic molecules.*

Robert Woodward used the principle of symmetry related to molecular orbits to develop his theory of how some molecules, through a series of addition-reactions, can be synthesized into many useful chemical products. He expanded organic synthesization to the formation of complicated molecules, some involving as many as fifty sequences or series of chemical reactions. Some examples of the products resulting from his addition-reactions are quinine, cholesterol, cortisone, lysergic acid (LSD), reserpine, strychnine, chlorophyll, and vitamin $B_{12}$.

*See also* Couper; Kekule

**WRIGHT'S THEORY OF GENETIC DRIFT (SEWALL WRIGHT EFFECT):** Biology: *Sewall Green Wright* (1889–1988), United States.

*In a small isolated population certain forms of genes may be randomly lost because they are not passed along to the next generation.*

The "Sewall Wright effect," also known as "the genetic sampling error," explores the changes in the gene pool of a small isolated and restricted population where there is a loss of particular genes and their characteristics that may lead to the emergence of new species. Within this small community natural selection does not usually take place due to inbreeding. He used statistics to determine the inbreeding coefficient as a way to compute the pedigrees within the local population. Sewall Wright and Sir Ronald A. Fisher (1890–1962), the British evolutionary biologist, computed the amount of inbreeding among members of populations as a result of random **genetic drift**. Together they developed the methods for computing the interactions of natural selection, mutation, migration, and genetic drift.

Sewall Wright's family lived in Melrose, Massachusetts, where as a young boy he deliberately dropped his middle name (Green). In 1892 his family moved to Galesburg, Illinois, where he attended high school and Lombard College. He then moved onto the University of Illinois where he received his PhD in biology in 1915. He spent the remainder of his career in research and teaching at other universities including the University of Chicago, University of California at Berkeley, and as a Fulbright Professor at the University of Edinburgh in Scotland (1949–1950), and finally at the University of Wisconsin–Madison.

**WRINCH'S CYCLOL THEORY OF PROTEIN STRUCTURE:** Biochemistry: *Dorothy Maud Wrinch* (1894–1976), England and United States.

*Chromosomes composed of sequences of amino acids are the only molecules with sufficient variety to permit the construction of complex molecules.*

Dorothy Wrinch had an eclectic academic career that included contributions in the areas of mathematics, biochemistry, philosophy, physics, as well as in sociology. She was born in Argentina in 1894 to a British couple who soon after her birth moved back to England. After graduation from high school, she received a scholarship to attend Girton College in Cambridge where she was influenced by Bertrand Russell to study philosophy and mathematics. She graduated with a first-class degree in mathematics in 1916, followed with a MSc and a DSc in mathematics from University College in London. After marrying John William Nicholson, who was the director of mathematics and physics at Balliol College at Oxford, she moved with him to Oxford where she taught mathematics at several women's colleges. During this time she earned her second master's degree in 1924 and her second doctorate degree in 1929. This was the first doctorate degree awarded to a woman by Oxford University. She published papers in the areas of applied mathematics and the philosophy of science. In the early 1930s Wrinch separated from her husband who by this time had become an alcoholic (they divorced in 1938). All the while she continued to receive fellowships in the new field of mathematics related to physics, chemistry, and biology.

She spent time at several European universities and in the mid 1930s she wrote five papers on the application of mathematics to chromosomes. This earned her a Rockefeller Foundation fellowship to study the application of mathematics to biological molecular structures. She traveled to several universities in the United States where she explained her theory of cyclol protein structure, which was based on concepts of mathematical symmetry.

Other scientists had suggested a hypothesis for the structure of fibrous protein by hydrogen bonding. Wrinch developed this suggested hypothesis in a viable model of protein structure. In 1936 her first cyclol model was presented in a paper that noted the possibility that polypeptides could cyclize and form closed rings that could form internal cross-links through what is known as cyclol reactions and thus could form stable peptide bonds. She figured out that such cyclol molecules would have a six-sided symmetry if the bonds were similar. This means that such rings can extend indefinitely to form what is known as "cyclol fabrics" that are proteins with no side chains. She presented this structure as a working hypothesis. After more research over the next few years, it was found that her cyclol hypothesis model was not accurate for globular proteins.

Although her theory was not entirely correct, it was useful when applied to chemical bonding and the study of organic compounds. During the early years of World War II in Europe she moved to the United States as a visiting lecturer in chemistry at Johns Hopkins University in Maryland. Dorothy became a visiting professor at Amherst and Smith Colleges and met Otto Charles Glaser who was a vice president at Amherst. They married in 1941. In 1943 she became a research professor of physics at Smith College in Massachusetts where she received a long-term fellowship in 1965. She retired from Smith in 1971 and moved to Woods Hole in Massachusetts. Her book on mathematical principles for the explanation of X-ray crystallography of complex crystal structures was published in 1946. It is titled *Fourier Transforms and Structure Factors*. Her theory of protein structure encompassed chemistry, physics, mathematics, as well as philosophy and contributed to molecular biology as a multidisciplinary study of life. Although her cyclol model for globular proteins was not completely accurate, it was a precursor for scientists to research the protein structure and develop a hypothesis for the DNA double-helix structure.

*See also* Crick–Watson

**WU'S THEORY OF BETA DECAY:** Physics: *Chien-shiung Wu* (1912–1997), United States.

> *The direction of the emitted beta particle is related to the direction of spin of the nucleus from which it originates.*

In 1934 Enrico Fermi verified Wolfgang Pauli's concept of beta decay, where a neutron disintegrates into an electron and neutrino, leaving behind a proton: neutron →

electron (β) + neutrino + proton (*see* Figure F2 under Fermi). This process is also known as the *nuclear weak force*, which is stronger than gravity but much weaker than the *strong nuclear force* that holds nuclei together. But there were problems with the Pauli/Fermi theory. In 1957, Chien-shiung Wu theorized that the problem was the direction of the beta decay. She demonstrated that the direction of emission of the beta particle was related to the spin orientation of the nucleus that was decaying. Thus, the emission process of the system is not identical to the mirror image of the system, and therefore parity (right–left symmetry) is not conserved during beta emission. Parity means that two systems that are mirror images of each other are the same in all respects except for this left–right or mirror image phenomenon and therefore should retain identical symmetry just as humans have a left-and-right side (mirror image) but also have bilateral symmetry.

*See also* Fermi; Feynman; Pauli; Wigner; Yang

**WURTZ'S THEORY FOR SYNTHESIZING HYDROCARBONS:** Chemistry: *Charles Adolphe Wurtz* (1817–1884), France.

*Hydrocarbons, including aromatic hydrocarbons, can be synthesized by reacting alkyl halides with sodium.*

A synthesizing reaction of hydrocarbons was used by Adolphe Wurtz in 1855 as a method for producing paraffin hydrocarbons by using alkyl halides and sodium in ether. Along with the German chemist Wilhelm Rudolph Fittig (1835–1910), he developed a similar type reaction for synthesizing aromatic hydrocarbons. He also developed a way of synthesizing chemicals from ammonia by substituting the carbon radical $C_2H_5$ for one of the hydrogen atoms in ammonia ($NH_3$). By using this technique Wurtz was able to produce a variety of ammonia-related hydrocarbons. In 1860 Wurtz, in cooperation with August Kekule, formed a conference of the International Chemical Congress where he was scheduled to read a paper by the Scottish chemist Archibald Couper who had anticipated Kekule's method of forming the ring structure of the carbon atom. Wurtz delayed his presentation of the paper, while in the meantime Kekule published his theory for the ring structure of the carbon atom containing six carbon atoms in a paper and thus received credit for the discovery of the ring structure of the benzene molecule ($C_6H_6$). Couper became so angry with Wurtz for not presenting his paper that it resulted in his discharge from Wurtz's laboratory. Couper became despondent and never did any serious chemical research again.

Wurtz's father was a Lutheran pastor who supported his son Adolphe's study of medicine rather than theology. A good student, Wurtz was more interested in the chemistry involved in medicine and was promoted to the faculty of medicinal chemistry at the local university in Strasbourg. He held several positions at various universities in France but found their laboratories inadequate for his chemical research thus he built his own laboratory in his home. He had some difficulty in convincing the French government to support chemical research in which Germany was the leader. Adolphe Wurtz (he never used his given name Charles) was the founder of the Paris Chemical Society and served as its president on three occasions. He conducted research at several institutions, published many papers in his lifetime, and made important contributions to organic chemistry.

*See also* Couper; Kekule

**WYNNE-EDWARDS' THEORY OF GROUP SELECTION:** Biology: *Vero Cooper Wynne-Edwards* (1906–1997), England.

*Different social behaviors are mechanisms for limiting a surplus of potential breeders beyond the quota that their habitat can carry.*

Wynne-Edwards was a keen observer of nature, particularly the roosting sites of the starling population in his home area. He noted that in addition to individual selection for breeding, there was group selection, which was an evolutionary construct. This group selection is based on the group's ability to control their rate of consumption of resources and to keep the breeding at a level that would benefit the group so it would not go extinct, whereas the individual selection of mates will generate populations of selfish individuals who overexploit the existing resources and will soon die out. He determined that one conflict with the idea of group selection is gene mutation in which the number of eggs laid may increase from two to six, and thus the increase in offspring may again exploit the available resources. Another conflict is immigration where new individuals who produce more than two eggs may upset the established balance of a two-egg group. He determined that other ecological factions could affect the established populations of a group such as weather (severe storms, freezing, drought, etc.), as well as human intervention.

Wynne-Edwards was one of the first ecologists, even before the discipline of ecology was a recognized field. He graduated with a degree in natural science (from which the science of ecology sprung) at Oxford University in 1927. He taught zoology at McGill University in Canada from 1930 to 1944. He returned to Britain and from 1946 to his retirement in 1974 was professor of natural history at Aberdeen University. His best-known book is *Animal Dispersion in Relation to Social Behavior* published in 1962 in which he expressed his theory of animal behaviors, such as territoriality, dominance hierarchies, groupings of flocks, and so forth as devices for controlling populations thus balancing the group and their resources. Others disputed his ideas. Some alternative explanations were altruism and population control as well as other ideas that led to the expansion of ecology and natural sciences related to sociobiology developed by Edward O. Wilson.

*See also* Buffon; Haeckel; Wilson (Edward Osborne)

# Y

**YALOW'S THEORY OF RADIOIMMUNOASSAY:** Physics: *Rosalyn Sussman Yalow* (1921–), United States. Rosalyn Yalow shared the 1977 Nobel Prize for Physiology or Medicine with the French American researcher Roger Guillemin and Andrew Schally, the Polish-born medical researcher.

> *Using the technique of radioimmunoassay (RAI) to detect small amounts of radioactive hormones plus a known amount of antibody, and then mixing these with an unlabeled hormone, it is possible to accurately measure the amount of the nonradioactive hormones.*

By mixing a small amount of radioactive hormone with an unknown amount of another nonradioactive hormone provided a means to accurately detect and measure amounts of the nonradioactive hormone in amounts as small as one pictogram which is $10^{-12}$ grams. Using this technique that was discovered by Rosalyn Yalow and the American physician and scientist Solomon Berson (1918–1972), it was then possible for Roger Guillemin (1924–) and Andrew Schally (1926–) to detect the various elusive hypothalamic hormones, which was a breakthrough in the field of endocrinology.

Yalow graduated from Hunter College in 1941 where she developed an interest in physics. From there she worked as a secretary at Columbia University's College of Physicians and Surgeons with the belief that no top graduate school in the United States would accept a woman. During World War II, when many young, college-age men went off to war, she accepted an assistantship at the University of Illinois where she was the only woman in a department with four hundred men. The University offered assistantships to women rather than close the campus due to the absence of qualified male teachers. After taking advanced physics course, she graduated with a PhD in 1945. She then moved to the Bronx Veterans Administration Hospital to establish a program in radioisotope services where she perfected the technique of using small quantities of radioisotopes to trace and measure substances in blood, particularly in the study of insulin levels in diabetic patients. Her collaborator in this endeavor was Solomon Berson whose death in 1972 precluded his receipt of the Nobel Prize. Their radioimmunoassay techniques proved to be very valuable in detecting small amounts of hormones, vitamins, and enzymes that could not be detected by other means. Although this discovery was a huge success, she and Dr. Berson refused to obtain a patent for their discovery.

**YANG'S THEORY OF NONCONSERVATION OF PARITY IN WEAK INTERACTIONS:** Physics: *Chen Ning Yang* (1922–), United States. Chen Ning Yang shared the 1957 Nobel Prize in Physics with Tsung-Dao Lee.

*The physical law of conservation of parity (symmetry) will break down during weak interactions of subnuclear elementary particles such as beta decay.*

Chen Ning Yang, a theoretical physicist, predicted in 1956 that the basic physical law of conservation of parity, first proposed by Eugene Wigner in 1927, would break down when the weak interactions (forces) related to the decay of elementary subnuclear particles were involved. *Parity* refers to the symmetrical quantum-mechanical nature of physical systems. *Parity conservation* refers to the basic physical concept of symmetry, which states that fundamental physical interactions cannot distinguish between right- or left-handedness, clockwise or counterclockwise, or mirror images of physical systems. In addition, for the conservation of parity, no distinction for the particle's orientation in space or the direction of time exists. Yang's theory predicted this concept was violated during the weak interactions of basic atomic elementary particles. These weak interactions are the fundamental forces that take place among elementary particles, including beta decay of nuclei, which produces neutrinos and electrons. Therefore, it is also known as *beta interactions*. These weak interactions are weaker than electromagnetic forces but stronger than gravitational interactions (the strong interaction involves the force that holds the nucleus together). Unlike electromagnetic and gravitational interactions, whose forces fall off as with the square of the distance and thus become less strong over long distances, the weak interactions fall off very rapidly, and thus are not effective beyond the size of the atom from which they originate. Yang and his collaborator, the Chinese-born American physicist Tsung-Dao Lee theorized that a subnuclear particle called a kaon would break down into two *pions*, which would maintain and conserve parity. But at times some of the kaons broke down in three pions, and thus, in this example of weak interactions, parity was not conserved (e.g., two odds and one even, or it could be described as two pions that spin clockwise while the other spins counterclockwise). This means the symmetry of left and right was not equal and electrons would exit the reactions in one direction more than in the other (nonsymmetrically). This resulted in the conclusion that parity would be conserved for electromagnetic and strong interactions but not for weak interactions. Chien-shiung Wu, who demonstrated that parity is not conserved in beta disintegrations, confirmed their theory. Physicists now believe there may be other antiparticles or energies that could account for this uneven symmetry. The concept of parity conservation for weak interaction of elementary subatomic particles is important for understanding the basic nature of matter.

*See also* Fermi; Feynman; Gell-Mann; Pauli; Wigner; Wu; Yukawa

**YANOFSKY'S THEORY FOR COLINEARITY OF DNA AND PROTEIN:** Biology: *Charles Yanofsky* (1925–), United States.

*Gene sequences and protein sequences are colinear and thus changes in DNA sequences can produce changes in protein sequences, thereby controlling alterations in RNA's structure that permits RNA to act as a regulatory molecule in both bacterial and animal cells.*

After graduating in 1948 from City College of New York, Charles Yanofsky entered Yale University's graduate PhD program in microbiology. He spent three years as a postdoctoral candidate working on gene mutations. He discovered that one gene's mutation's effects are compensated by another so-called suppressor gene's mutation that will supply the missing enzyme in the first mutated gene, thus making the first mutated gene's harmful influence ineffective. Charles Yanofsky moved to Stanford University in California in 1958 as an associate professor in microbiology where he demonstrated that the linear sequence of amino acid molecules found in proteins are determined by the arrangement of nucleotide molecules found in DNA material. This provided the evidence for the assumption for the double helix structure of DNA proposed by James Watson and Francis Crick. Yanofsky received many awards for his research and has served as a professor of biology since 1961 at Stanford University.

*See also* Crick–Watson

**YOUNG'S WAVE THEORY OF LIGHT:** Physics: *Thomas Young* (1773–1829), England.

*Light is transmitted through the aether as a wave front of beams that are both identical and singular.*

Thomas Young studied the functioning of the human eye. His theory that the lens of the eye changed shape to adjust to light and distance led him to explore the nature of light and how it traveled from one object to another. In the early nineteenth century, there were two conflicting theories of the nature of light. One claimed light was a stream (emission) of particles sent out by objects, which were received by the eye; the other stated that light consisted of minute standing waves and was transmitted by the aether. But the prevailing concept was the corpuscular theory for the emission of light, which claimed the polarization of light was possible only if light was a collection of single tiny particles originating from an object. This concept did not conform to Young's experimental evidence or to the mathematics of that time. Young proposed the transmission theory of light as a wave front of identical "beams" (not corpuscles), passing through a medium that he and others referred to as aether. Young experimented with a beam of light that he focused through two pinholes in a barrier to the path of a light beam. This produced two separate beams of light emanating from the pinholes on the other side of the barrier. These new standing wavelets exhibited two curved wave fronts whose matching crests showed up as alternate areas of light on a back screen (*see* Figure Y1).

Later, two narrow slits were used instead of pinholes. This phenomenon is known as *diffraction*, where the waves spread and bend as they pass through the small openings in

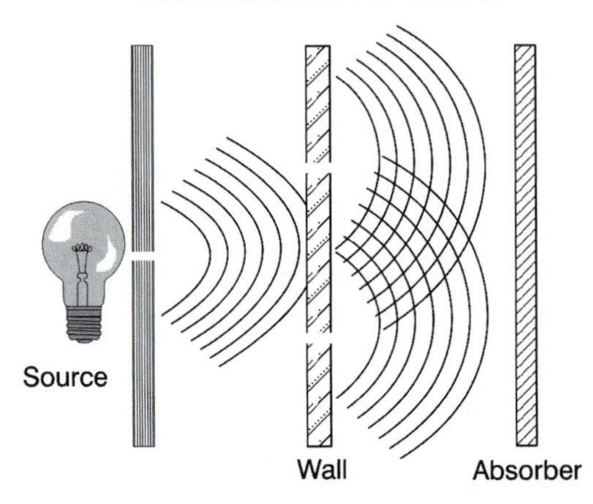


Figure Y1. Young's experiment demonstrated the wave nature of light. The light source passes through the hole in the first barrier, proceeds as a wave front to the second wall, where it passes through two holes, and emerges as two wave fronts that are recorded on the absorber wall. The light areas on the third wall will occur when the crest of the light waves are "in phase" and add to their brightness, while the dark areas are where the waves are "out of phase" and interfere with each other.

the barrier. At the point where the crests of the light wavelets were "matched," they intensified each other to form bright strips of light. Conversely, where the crests of the waves were not matched (interfere), they counteracted or blocked each other to form dark images. Diffraction (the splitting of the light beam into wavelets) and interference (the matching, or not matching, of the wavelet's crests) are basically the same phenomenon and now apply to all forms of electromagnetic radiation. Young's interference experiment was a classic demonstration proving the wave nature of light. It took some time for other physicists to understand the importance of Young's wave theory, but once accepted, it was used to exhibit why the different colors of the spectrum have different wavelengths. Using the wave front theory of light, Young and other physicists explained transverse wave propagation, the mechanical quality of the light medium, polarization, reflection and refraction, and other optical phenomena. His theory later assisted in determining the speed of light in air and water. His wave theory was augmented by the "quantum/photon" theory of light proposed by Schrödinger and Einstein, which resulted in the wave–particle duality of light.

*See also* Einstein; Fresnel; Hertz; Huygens; Maxwell; Schrödinger

**YUKAWA'S MESON THEORY FOR THE "STRONG INTERACTION":** Physics: *Hideki Yukawa* (1907–1981), Japan. Hideki Yukawa was the first Japanese citizen to be awarded the Nobel Prize for Physics in 1949.

*Nuclei containing more than one positive proton must be held together by a force stronger than that of the protons' opposing positive charges.*

Hideki Yukawa knew the "electroweak" force or the "weak interaction" that was involved in beta decay was much weaker than the force that binds nucleons of the nucleus together. Beta decay is the simplest type of radioactivity: a neutron decays into a proton, an electron, and what was later discovered to be a neutrino, which is considered massless (*see* Figure F2 under Fermi). Yukawa believed there must be a heavier particle that could fuse protons and neutrons within the nucleus of atoms. Using electromagnetic forces as an analogy, he applied quantum theory to predict that a stronger force was responsible for "binding" protons and neutrons in nuclei. The difference was that electromagnetic photons (visible light), which are considered massless, interact over infinite distances, whereas Yukawa's predicted nuclear binding "strong interaction" particle would be many times heavier than an electron and could react only over a distance less than the diameter of an atom (about $10^{-12}$). In 1935 Yukawa predicted a new particle would bind nucleons in a nucleus. A few years later Carl Anderson confirmed the discovery of this new elementary particle. It was named the meson, and later *mu-meson*, which is now called a *muon*. Muons did not interact frequently enough with the nucleons (quarks, neutrons, and protons) to "glue" them together adequately. It was later discovered that the muon was a decay product of another particle with 265 times the mass of an electron. This heavier particle discovered by the British physicist Cecil Powell (1903–1969) in 1947 was first called the pi-meson and was later named the *pion*. The decay of the pion confirmed Yukawa's prediction for the "strong interaction" (force) that binds particles in nuclei.

*See also* Anderson (Carl); Fermi; Feynman; Gell-Mann; Pauli; Wigner; Wu; Yang

# Z

**ZEEMAN'S THEORY OF THE MAGNETIC EFFECT ON LIGHT:** Physics: *Pieter Zeeman* (1865–1943), Netherlands. Pieter Zeeman shared the 1902 Nobel Prize in Physics with Hendrik Lorentz.

*The spectral lines of light emitted from atoms are split into either two or three lines when the atoms emitting the light are subjected to a magnetic field.*

It had been known for some time that the light given off from burning chemical elements, when viewed through a spectroscope, would form distinct patterns of colors and dark lines. Sodium was the commonly used element for spectroscopic viewing. Its spectral lines are referred to as the "D-lines" due to their position in the electromagnetic spectrum. Pieter Zeeman, who undertook to verify Hendrik Lorentz's theory on atomic structure, set up a spectroscope to view these D-lines, placing an **electromagnet** between the scope and the sodium light source. He noticed that when the magnetic field was oriented perpendicular to the path of the light, the spectral lines were split into three distinct lines. When the magnetic field was oriented parallel to the light path, the lines were split into two images. This phenomenon, which is the splitting of spectral lines of a light source when passing through a magnetic field, became known as the *Zeeman effect*. Zeeman calculated the ratio of the electrical charge to the mass of the vibrating sodium ions, which proved it had a negative charge.

*See also* Bohr; Lorentz; Maxwell

**ZENO'S PARADOXES:** Physics: *Zeno of Elea* (c.490–430 BCE), Greece.

*If space can be continually divided into an infinite number of units, it will take an infinite amount of time to pass through all these units of space. Therefore, motion is an illusion.*

Zeno of Elea, a pre-Socratic philosopher, devised paradoxes as arguments to contradict his philosophical opponents. The "theory" that motion cannot exist is only one of several of Zeno's paradoxes, based on the "dichotomy" that motion cannot exist because before it can reach where it is going, it must first reach a midpoint (half of its destination). He continued by stating that before this midpoint could be reached, it must reach one-fourth of its course, and before this, its one-eighth point, its one-sixteenth point, and so on. If one continues with this concept of motion, it can never proceed from where it starts. A similar but flip-side paradox is best explained by Zeno's story of the race between Achilles and the tortoise. Achilles and the tortoise start from the same point, but the tortoise is allowed to start first and reaches point A (half the distance of the race). Before Achilles can pass the tortoise, he must also reach point A, but by this time the tortoise has proceeded to point B. Now Achilles must run to point B, but the tortoise has proceeded to point C and so on. In a race so designed, Achilles will never catch the tortoise because as hard as he tries, he can cut the remaining distance only in half each time; thus the tortoise is always ahead and Achilles cannot win. This is an example of dividing the race into an infinite number of tasks, just as Zeno also stated that a line or space could be divided into an infinite number of units. This argument was used by Democritus to determine the atomic nature of matter by continually dividing a handful of soil into halves, over and over again, into an almost infinite number of times, until one tiny piece of matter so small it cannot be further divided remains—thus the atom. Zeno's paradox remained unsolved for two thousand years until it was explained by the use of calculus as the *convergence series*, an infinite series with a finite sum.

*See also* Atomism Theories

**ZIEGLER'S THEORY OF STEREOSPECIFIC POLYMERS:** Chemistry: *Karl Waldeman Ziegler* (1898–1973), Germany. Karl Ziegler shared the 1963 Nobel Prize in Chemistry with Giulio Natta.

*The stereoregularity of a polymer depends on the catalyst used to prepare it, and once prepared, the polymer's stereochemistry does not change.*

Ethylene consists of long chains of thousands of ethylene molecules which, in turn, make up of thousands of polymer units of ethylene. Therefore, the stability of this long chain was unstable and tended to break causing the formation of branches that weakened the ethylene polymer plastics that had a boiling point just above 212° F. Ziegler used a group of catalysts (usually metallic ions of titanium or aluminum) that prevented branching of the chains. By using these catalysts, much stronger plastics, including ones that could be kept in water without softening, can be manufactured. Ziegler mainly studied the polymer known as polyethylene, while Giulio Natta used a similar system to study other polyalkenes, such as isotactic, syndiotactic, or atactic forms of methyl polymers. The $TiCl_4$ and $VCl_4$ catalysts are used to convert propene to isotactic polypropylene and syndiotatic polymers. The Ziegler–Natta catalyst was a major development in understanding the chemistry and the production of various polymerization processes and their products, such as, plastic bottles for milk and beverages, and household cleaners, toys, components for appliances, moulds, to name a few.

**ZINN'S CONCEPT OF A "BREEDER REACTOR":** Physics: *Walter Henry Zinn* (1906–2000), United States.

*When irradiated by neutrons, uranium-238 can be converted into fissionable plutonium-239.*

It was known for some time that when neutrons were slowed down, they could penetrate the nuclei of uranium, thus causing the uranium nuclei to split into nuclei of lighter elements while collectively giving off great amounts of energy. Walter Zinn and his mentor, Leo Szilard, demonstrated that a small mass of the split uranium nucleus was converted into energy, as predicted by Einstein's formula, $E = mc^2$. At the beginning of World War II, Zinn worked with Enrico Fermi on the Manhattan Project to build the first nuclear pile. Zinn was the person who slowly pulled the control rods from the pile to allow more neutrons to interact with purified uranium. At the time, no one knew if it would work or blow up. This pile was successful and demonstrated that a sustainable fission reaction was possible, which led to the first "atomic" bombs. Zinn was also in charge of dismantling the reactor. In 1951 Zinn developed the first breeder reactor that used neutrons emitted from the core of an atomic reactor to change a blanket of U-238 surrounding the core into plutonium-239 (Pu-239). Pu-239, first identified in 1940 by Glenn T. Seaborg, who used a cyclotron, is fissionable with a long half-life. About 0.66 of a pound is needed to reach a critical mass and become a nuclear bomb, which is about one-third as much required of the less plentiful U-238. Pu-239 is highly radioactive but relatively easy to produce and can be used in lightweight reactors to produce heat and electricity.

*See also* Fermi; Seaborg; Szilard

**ZUCKERANDL'S THEORY FOR MEASURING THE RATE OF EVOLUTION:** Biology: *Emile Zuckerandl* (1922–), United States.

*The differences in the hemoglobin chains in mammals can be used as a "clock" to measure the time spans of the evolution of species.*

While comparing the amino acids in the hemoglobin of the blood of different animals, Emile Zuckerandl discovered that out of the 146 amino acids in one of the human hemoglobin chains, only one was different from the gorilla. However, there were more differences in the amino acids for other mammals. He used this concept to formulate a "mean" difference of twenty-two hemoglobin chains for all mammals. He then considered the time during which these animals "split" off from a common ancestor to be about eighty or ninety million years, surmising it takes approximately seven or eight million years for evolution to change one pair of amino acids. This theory was improved and became valuable for future biologists to estimate the rate of organic evolution.

*See also* Darwin; Waddington; Wilson (Allan)

**ZWICKY'S THEORY FOR SUPERNOVAS AND NEUTRON STARS:** Astronomy: *Fritz Zwicky* (1898–1974), United States.

*Supernovas, which are distinct from novas, are brilliant stellar explosions that collapse into neutron stars under their own gravitational force.*

Fritz Zwicky was the first to theorize that supernovas were different from other bright objects in the sky (*nova* means "new" in Latin). His theory stated that supernovas were stellar explosions that produced great brightness. Zwicky claimed that only about two or three supernovas are ever discovered during each thousand-year period. He determined that supernovas have a brightness that is about fourteen to fifteen times that of the sun, making them visible at the great distances where galaxies are found. He also calculated that when a supernova burned out and there was no longer radiation to maintain its size and brilliance, it would collapse. According to the law of gravity, it would attract all of its mass into a dense core and end its existence as a neutron star. A neutron star is as massive as a regular star, but it is only 7 to 9 miles in diameter. In other words, its density is so great that a teaspoonful would weigh many tons. Years later Zwicky's prediction for supernovas was borne out when the existence of superdense neutron stars was discovered. His work with supernovas resulted in theories of galaxy evolution, including galaxy clusters and super-clusters of many galaxy clusters, which indicate that distant matter in the universe may not be evenly distributed.

*See also* Baade; Hubble; Rossi

# Glossary

**absolute temperature.** In theoretical physics and chemistry, refers to the Kelvin scale, specifically for absolute zero which is −273.13° Celsius or −459.4° F. This is the temperature at which all matter possesses no thermal energy and at which all molecular motion ceases, with the exception of molecules vibrating in place without moving. It has never been reached.

**AC.** Abbreviation for *alternating current.* Electric current in a circuit that reverses its direction at repeated intervals and that was discovered by Nikola Tesla.

**acid.** A substance that releases hydrogen ions when added to water. Strong acids are sour tasting, turn litmus paper red, and react with some metals to release hydrogen gas.

**adiabatic.** Refers to a reversible thermodynamic process in which there is no transfer of heat into or out of a closed (isolated) system.

**adsorption.** Adherence or collection of atoms, ions, or molecules of a gas or liquid to the surface of another substance, called the *adsorbent*; for example, hydrogen gas collects or adsorbs to the surface of several other elements, particularly metals. An important process in the dyeing of fabric.

**aether.** Early scientists assumed a "medium" called aether, which in Greek mythology typifies the upper air, occupied all space, and thus it was believed to be required for the transmission of electromagnetic waves. Also referred to as *ether*.

**agar.** A gelatinous substance extracted from a specific species of red marine algae. Used mainly as a gelling agent in bacterial culture media. Also an ingredient in creams, ointments, and commercial laxatives.

**AIDS.** Acquired Immune Deficiency Syndrome; disease that compromises the immune system through persistent opportunistic infections and malignancies and is believed to be caused by the human immunodeficiency virus (HIV).

**albedo.** In astronomy, *bond albedo* is the fraction of the total light incident that celestial bodies (e.g., planets, asteroids, satellites) reflect back to space in all directions. In the field of optics, *normal albedo* (also called normal reflectance) is the measurement of the fraction of light or electromagnetic radiation that is reflected by any surface that is viewed vertically.

**alchemist (alchemy).** A forerunner of modern chemistry (chemists) practiced from approximately 500 BCE thru the sixteenth century. It had a twofold philosophy: the search for and use of the philosophers' stone to transmute base metals into gold and prepare and perfect medicine for people, called the *elixir vitae*.

**alkaloid.** In organic chemistry, a basic nitrogenous compound obtained from plants, soluble in alcohol and insoluble in water (e.g., morphine, nicotine, caffeine, cocaine).

**allele.** The shortened form of the term *allelomorph*. An alternative, and possibly a mutational, form of the same gene. For instance, in a diploid cell, there are two alleles (from each parent) of one gene, one of which is dominant, the other recessive. The dominant gene determines the particular characteristics displayed by the organism. Each allele has a unique nucleotide sequence.

**alpha particle.** A nucleus of a helium atom ($H^{++}$)—that is, two positive protons and two neutrons, without any electrons. Alpha particles, along with beta and gamma particles, constitute the three basic forms of radiation resulting from nuclear decay.

**altimeter.** An instrument that is used to measure the altitude of an object above a fixed level (e.g., sea level or at the top of a mountain). A *pressure altimeter* (aneroid barometer) measures air pressure from a stationary position outside an aircraft. A *radar or radio altimeter* measures the height of the aircraft above ground level during the landing process.

**amino acid.** An organic compound comprising both an amino group ($NH_2$) and a carboxylic acid group (COOH). They are polymerized to form proteins and peptides. Amino acids occur naturally and also have been synthesized in laboratories. It is believed that products of a naturally occurring synthesis of amino acids may be the buildings blocks of life.

**amniocentesis.** The surgical removal of a sample of amniotic fluid from a pregnant woman. The chemical analysis of the fluid can determine the sex of the fetus, as well as genetic disorders such as Down's syndrome, developmental disorders such as spina bifida, and a number of biochemical and/or chromosomal abnormalities.

**amplifier.** A device capable of increasing the level of power or the magnitude, for example, an electric current that uses a transistor or an electron tube with an electric signal that varies with time and does not distort the shape of the electrical waves.

**anode.** The positively charged electrode in an electrolytic cell, electron tube, or storage battery; the collector of electrons.

**antibody.** A blood serum protein, sometimes occurring normally or generated in response to an invading antigen, that specifically reacts with a complimentary antigen to produce immunity from a number of microorganisms and their toxins.

**antimatter.** *See* antiparticle.

**antineutrino.** The antiparticle to the neutrino. *See also* antiparticle.

**antiparticle.** A subatomic particle—a positron (positive electron), antiproton, or antineutron—with the identical mass of the ordinary particle to which it corresponds but opposite in electrical charge or in magnetic moment. Antiparticles make up antimatter, the mirror image of the particles of matter that make up ordinary matter as we know it on Earth. This is a theoretical concept devised to relate relativistic mechanics to the quantum theory.

**antiproton.** The antiparticle to the proton. *See also* antiparticle.

**aperture.** An opening (i.e., hole or slit) through which light waves, radio waves, electrons, or radiation can pass. This may be the adjustable opening on optical instruments (e.g., cameras and telescopes).

**aplanatic lens.** A lens whose surfaces are not segments of a sphere. It is used to correct imperfect focusing called *spherical aberration*.

**asteroid.** Derived from the Greek word *asteroeidēs*, which means "starlike." They are small bodies that revolve around the sun. They are sometimes called small "planetoids," and when they become fragmented and land on Earth they are considered meteorites. Most asteroids are found in a planetary orbit called the *asteroid belt*, located between the orbits of Mars and Jupiter.

**atmosphere.** Also known as *ATM* for standard atmosphere. A unit of pressure that is equal to 101.325 pascals which is the air pressure that is measured at mean sea level. The actual atmospheric pressure fluctuates around this number, but the unit is used to express pressures that are in excess of standard atmospheric pressure, such as those employed during high-pressure chemical transactions. (The *pascal* is named after the French mathematician Blaise Pascal and refers to a unit of pressure equal to one newton per square meter.)

**atomic number (proton number).** The number of positively charged protons found in the nucleus of an atom, upon which its structure and properties depend. This number determines the location of an element in the Periodic Table. For a neutral atom the number of electrons equals the number of protons.

**atomic weight (atomic mass).** The total number of protons plus neutrons in an atom.

**AU (astronomical unit).** The average distance of Earth from the center of the sun—approximately 93 million miles, or 150 million kilometers.

**autocatalytic.** The theoretical process where primordial organic molecules may have replicated themselves in early prebiotic environments.

**autonomic.** Independent, spontaneous, or involuntary. Relates to the autonomic nervous system and autonomic reflex system in vertebrates and other animals, as well as the autonomic movement in plants.

**autopoiesis.** The self maintenance of an organism.

**axiom.** An assumption upon which a mathematical theory is based.

**baryons.** Also known as heavy particles. They are the family of heavy subatomic particles that are made up of three quarks and that include protons and neutrons. Baryons are strongly interacting fermions that experience the strong nuclear force.

**base.** An alkali substance that reacts with (neutralizes) an acid to form a salt, for example, $4HCl + 2Na_2O \rightarrow 4NaCl + 2H_2O$ [hydrochloric acid + sodium hydroxide yields sodium chloride (table salt) + water].

**biosynthesis.** The natural synthesis (fusion) of an organic chemical compound by living organisms.

**birefringence.** Also known as double refraction. The splitting of a beam of ordinary light into two beams of light that travel at different velocities by a medium, such as calcite or quartz. Also defined as the difference in the indices of refraction of a crystal.

**black holes.** Theoretically, they are thought to be vortex areas in space where massive stars have collapsed, creating such great gravity that not even light can escape into space.

**bolides.** Two or more parts of a large meteor formed when the meteor, usually called a *fireball*, produces bright streaks of light and splits. Called bolides after the Greek word *bolis*, which means "missile." A loud hissing noise can sometimes be heard as one of these large meteors passes through the atmosphere.

**bonding (chemical).** Electrostatic force that holds together the elements that form molecules of compounds. This attractive force between atoms is strong enough to hold the compound together until a chemical reaction causes the substance to either form new bonds or break the bonds that form the molecule. *See also* covalent bond; ionic bonding.

**bosons.** One of two main classifications of subatomic particles, they are weak "force" particles, (photons, pi mesons, gluons, positive W and negative W particles, neutral Z particles, gravitons). The other main classification of subatomic particles is the fermion.

**calorimeter.** An instrument that measures thermal activity (heat) generated in or emitted by chemical reactions that result in a change of state of the involved chemical(s).

**calx.** Calcium oxide (CaO). The crumbly, white, water-soluble solid residue that is left after the calcination of calcium carbonate limestone that results in the removal of all the carbon dioxide from the mineral. Also called caustic or burnt lime; chalk. It is used in the pulp and paper industries and as a flux in the manufacture of steel.

**capacitor.** A storage device (condenser) for static electricity, consisting of two or more metal surfaces (conductors) separated from each other by a dielectric. It stores the electrical energy and impedes the flow of direct current. *See also* Leyden jar.

**catalyst.** Any substance that affects the rate of a chemical reaction without itself being consumed or undergoing a chemical change. Platinum/palladium pellets in automobile catalytic converters are chemical catalysts. A biological catalyst (e.g., an enzyme) affects chemical reactions in living organisms.

**catastrophism.** In biology and geology, the idea that catastrophic events that have occurred in the past (e.g., earthquakes and volcanoes, meteor impacts, and major climate changes) radically altered Earth's surface and/or biological processes.

**cathode.** A negatively charged electrode or plate, as in an electrolytic cell, storage battery, or an electron tube similar to a TV. Also, the primary source of electrons in a cathode ray tube such as the Crookes tube.

**Cepheids.** A population of giant yellow stars that pulses regularly by expanding and contracting due to the changes in their surface temperatures, resulting in an oscillation of their luminosity that ranges from $10^3$ to $10^4$ greater than the sun. The importance of a Cepheid variable is its function as a "standard candle" or candela (indicator) to determine its distance from Earth, thus an essential component in celestial mapping. Also called *Cepheid variables*.

**chemical reduction.** A chemical reaction in which the oxidation number (oxidation state) of atoms is changed. Also called reduction/oxidation reaction or redox for short.

**chiral bag model.** In the field of quantum chromodynamics (QCD), it refers to one of several models of the nucleon, which is a general term for either the neutron or proton as a constituent part of the nucleus. The nucleon is made up of three quarks, but the actual equations of motion for QCD are unknown. The chiral bag model, a composite of two other models, is a theoretical attempt to address the asymmetry of the nucleon.

**chloroplasts.** Chlorophyll-containing organelles (cell plastids) found in abundance in plant cells that undergo photosynthesis.

**chromatography.** Any of a group of techniques used to separate complex mixtures (i.e., vapors, liquids, or solutions) by a process of selective adsorption (not to be confused with absorption), the result being that the distinct layers of the mixture can be identified. The most popular techniques are liquid, gas, column, and paper chromatography.

**chromosomes.** The complex, DNA-containing, threadlike material inside the nuclei of the cells of living organisms that determines hereditary characteristics of that organism.

**chromosphere.** The transparent, gaseous (mainly hydrogen) layer of the sun's atmosphere that rests on and completely surrounds the photosphere. It is approximately several thousand miles thick and acts as a thermal buffer zone between the photosphere and the coronal layer. Temperatures range from 6,000°C where it merges with the photosphere to 20,000°C at the region below the corona.

**cloud chamber.** A device for detecting the paths of high-speed particles as they move through a chamber filled with air or gas which is saturated with water vapor. The device is fitted with a piston that, when moved outwardly, affects the expansion of the gas and the cooling of the vapor. A fog or cloud of minute droplets then forms on any nuclei or ions present in the chamber. Also known as the *Wilson cloud chamber*.

**codons.** A sequence of three adjacent nucleotides within a molecule of $_{m}$RNA (messenger RNA) that carries the genetic code (*triplet code*) of one amino acid during protein synthesis. It is the basic unit of the genetic code.

**coke.** The residue produced after bituminous coal or other carbonaceous materials, such as petroleum or pitch, are heated to extremely high temperatures in the absence of air. Primarily consisting of carbon, it is used as a fuel in blast furnaces.

**comet.** A nebulous celestial formation consisting of rocks, ice, and gases. Comets are composed of three main parts: the *nucleus*, which is the center made of rock and ice; the *coma*, which is composed of the gases and dust that form around the nucleus as it evaporates; and the *tail*, which is made up of the gases and spreads out from the coma.

**compound.** A substance in which two or more elements are joined by a chemical bond to form a substance different from the combining elements. The combining atoms do not vary their ratio in their new compound and can only be separated by a chemical reaction, not a physical force. *See also* bonding.

**conductor.** Substances that allow heat or electricity to flow through them.

**cosmogony.** The astrophysical study of the origin and evolution of the universe.

**cosmology. (cosmos, cosmological).** The study of universe on the smallest and largest of scales in terms of time, space, and the makeup of the universe. It includes theories about the origin of the universe and everything in it, the evolution of the universe from past to present to future, and the structure of the universe and its celestial bodies at various stages of their evolution.

**covalent bond.** Sharing of electrons by two or more atoms to form a pair of electrons. This type of bonding always produces a molecule. Also known as *electron pair bond*. *See also* bonding.

**critical mass.** The minimum mass of fissionable material (U-235 or PU-239) that will initiate an uncontrolled fission chain reaction, as in a nuclear (atomic) bomb.

**critical temperature.** The temperature above which a substance cannot be converted from the liquid to the gaseous state or vice versa, regardless of the pressure applied. Also, the temperature at which a magnetic material will lose its magnetism.

**cryogenics.** Study of the behavior of matter at very low temperatures below −200°C. The use of the liquefied gases (oxygen, nitrogen, hydrogen) at approximately −260°C is standard industrial practice.

**cyclotron.** A particle accelerator made up of two hollow cylinders (similar to two opposing D structures) that are connected to a high frequency alternating voltage source in a constant magnetic field. The charged particles, which are injected near the midpoint of the gap between these two hollow cylinders, are then accelerated in a spiral path of increasing expanse so that the path traveled by these accelerated particles increases with their speed where a deflecting magnetic field deflects them to a target. *See also* particle accelerator.

**daltons.** Named after English chemist John Dalton, it is an arbitrarily defined unit that is used to express the masses of atoms, molecules, and nuclear particles. The standard (dalton) is the unit of mass equal to one-twelfth the mass of $^{12}C$. Also known as atomic mass unit (*AMU*).

**dark matter.** Nonluminous matter that is assumed to be present in the Milky Way and other galaxies that explains the motions of the stars and clouds of gases in those galaxies. Cosmological theory states that such dark matter makes up over 90% of all matter in the universe and must exist to achieve the critical density necessary to close the universe.

**DC.** Abbreviation for *direct current*. Electric current that flows in only one direction.

**declination.** In astronomy, it is the angular distance north (positive) or south (negative) of the celestial equator, that is, the circle formed on the celestial sphere in the same plane as Earth's equator. In navigation, it is the arc between the equator and the point measured on a great circle perpendicular to the equator.

**dendrochronology.** The study of tree rings as a dating method for events and conditions over a limited period time. It is based on the number, width, and density of annual rings of older trees that have been cut into cross sections. Using trees such as Douglas fir and white pine enables scientists to establish a master tree index that can date, rather accurately, both events and climatic conditions over the past several thousand years.

**deterministic/determinism.** The doctrine that espouses that all phenomena are causally determined by prior events. It has also been stated as the relationship between a cause and its effect, particularly natural phenomena, or as the hypothesis stating a set of precisely determined conditions will always repeat the same effect, or that an event cannot precede its cause. Also known as *causality*.

**diastolic.** Refers to the rhythmic relaxation and dilation of the heart's chambers, particularly the ventricles. The diastolic reading on a blood pressure monitor that records the lowest arterial blood pressure during the time the ventricles fill with blood.

**diphtheria.** An acute infectious disease of humans that is caused by the growth of the *corynebacterium diphtheriae* bacillus on a mucous membrane, especially in the throat and nose and characterized by respiratory difficulty and high fever. Prior to the discovery of a vaccine in 1923, and the development of various antibiotics that are successful in treating the disease, diphtheria had a high mortality rate, especially among children.

**dipole.** A pair of magnetic poles or electric charges of equal magnitude but with opposite polarity that are separated by a small distance.

**DNA.** Abbreviation for *deoxyribonucleic acid*. The complex ladder-like, double-stranded nucleic acid molecule present in chromosomes that forms a double helix of repetitive building blocks and shapes the inherited genetic characteristics of all living organisms, with the exception of a small number of viruses.

**Doppler effect.** The apparent change or shift in the observed frequency of a sound or electromagnetic wave due to the relative movement between the source and the observer. The same principle applies when determining the distance of stars in the galaxy. The Doppler frequency or shift is based on the color shift (frequency of light) related to the star's velocity. The light frequency for a star receding from Earth is redder (longer wave lengths) than a star approaching Earth, which emits a blue light (shorter wave length).

**ecology.** The scientific study of the interrelationships of organisms to each other and their physical, chemical, and biological environments.

**electrolysis.** A process in which an electric current is passed through a liquid, known as an electrolyte, producing chemical changes at each electrode. The electrolyte decomposes, thus enabling elements to be extracted from their compounds. Examples are the production of chlorine gas by the electrolysis of sodium chloride and the electrolysis of water to produce oxygen and hydrogen.

**electrolyte.** A compound that, when molten or in solution, will conduct an electric current. The electric current decomposes the electrolyte.

**electromagnet.** A strong magnet composed of a wire coil that is wrapped around a soft-iron core through which a current of electricity is passed and which becomes demagnetized when the flow of electric current is suspended.

**electron.** An extremely small, negatively charged particle that moves around the nucleus of an atom. The interaction of the electrons of atoms is the chemistry of Earth's elements.

**electrophoresis.** A method for separating and analyzing colloidal particles in a stable liquid which is under the influence of an electric field. The movement of the colloids is the result of Coulomb's law: *two bodies charged with the same sort of electricity will repel each other in the inverse ratio of the square of the distance between the centers of the two bodies*.

**electroscope.** An apparatus that detects the presence and signs of minute electrical charges using a process of electrostatic attraction and repulsion.

**empirical.** Relates to actual observation, practical experience, and experimentation rather than scientific theory.

**endosymbiosis.** Refers to the process responsible for the origination of a new organism, namely the fusion of two independently evolved organisms. One, called the *host*, and the other, called the *endosymbiont*, become a tightly joined system that eventually evolves into just one organism. In many cases, but not all, endosymbiosis is obligate, that is, neither organism can survive without the other. Often there is no benefit to the host organism or the endosymbiont is harmful to the host or the host to the endosymbiont. It is a controversial theory for the formation of life, specifically that some inorganic chemicals combined to form organic molecules.

**energy.** The capacity to do work. Heat and work are forms of energy and are interchangeable. Some examples of energy are heat, light, sound, radioactive, and mechanical.

**entropy.** Disorganization, randomness. In thermodynamics, it is the function of the system where the amount of heat transfer introduced in a reversible process is equal to the heat that is absorbed by the system from its surroundings, divided by the absolute temperature of the thermodynamic system.

**enzyme.** Any of a number of proteins or conjugated proteins which are produced by living organisms that act as biochemical catalysts in those organisms.

**equilibrium.** A state or condition in which the influences of energy forces and related reactions are canceled by each other, the result of which is a balanced stable, and unchanging system. *Thermal equilibrium* is said to occur when no heat exchange has taken place within a body or between the body and its surroundings.

**equinox.** One of two points or moments on the celestial sphere when the center of the sun intersects the celestial equator, either in a north or southbound direction.

**ether.** *See* aether.

**eugenics.** The genetic principles of heredity to improve a species, most often associated with breeding or engineering of a "superior" race of humans while discouraging the breeding of those considered "inferior." Animal and plant breeding as well as genetic counseling might be considered less extreme applications of eugenics.

**eukaryotic.** Describes the state of a cell (eukaryote that makes up all living things except bacteria and cyanobacteria) containing a definitive nucleus, in which nuclear material is surrounded by a membrane and cytoplasm-containing organelles. Along with prokaryotes, they are the two major groups into which organisms are divided.

**fermion.** A subatomic particle (electron, proton, or neutron) having odd half-life integral angular momentum, which obeys the Pauli exclusion principle: *no more than one in a set of identical particles may occupy a particular quantum state.*

**fission.** The splitting of an atom's nucleus with the resultant release of enormous amounts of energy and the production of smaller atoms of different elements. Fission occurs spontaneously in the nuclei of unstable radioactive elements, such as U-235 and Pu-239, and is used in the generation of nuclear power, as well as in nuclear bombs.

**fluorescent.** Consisting of a gas-filled tube with an electrode at each end. Passing an electric current through the gas produces ultraviolet radiation which is converted into visible light by a phosphor coating on the inside of the tube. This emission of light by the phosphor coating is called fluorescence.

**forensics.** Relates to public discussion or debate, particularly in legal proceedings, concerning engineering practices, medical evidence, chemical studies, and so forth where the findings are presented as legal evidence in a court of law.

**fractal.** An irregular or fragmented geometrical shape whose intricate structure is such that, when magnified, the original structure is reproduced (self-similarity). Fractals are important in the study of certain branches of physics, as well as in chaos theory and computer-generated graphics.

**fusion.** An endothermic nuclear reaction yielding large amounts of energy in which the nuclei of light atoms (e.g., forms of heavy hydrogen, such as deuterium or tritium) join or fuse to form helium (e.g., energy of our sun or the hydrogen bomb). The opposite of fission.

**galaxy.** A huge grouping of millions, or even billions, of stars held in one of several shapes by their mutual gravity. There are elliptical, irregular, and spiral galaxies.

**galvanometer.** An instrument that measures a small electrical current using mechanical motion derived from the electrodynamic or electromagnetic forces produced by the current.

**gene.** The basic unit of hereditary material that is composed of a sequence of nucleotides of a section of DNA or RNA molecules. The sequence of nucleotides determines the structure of amino acids in proteins, which is fundamental to all other biological processes. Genes, individually or in groups, determine inherited characteristics.

**genetic drift.** In population genetics, it is the statistical effect that results from the random fluctuations of gene frequencies from generation to generation, primarily in small populations. In other words, chance alone can have a profound effect, thus the concept of genetic drift.

**genetics.** The science of biological heredity and the mechanisms by which characteristics are passed along to succeeding generations.

**genomes.** The complete hereditary information encoded in the DNA (or for some viruses, in the RNA) of an organism of species. In other words, all the genes are contained in a single set of haploid chromosomes. During reproduction, each parent contributes its genome to its offspring.

**geomagnetism.** Refers to Earth's magnetism and, in a broader sense, the magnetic phenomena of interplanetary space.

**glaciation.** The alteration of the surface of Earth by passage of glaciers, mainly by erosion or deposition.

**global warming.** The increase in global temperatures reportedly augmented by the emission of industrial gases, along with other natural air pollutants, that traps heat from the sun. A natural cloud cover acts as an "insulating blanket" which keeps the heat of Earth and the lower atmosphere from radiating into the outer atmosphere and on into space. Scientists on both sides of the issue continue to debate whether there is increasing evidence that the addition of pollutant clouds into the atmosphere has increased Earth's temperature with the potential to cause climatic, often catastrophic, changes in the environment. Also referred to as the *greenhouse effect*.

**gluon.** A hypothetical, massless, neutral elementary particle that carries the strong force (interactions) that binds quarks, neutrons, and protons together. Gluons can also interact among themselves to form particles that consist only of gluons without quarks and are called *glueballs*.

**graviton.** A hypothetical (not yet discovered) carrier particle presumed to be the quantum of gravitational interaction, having a mass and charge of zero and a spin of 2.

**ground state.** The lowest stable energy state of a system of interacting elementary particles.

**gutta-percha.** It is a natural polymer derived from the milky, thermoplastic substance that is obtained from gutta-percha trees (genera *Palaquium* and *Payena*) found in Malaysia. It is an excellent electrical insulator, particularly in submarine cables. Also used in golf balls and waterproofing products.

**hadron.** An elementary particle, part of the largest family of elementary particles, that has strong interactions, usually producing additional hadrons during high energy collisions.

**half-life.** The time required for one-half of the atoms of heavy radioactive elements to decay or disintegrate by fission into lighter elements.

**halogens.** Electronegative monovalent nonmetallic elements of Group 17 (VIIA) of the Periodic Table (fluorine, chlorine, iodine, bromine, astatine). In pure form, they exist as diatomic molecules (e.g., $Cl_2$).

**heliocentric.** Refers to belief that the sun is the center of the solar system or universe.

**homeostasis.** The physiological state of equilibrium within an organism. In other words, its chemical composition, as well as other internal functions, are in balance (e.g., body temperature, acid-base balance).

**hominid.** Member of the mammal family of which homo sapiens is the only surviving species.

**hominoid.** Manlike; an animal that resembles a human. (Humans and anthropoid apes are usually included in the superfamily commonly referred to as hominoids.)

**homologous.** In evolutionary theory, refers to the structural relationship between the physical parts of different species or organisms due to evolutionary development (e.g., the wing of a bird and the pectoral fin of fish; the flipper on a sea lion and the arm on a primate).

**humoral theory.** Pertains to the practice of medicine, primarily in the Middle Ages, whereby the body was governed by four principle humors or fluids (blood, phlegm, choler, and black bile). These were present in varying proportions in each person, the balance of which was essential for continued good health. If any of these four "humors" were out of balance, a procedure (e.g. blood letting) was performed by the physician in an effort to restore "balance."

**hydrostatic.** The study of liquids at rest (e.g., liquids contained in dams, storage containers, and hydraulic machinery).

**hypotenuse.** In a right triangle, the side opposite the right 90° angle.

**impedance.** In electronics, it is a term that describes a portion of the overall opposition of a circuit to a sine wave of alternating current, that is, how much the circuit impedes the flow of current. The term, which is measured in ohms Ω, is often used interchangeably with the term "resistance" when referring to simple circuits that have no capacitance or inductance. However, impedance is more complex and includes the effects of capacitance and inductance. Impedance varies with frequency, whereas the effect of resistance is constant regardless of frequency.

**incidence.** In optics, it refers to the incidence angle (also angle of incidence) that is formed between a beam or a ray on a surface and the perpendicular line at the point of incidence (arrival).

**inclination.** The angle between a reference plane and the axis of direction, that is, the deviation from the vertical to the horizontal. In astronomy, it is also called *magnetic inclination* or *magnetic dip*. The dip angle of Earth's magnetic field.

**inductor.** A passive electrical device, such as a coil of copper wire wrapped around a ferromagnetic material, that introduces electromagnetic force (inductance) into an electrical circuit.

**in vitro.** Meaning "in glass" in Latin. Refers to an observable biological reaction that occurs under artificial conditions outside of a living organism, usually in a test tube or a petri dish.

**ion.** An atom or a group of atoms that have gained or lost electron(s) and thus have acquired an electrical charge. The loss of electrons gives positively charged ions. The gain of electrons results in negatively charged ions. If the ion has a net positive charge in a solution, it is a *cation*. If it has a net negative charge in solution, it is an *anion*. An ion often has different chemical properties than the atoms from which it originated.

**ionic bonding.** Donating of electrons from one element to another element, forming positively and negatively charged ions respectively. The electrostatic attraction between the oppositely charged ions constitutes the bond. Also known as *electrovalent bond*.

**ionization.** The chemical process for producing ions in which a neutral atom or molecule either gains or loses electrons, giving it a net charge, thus becoming an ion.

**irrational numbers.** Any real number that is not the quotient of two integers. They are usually algebraic (roots of algebraic equations) or transcendental numbers.

**isomer.** In chemistry, chemical compounds with the same molecular composition but with different chemical structures. For example, butane has two isomers, $C_4H_{10}$ *and* $C_2H_4(CH_3)_2$. In nuclear physics, isomers refer to the existence of atomic nuclei with the same atomic number and the same mass number but different energy states.

**isomerism.** Refers to the condition whereby certain chemical compounds have the same molecular formulae but different molecular structures.

**isostasy.** The theoretical gravitational equilibrium existing in the earth's crust. If there is a disturbance on the surface of Earth (e.g., erosion or glacier movement, which is also referred to as deposition), there are counterbalancing movements in Earth's crust. The areas of deposition will sink, whereas the areas of erosion will rise. The same counterbalancing effect also occurs in Earth's oceans as the lack of density in ocean water is compensated by an excess density in the material under the ocean's floor.

**isotopes.** Atoms of the same element with different numbers of neutrons in their nuclei. All atoms of an element always contain the same number of protons in their nuclei. Thus, their proton (atomic) number remains the same. However, an atom's nucleon number, which denotes the total number of protons and neutrons, can be different. These atoms of the same element with different atomic weights (mass) are called *isotopes*. Isotopes of a given element all have the same chemical characteristics (electrons and protons), but they may have slightly different physical properties.

**kaon.** An elementary particle that is a subclass of the hadrons. Mesons consist of quark—antiquark pairs. They have zero spin, a nonzero strangeness (quantum) number, and a mass of approximately 495 MeV. It is the lightest hadron to contain a strange quark. Also known as a *K meson* in particle physics.

**kinetic energy.** Energy association with motion.

**latitude.** Angular distance of a point on Earth's surface measured along a meridian from the equator (zero latitude) north or south to the poles, which are at 90°N or 90°S.

**lepton.** In particle physics, any light particle. Leptons have a mass smaller than the proton mass and do not experience the strong nuclear force. They interact with electromagnetic and gravitational fields and essentially interact only through weak interactions.

**Leyden jar.** An early and improved form of *capacitor* (condenser). Metal foil was placed on both the inside and outside of the glass jar, allowing the glass to act as a dielectric or nonconducting substance to separate the electrical charges. A charge of stored static electricity occurred as the wire touched the inside foil, which was fed through the insulating cork on the top of the jar. A circuit was completed when the wire conducted the electricity to the foil on the outside of jar, or a spark jumped to a finger if it was brought near the wire exiting the jar. *See also* capacitor.

**libration.** The very slow oscillatory rotation, either real or apparent, of a satellite that does not possess enough energy to make a full rotation as seen from a larger celestial body around which it revolves. An example: the libration's of Earth's moon that enables 59% of its surface to be observed on Earth despite its synchronous rotation.

**logarithm.** A mathematical method developed in the sixteenth century that simplified the multiplication and division processes for large sums by using exponents of the number 10, which are called *logarithms* (shortened to *logs*). Multiplication is reduced to addition; division reduced to subtraction. For example, the log of 100 written as ($10^2$) is 2; the log of 1000 written as ($10^3$) is 3. Thus, multiplying 100 × 1000 can be facilitated by adding their logs which since the sixteenth century have been recorded in a series of tables of logarithms. The answer is 100,000 or log ($10^5$). In advanced mathematics, logarithmic tables have been formulated to deal with computations involving far more complex infinite numbers.

**macromolecules.** Very large molecules composed of many relatively simple structural units, each of which consists of several atoms that have bonded together. Examples are polymers (natural and synthetic) and proteins.

**magnet.** A body or an object that has the ability to attract certain substances (e.g., iron). This is due to a force field causing the movement of electrons and the alignment of the magnet's atoms.

**magnetic moment.** In physics, the ratio between maximum torque which is exerted on a magnetized body, electric current-carrying coil, or magnetic domain, including nuclei, and the strength of that magnetic domain or field. Also called *magnetic dipole moment.*

**magnetohydrodynamics.** In physics, the study of motion or dynamics of electrically conducting fluids (plasmas, ionized gases, liquid metals) and their interactions with magnetic fields. Also known as *hydromagnetics* or *magnetofluid dynamics.*

**magnetosphere.** The comet-shaped regions surrounding Earth and the other planets where the charged particles are controlled by the planet's own magnetic field rather than the sun's. Earth's geomagnetic field is believed to begin at an altitude of about 100 kilometers and extends to the far-away borders of interplanetary space.

**manometer.** A double-leg (U-tube) instrument designed to measure the difference between two fluid pressures near to normal atmospheric pressure (14.7 psi). The *barometer* and the *sphygmomanometer* that measures arterial blood pressure are common forms of the manometer.

**maser.** (microwave amplification by stimulated emission of radiation). A device that converts incident electromagnetic radiation from a wide range of frequencies to one or more discrete frequencies of highly amplified microwave radiation.

**mass.** The quantity (amount) of matter contained in a substance. Mass is constant regardless of its location in the universe. Mass should not be confused with weight.

**mean.** Determined by adding all the values or a set of numbers and dividing the sum by the total numbers or values. Usually associated with the term "average."

**meiosis.** A type of division of the nuclei of cells during which the number of chromosomes is reduced by half.

**membranes.** In the field of astrophysics, membranes are multidimensional objects that are components of M-theory which is a proposed "master theory" that unites five superstring theories within a single dominant framework to explain the universal forces. M-theory purports to involve eleven space-time dimensions, of which membranes, or branes or p-branes, are a theoretical ingredient that helps explain the concept of strings as the model for the universe.

**memes.** Units of cultural information that flourish from one individual to another in much the same manner as genes propagate from one organism to another during biological evolution, that is by the process of natural selection. Examples of memes are belief systems, clothing fashions, pottery styles, music, slogans. The concept of memes has spawned its own abstract scientific theory called "memetics."

**meson.** An elementary particle with strong nuclear interactions, having a baryon number zero. Mesons are unstable and decay to the lowest accessible mass states.

**metabolism.** A chemical transformation that occurs in organisms when nutrients are ingested, utilized, and finally eliminated (e.g., digestion, absorption, followed by a complicated series of degradations, syntheses, hydrolysis, and oxidations utilizing enzymes, bile acids, and hydrochloric acid). Energy is an important by-product of the metabolizing of food.

**meteorite.** A small portion of a larger meteor, meteoroid, or a disintegrated chunk of an asteroid that has not completely vaporized as it entered and passed through Earth's atmosphere and that eventually lands on Earth's surface.

**mitochondria.** In cell biology, mitochondria (singular: mitochondrion) are the membrane-enclosed organelles that are found in most eukaryotic cells. They contain enzymes responsible for the conversion of food into usable energy in the cytoplasm of cells. They have their own DNA (mitochondrial DNA or mtDNA), as well as their own independent genomes.

**mole.** The SI base unit that measures the amount of substance of a system with a weight in grams numerically equal to the molecular weight of the substance. It is equal to the amount of the substance that contains as many elementary units as there are atoms in 0.012kg of carbon-12. This is known as Avogadro's constant. The mole's use is usually limited to the measurement of subatomic, atomic, and nuclear particles. Symbol: mol.

**molecule.** The smallest particle of a substance containing more than one atom (e.g., $O_2$) or a compound that can exist independently. It is usually made up of a group of atoms joined by covalent bonds.

**morphology.** In biology, the study of the form and structure of living organisms, primarily their external structure.

**Mössbauer effect.** A physical phenomenon involving the resonant and recoil-free emission and absorption of gamma rays by atoms bound in solid form.

**muon.** The semistable second generation lepton, with a mass 207 times that of an electron. It has a spin of one-half and a mass of approximately 105 MeV, and a mean lifetime of approximately $2.2 \times 10^{-6}$ second. Also known as *mu-meson*.

**nanotechnology.** A field of applied science and technology dealing with the control of matter on the atomic and molecular scale (1 to 100 nanometers), and the fabrication of devices and products within that size. Examples: computer chips; polymers based on molecular structure.

**nebula.** An immense and diffuse cloudlike mass of gas and interstellar dust particles, visible due to the illumination of nearby stars. Examples are the Horsehead Nebula in Orion and the Trifid Nebula in Sagitarius.

**neutrino.** An electrically neutral, stable fundamental particle in the lepton family of subatomic particles. It has a spin of one-half and a small or possibly a zero at-rest mass, with a weak interaction with matter. Neutrinos are believed to account for the continuous energy distribution of beta particles and are believed to protect the angular momentum of the beta decay process.

**neutron.** A fundamental particle of matter with a mass of 1.009 (of a proton) and having no electrical charge. It is a part of the nucleus of all elements except hydrogen.

**nucleon.** A general term for either the neutron or proton, in particular as a constituent of the nucleus.

**nucleosynthesis.** The process of creating new atomic nuclei from pre-existing nucleons (protons and neutrons), or the synthesis of chemical elements by nuclear processes. *Primordial nucleosynthesis*, also call nucleogenesis, occurred within a few minutes after the "big bang" when the universe was extremely hot and was responsible for the abundance of lighter elements, such as helium, in our cosmos. *Stellar nucleosynthesis*, the principal form of nucleosynthesis today, takes place in stars by either nuclear fusion or nuclear fission.

**nucleotide.** The structural unit of nucleic acid found in RNA and DNA.

**nucleus.** The core of an atom which provides almost all of the atom's mass. It contains protons, neutrons, and quarks held together by gluons (except hydrogen's nucleus which is a single proton) and has a positive charge equal to the number of protons. This charge is balanced by the negative charges of the orbital electrons.

**organelle.** A distinct subcellular structure, with a specific function and defined shape and size, found in the cytoplasm of the cell (e.g., mitochondria).

**oxide.** A compound formed when oxygen combines with one other element—a metal or nonmetal (e.g., magnesium oxide).

**ozone layer.** The layer is found in the upper atmosphere, between 10 and 30 miles in altitude. This thin layer of gases contains a high concentration of ozone gas ($O_3$) which partially absorbs solar ultraviolet (UV) radiation and prevents it from reaching Earth. It is mostly formed over the equator and drifts toward the North and South Poles. It seems to have a cyclic nature. Also called the *ozonosphere*.

**pangenes.** In evolutionary theory, they are hypothetical protoplasmic particles inside the nuclei of the cells of living organisms that control heredity. Originally coined by Darwin, it is a term no longer considered to be accurate, nor is it used by credible biologists.

**parallax.** The apparent change in direction and/or position of an object viewed through an optical instrument (e.g., telescope), which occurs by the shifting position of the observer's line of sight.

**particle.** A very small piece of a substance that maintains the characteristics of that substance. Also known as fundamental particles found in atoms.

**particle accelerator.** A machine designed to speed up the movement of electrically charged subatomic particles that are directed at a target. These subatomic particles, also called *elementary particles*, cannot be further divided. They are used in high-energy physics to study the basic nature of matter, as well as the origin of life, nature, and the universe. Particle accelerators are also used to synthesize elements by "smashing" subatomic particles into nuclei to create new, heavy, unstable elements, such as the superactinides. *See also* cyclotron.

**parton.** In particle physics, a termed originated by Richard Feynman in the late 1960s. It is a theoretical point-like fundamental particle that is a constituent of the proton, neutron, and other baryons. Today these particles are called quarks and gluons. The parton model aids in the interpretation of very high-energy experiments on nucleons as well as short-distance interactions.

**Periodic Table of the Chemical Elements.** An arrangement of the chemical elements in sequence in the order of increasing atomic numbers. It is arranged in horizontal rows for periods and in vertical columns for groups and illustrates the similarities in properties of the chemical elements.

**phage.** A parasitic virus in a bacterium that has been isolated from a prokaryote. Also called *bacteriophage*.

**phlogiston.** The hypothetical substance believed to be the volatile component of combustible material. It was used to explain the principle of fire before oxidation and reduction were known and prior to the discovery of the principle of combustion.

**photon.** The quantum unit of electromagnetic radiation or light that can be thought of as a particle. Photons are emitted when electrons are excited and move from one energy level (orbit) to another.

**photosynthesis.** Process by which chlorophyll-containing cells in plants and bacteria convert carbon dioxide and water into carbohydrates, resulting in the simultaneous release of energy and oxygen.

**phyla.** (plural of **phylum**). A primary taxonomic ranking of organisms into groups of related classes. Phyla are grouped into kingdoms, except in most plants where kingdom is replaced by division.

**pi.** The transcendental number 3.141592 for the ratio of the circumference of any circle to its diameter, using the symbol $\pi$.

**pion.** A short-lived elementary particle classified as a meson which is primarily responsible for the strong nuclear force. It exists in three forms: neutral, positively charged, and negatively charged. The charged pions decay into muons and neutrinos, and the neutral pion decays into two gamma ray photons. Also called *pi-meson*.

**polygon.** A simple closed curve in the plane that is bounded by three or more line segments.

**positron.** The positively charged antiparticle of an electron: $e^+$ or $p^+$.

**prebiotic.** Refers to the period on Earth before the existence of organic life.

**precession.** Refers to the wobbling or circling of Earth's orbit. It is a complex motion of a rotating body (Earth) subject to a torque acting upon it as a result of gravity.

**primordial.** The original or first in a sequence, usually referring to the earliest stage of development of an organism or its parts.

**prism.** A homogeneous, transparent solid, usually with a triangular base and rectangular sides, used to produce or analyze a continuous spectrum of light.

**prokaryote.** Any organism of the Prokaryote kingdom in which the genetic material is not enclosed within the cell nucleus and possesses a single double-stranded DNA molecule. Only bacteria and cyanobacteria are prokaryotes. All other organisms are eukaryotes.

**proton.** A positively charged particle found in the nucleus of an atom.

**quantum.** The basic unit of electromagnetic energy that is not continuous, but occurs in discrete bundles called "quanta." For example, the photon is a small packet (quantum) of light with both particle and wave-like characteristics. A quantum unit for radiation is the frequency $v$ to the product $\hbar v$, where $\hbar$ is Planck's constant. The quantum number is the basic unit used to measure electromagnetic energy. To simplify, it is a very small bit or unit of something.

**quark.** A hypothetical subnuclear particle having an electric charge one-third to two-thirds that of the electron. Also known as the *fundamental subatomic particle*, which is one of the smallest units of matter.

**radical.** Also known as "free radical." A group of atoms having one unpaired electron. Also, in mathematics, a given root of a quantity.

**radioisotope.** The isotopic form of a natural or synthetic element that exhibits radioactivity. The same as a radioactive isotope of an element.

**rectifier.** A device (diode) that converts alternating current (AC) to direct current (DC).

**reduction.** The acceptance of one or more electrons by an atom or ion, the removal of oxygen from a compound, or the addition of hydrogen to a compound.

**resistors.** Two-terminal devices used in electric and electronics circuits that are designed to resist an electric current, thus limiting the flow of the current or causing a drop in voltage.

**retrovirus.** An animal virus containing RNA in which the genome replicates through reverse transcription and which has two proteinaceous structures, enabling it to combine with the host's DNA. Retroviruses contain oncogenes, which are cancer-causing genes that become activated once the virus enters the host's cell and begins to reproduce itself.

**RNA.** Abbreviation for *ribonucleic acid*. The linear, single-stranded polymer of ribonucleotides, each of which contains sugar (ribose) and one of four nitrogen bases (adenine, guanine, cytosine, uracil). It is present in all living cells (prokaryotic and eukaryotic) and carries the genetic code, which is transcribed from the DNA to the ribosomes within the cell where this genetic information is reproduced.

**semiconductor.** Usually a "metalloid" (e.g., silicon) or a compound (e.g., gallium arsenide), which has conductive properties greater than those of an insulator but less than those of a conductor (metal). It is possible to adjust their level of conductivity by changing the temperature or adding impurities.

**sidereal period.** The actual period (length of time) of revolution of a planet in its orbit around the sun using the stars as reference points.

**singularity.** Often referred to as a space-time singularity, it is a region of space-time where one or more components of curved spaces become infinite. Also defined as the location at which the fabric of space-time experiences a "devastating rupture." Examples of space-time singularities: the big bang, black holes.

**solar system.** Consists of the sun and the other celestial objects that are bound to it by gravitational forces, including the eight planets and three dwarf planets (Ceres, Eris, and Pluto) that orbit the sun along with their satellites (moons), asteroids, comets, meteors, as well as the remnants from the formation of the solar system that are located in the region called the Kuiper belt.

**solenoid.** An electromagnetic coil of insulated wire that produces a magnetic field within the coil. Most often it is shaped like a spool or hollow cylinder with a movable iron core that is pulled into the coil when electric current is sent through the wire. It then is able to move other instruments, for example, relay switches, circuit breakers, automobile ignitions.

**soma.** The entire physical body of an organism, with the exception of its germ cells and tract.

**species.** The lowest ranking in the classification of organisms. It is the distinguishable group with a common ancestry, able to reproduce fertile offspring, and that are geographically distinct. (Related species are grouped into a genus.)

**spectrophotometry.** The quantitative analysis of radiant energy, specifically visible, ultraviolet, and infrared light, as well as X-rays. With a spectrophotometer, an instrument designed to measure light intensity, it is possible to measure light intensity as a function of color, in other words, the wavelength of light.

**spectroscopy.** The analysis of chemical elements that separates the unique light waves either given off or absorbed by the elements when heated.

**Standard Model.** In particle physics, a collection of established experimental knowledge and theories that summarize the field. It includes the three generations of quarks and leptons, the electroweak theory of weak and electromagnetic forces, and quantum chromodynamic theory of strong forces.

**steroid.** A class of lipid proteins, such as sterols, bile acids, sex hormones, or adrenocortical hormones, that are derived from cyclopentanoperhydrophenanthrene. A shorter term for *anabolic steroid.*

**stoichiometry.** The calculation of measurable numerical relationships of chemical elements and chemical compounds as reactants and products in chemical reactions.

**subatomic particle.** A component of an atom whose reactions are characteristic of the atom, for example, electrons, protons, and neutrons.

**superconductivity.** A property of a metal, alloy, or compound that at temperatures near absolute zero loses both electrical resistance and magnetic permeability (is strongly repelled by magnets), thereby having infinite electrical conductivity.

**supernova.** A great explosion of a large star that collapses because of its gravitational force, sending great bursts of electromagnetic radiation (light) into space.

**superstrings.** In physics, a component of superstring theory incorporating supersymmetry that is an attempt to explain the fundamental forces of nature and all the particles into one theory of general relativity. Superstrings are one-dimensional, closed curves or loops of vibrating energy with zero thickness and length that is measured as the Planck length, namely, $10^{-35}$ m.

**syllogism.** In broad terms, it is a form of a "logical argument" in which the conclusion is inferred from a major and a minor premise. It is the basis of *deductive reasoning*, that is, reasoning from the general to the specific, and is sometime referred to as specious reasoning.

**symbiosis.** The interrelationship between two different organisms or two different species in which one but not always both benefit. For instance, *parasitism* is a form of symbiosis.

**systolic.** Refers to *systole* which is the rhythmic contraction of the heart, particularly the ventricles, by which blood is driven through the aorta and pulmonary artery after each dilation or diastole.

**tetanus.** An infectious disease of both humans and animals caused by the *Clostridium tetani* bacteria. Infection can occur after a deep wound is contaminated by dirt. Symptoms include violent and involuntary muscle spasms and contractions, including those of the jaw. Hence, the common term for tetanus—*lockjaw*. A vaccine for tetanus is routinely administered whenever an injury involves an open wound that has been exposed to dirt or debris.

**thermionic emission.** The emission of electrons or ions, usually into a vacuum, from a heated object, such as a cathode of a thermionic tube.

**thermodynamics.** The study of energy and laws governing transfer of energy from one form to another, particularly relating to behavior of systems where temperature is a factor (i.e., direction of the flow of heat and availability of energy to perform work).

**thermonuclear.** Release of heat energy when the nuclei of atoms split (fission, atom bomb, or nuclear power plant) or when nuclei combine (fusion, hydrogen bomb).

**tincture.** An alcoholic extract (e.g., vegetable or herb) or a solution of a nonvolatile substance (e.g., iodine). They are more dilute, usually only 10%, than fluid extracts and less volatile than spirits.

**transistor.** A device that overcomes the resistance when a current of electricity passes through it, used widely in the electronics industry.

**ultraviolet (UV).** The radiation wavelength in the electromagnetic spectrum from 100 to 3,900 angstroms (Å), between the X-ray region and visible violet light.

**universe.** All the space, matter, and energy that exists, including that which existed in the past and is postulated to exist in the future.

**valence.** The whole number that represents the combing power of one element with another element. Valence electrons are usually, but not always, the electrons in the outermost shell.

**vector.** A quantity specified by magnitude and direction whose components convert from one coordinate system to another in the same manner as the components of a displacement. Vector quantities may be added and subtracted.

**velocity.** The time rate at which an object is displaced. Velocity is a vector quantity whose quantity is measured in units of distance over a period of time.

**weight.** The measure of the mass or heaviness of an object. It is determined by the gravitational force exerted on an object.

**zygote.** A fertilized egg that develops into an embryo. It is the product of the union of two gametes.

# APPENDIX A

# Alphabetical Listing of Entries by Scientific Discipline

**ANTHROPOLOGY**

Johanson; Leakey

**ASTRONOMY**

Adhemar; Airy; Al-Battani; Ambartsumian; Ångstrom [also *Physics*]
Baade; Bahcall; Bessel [also *Mathematics*]; Bode; Bok; Bradley; Brahe; Burbidge (aka B2FH)
Cassini; Chandrasekhar; Chang (Heng) [also *Mathematics*]; Copernicus
Douglass; Drake
Eddington, Eratosthenes; Eudoxus
Fairbank; Friedmann
Geller; Gold; Guth
Hale; Halley; Hawking; Herschels; Hertzsprung; Hewish; Hoyle; Hubble; Huggins
I-Hsing [also *Mathematics*]
Janssen; Jeans; Jeffreys (Harold)
Kapteyn; Kepler; Kimura (Hisashi); Kirkwood; Kuiper
Lambert [also *Mathematics, Physics*]; Leavitt; Lemaître; Lockyer; Lovell; Lowell
Maunder; Misner
Nicholas
Olbers; Oort
Penzias; Pogson; Poseidonius; Ptolemy
Roche; Rubin; Russell; Ryle
Sagan; Saha; Sandage; Schiaparelli; Schmidt; Schwarzschild; Shapley; Slipher; Smoot; Spencer-Jones; Struve
Taylor
Whipple; Wolf
Zwicky

## BIOCHEMISTRY

Elion
Fox
Kendall; Khorana
Merrifield; Mullis [also *Biology*]
Porter
Theorell; Todd
Wrinch

## BIOLOGY

Baer; Bakker; Baltimore; Behring; Bonnet; Buffon
Candolle; Chambers; Chang (Min Chueh); Chargaff; Clarke; Cohn; Crick–Watson; Cuvier
Dale; Darlington; Darwin; Dawkins; De Beer; Delbruck/Luria; De Vries; d'Herelle; Dobzhansky; Dulbecco
Einthoven; Eldredge; Elton; Erasistratus
Fabricius; Fallopius; Fleming (Alexander); Florey; Fracastoro [also *Medicine*]
Galen; Gallo; Galton; Garrod [also *Medicine*]; Gilbert (Walter); Goldstein [also *Medicine*]; Gould
Haeckel; Haldane; Harvey
Ingenhousz; Ingram; Isaacs
Jacob-Monod; Jeffreys (Alec); Jenner; Jerne
Kimura (Motoo); Koch
Lamarck; Landsteiner; Lederberg; Leeuwenhoek; Leishman; Levene; Levi-Montalcini; Linnaeus; Lister; Lysenko
Malpighi; Malthus; Margulis; Maynard-Smith; McClintock; Mendel; Meselson–Stahl; Miescher; Muller; Mullis [also *Biochemistry*]
Neher [also *Physics*]; Nicolle
Ochoa; Oken; Oparin
Pardee; Pasteur; Perutz; Pfeiffer
Ramón y Cajal; Raup; Ray; Redi; Reed; Reichstein [also *Chemistry*]; Robbins; Roberts
Sabin; Sachs; Sarich; Schleiden; Schwann; Sharp; Spallanzani; Strasburger; Swammerdam
Tatum; Temin; Theophrastus; Tonegawa
Vesalius; Virchow
Waddington; Waldeyer-Hartz; Wallace; Watson–Crick; Weismann; Wilson (Allan); Wilson (Edward); Wright (Sewall); Wynne-Edwards
Yanofsky
Zuckerandl

## CHEMISTRY

Abegg; Adams; Arrhenius; Aston [also *Physics*]; Avogadro
Babo; Baekeland; Baeyer; Berzelius; Black; Boyle; Bunsen
Calvin; Cannizzaro; Caspersson; Cavendish; Charles; Chevreul; Claude; Corey; Cori; Couper; Crutzen; Curie; Curl
Daguerre; Dalton; Daniell; Davy; Debye-Hückel; Democritus; Dewar [also *Physics*]; Djerassi; Döbereiner; Domagk; Draper, Dumas

Ehrlich; Eigen; Elion; Ernst; Eyring
Fajans; Fischer; Fleischmann; Flory; Frankland
Gay-Lussac; Gerhardt; Giauque [also *Physics*]; Graham
Haber; Hahn; Harkins; Haworth; Helmont; Higgins; Hodgkin; Hoffman; Hückel
Ideal Gas Law; Ingold; Ipatieff
Kekule; Kipping; Krebs; Kroto
Langmuir; Laurent; Lavoisier; Le Bel; Le Chatelier; Lewis; Liebig
Martin; Mendeleev; Meyer; Miller; Mitscherlich; Mulliken
Natta; Nernst; Newlands; Noddack; Norrish; Northrop
Odling; Ostwald
Paracelsus; Parkes; Pauling; Ponnamperuma; Priestley; Prigogine; Proust
Ramsay; Reichstein [also *Biology*]; Revelle; Rowland
Sanger; Scheele; Seaborg; Sidgwick; Soddy; Sorensen; Spedding; Stahl
Tiselius; Turner
Urey
Van't Hoff
Wallach; Werner (Alfred); Wilkinson; Williamson; Wohler; Woodward; Wurtz
Ziegler

**COMPUTER SCIENCE**

Amdahl; Metcalfe; Minsky [also *Mathematics*]; Moore [also *Mathematics*]

**GEOLOGY**

Agassiz; Agricola; Airy [also *Astronomy*]
Barringer; Beaumont
Charpentier
Dana
Ekman; Ewing
Haüy; Hess (Harry)
Lyell
Mohorovicic
Richter
Shepard; Steno; Suess
Wegener; Werner (Abraham)

**MATHEMATICS**

Abel; Archimedes
Babbage; Balmer [also *Physics*]; Banach; Bernoulli (Daniel) [also *Physics*]; Bernoulli (Jakob); Bessell [also *Astronomy*]; Boole
Cantor; Cardano; Chang (Heng) [also *Astronomy*]; Chapman-Enskog [also *Physics*]; Conway
D'Alembert [also *Physics*]
Euclid; Euler
Fermat; Fibonacci
Gauss; Gibbs [also *Physics*]; Godel
Hadamard; Hamilton; Hardy
I-Hsing [also *Astronomy*]

Lagrange; Lambert [also *Astronomy*, *Physics*]; Liebniz; Lindemann; Lorenz
Minsky [also *Computer Science*]; Moore [also *Computer Science*}
Nash; Noether
Peano; Pearson; Pythagoras
Regiomontanus; Reichenbach [also *Philosophy*]; Riemann
Snell [also *Physics*]
Tartaglia; Turing
Ulam
Von Neumann
Whitehead [also *Physics*]

**MEDICINE**
Banting; Bell
Fracastoro
Garrod; Goldstein
Mesmer; Montagnier

**METEOROLOGY**
Bergeron; Bjerknes [also *Physics*]
Charney [also *Physics*]
Hadley
Schneider

**MICROBIOLOGY**
Arber; Enders; Nathans

**PHILOSOPHY**
Anaximander
Bacon
Descartes
Reichenbach [also *Mathematics*]
Thales

**PHYSICS**
Abbe; Alvarez; Ampère; Anderson (Carl); Anderson (Philip); Ångström [also *Astronomy*]; Arago; Aristotle; Aston [also *Chemistry*]; Atomism Theories; Auger
Babinet; Balmer [also *Mathematics*]; Bardeen; Becquerel; Beer; Bernoulli (Daniel) [also *Mathematics*]; Bethe; Biot-Savart; Birkeland; Bjerknes [also *Meteorology*]; Bohm; Bohr; Boltzmann; Born-Haber
Cagniard de La Tour; Cailletet; Carnot; Casimir; Celsius; Chadwick; Chapman-Enskog [also *Mathematics*]; Charney [also *Meteorology*]; Charpak; Chu; Clausius; Cockroft-Walton; Compton; Coriolis; Coulomb; Crookes
D'Alembert [also *Mathematics*]; Davisson; De Broglie; Dehmelt; Dewar [also *Chemistry*]; Dicke; Diesel; Dirac; Doppler; Dyson
Edison; Einstein; Eötvös; Esaki; Everett

Fahrenheit; Faraday; Fermi; Fessenden; Feynman; Fick [also *Physiology*]; Fitzgerald; Fizeau; Fleming (John); Flerov; Foucault; Fourier; Fowler; Franck; Franklin (Benjamin); Franklin (Rosalind); Fraunhofer; Fresnel; Friedman; Frisch

Gabor; Galileo; Galvani; Gamow; Gassendi; Geiger–Nutter; Gell-Mann; Giauque [also *Chemistry*]; Gibbs [also *Mathematics*]; Gilbert (William); Glaser; Glashow

Hall; Heisenberg; Helmholtz; Henry; Hertz; Hess (Victor); Higgs; Hooke; Huygens

Jansky; Joliot-Curie; Josephson; Joule

Kamerlingh-Onnes; Kapitsa; Karle; Kelvin; Kerr; Kerst; Kirchhoff; Klitzing; Kohlrausch; Kusch

Lamb; Lambert [also *Astronomy*, *Mathematics*]; Landau; Landauer; Langevin; Laplace; Larmor; Lawrence; Lederman; Lee; Lenard; Lenz; Lorentz

Mach; Maiman; Malus; Mansfield; Marconi; Matthias; Maupertuis; Maxwell; McMillan; Meissner; Meitner; Michelson; Millikan; Minkowski; Moseley

Nambu; Néel; Neher [also *Biology*]; Newcomb; Newton; Noyce

Oersted; Ohm; Oliphant; Oppenheimer

Pascal; Pauli; Peierls; Penrose; Perl; Perrin; Planck; Poynting; Prévost; Purcell

Quantum Theories

Rabi; Raman; Ramsey; Raoult; Rayleigh; Reines; Ricciolo; Richardson; Röentgen; Romer; Rossi; Rubbia; Rumford; Rutherford; Rydberg

Sakharov; Salam; Schrödinger; Schwinger; Seebeck; Segre; Shockley; Siemens; Simon; Snell [also *Mathematics*]; Stark; Stefan; Steinberger; Stern; Stokes; Stoney; Szilard

Tamm; Teller; Tesla; Thomson; Ting; Tomonaga; Torricelli; Townes; Townsend; Tyndall

Uhlenbeck

Van Allen; Van de Graaf; Van der Meer; Van der Waals; Van Vleck; Volta; Von Laue

Walton; Watson (William); Watson-Watt; Weber; Weinberg; Weizsäcker; Wheeler; Whitehead [also *Mathematics*]; Wien; Wigner; Wilson (Charles); Witten; Wolfram; Wu

Yalow; Yang; Young; Yukawa

Zeeman; Zeno; Zinn

**PHYSIOLOGY**

Fick [also *Physics*]; Langley; Pavlov

# APPENDIX B

# Nobel Laureates in Chemistry (1901–2007)

Nobel laureates are listed consecutively by year starting with 1901 and ending with 2007. The country indicates where the chemist did major work. If the chemist's country of birth is different than where major work was, or is, done, this is indicated after the "b." In many years, several chemists are recognized. When this occurs, it may be for work in similar areas of chemistry for working collaboratively, or independently. When the prize has been shared for work in an entirely different area, the name, country, and work are shown as a separate entry.

| Year | Recipient(s) | Country | Work |
|---|---|---|---|
| **1901** | Jacobus van't Hoff | Germany, b. Netherlands | laws of chemical dynamics and osmotic osmotic pressure in solutions |
| **1902** | Herman Emil Fischer | Germany | sugar and purine syntheses |
| **1903** | Svante August Arrhenius | Sweden | electrolytic dissociation theory |
| **1904** | William Ramsay | United Kingdom | discovery of noble gases |
| **1905** | Adolf von Baeyer | Germany | organic dyes and hydroaromatic compounds |
| **1906** | Henri Moissan | France | isolation of fluorine and electric furnace |
| **1907** | Eduard Buchner | Germany | fermentation in absence of cells and biochemistry |
| **1908** | Ernest Rutherford | United Kingdom, b. New Zealand | radioactive decay |
| **1909** | Wilhelm Ostwald | Germany, b. Russia | chemical equilibrium, kinetics, and catalysis |

*(Continued)*

| Year | Recipient(s) | Country | Work |
|---|---|---|---|
| **1910** | Otto Wallach | Germany | pioneering work with alicyclic compounds |
| **1911** | Marie Curie | France, b. Poland | discovery of radium and polonium |
| **1912** | Victor Grignard | France | discovery of Grignard's reagent |
| | Paul Sabatier | France | hydrogenation with metal catalysts |
| **1913** | Alfred Werner | Switzerland, b. Germany | bonding of inorganic compounds |
| **1914** | Theodore Richards | United States | determination of atomic weights |
| **1915** | Richard Willstätter | Germany | studies of plant pigments, especially chlorophyll |
| **1916*** | | | |
| **1917*** | | | |
| **1918** | Fritz Haber | Germany | synthesis of ammonia |
| **1919*** | | | |
| **1920** | Walter H. Nernst | Germany | thermochemistry |
| **1921** | Frederick Soddy | United Kingdom | radioactive substances and isotopes |
| **1922** | Francis W. Aston | United Kingdom | mass spectrography and discovery of isotopes |
| **1923** | Fritz Pregl | Austria | organic microanalysis |
| **1924*** | | | |
| **1925** | Richard A. Zsigmondy | Germany, b. Austria | colloid chemistry |
| **1926** | Theodore Svedberg | Sweden | disperse systems |
| **1927** | Heinrich Otto Wieland | Germany | bile acids |
| **1928** | Adolf Windaus | Germany | sterols relationship with vitamins |
| **1929** | Arthur Harden | United Kingdom | fermentation of sugar and sugar enzymes |
| | Hans von Euler-Chelpin | Sweden, b. Germany | |
| **1930** | Hans Fischer | Germany | synthesis of haemin |
| **1931** | Carl Bosch | Germany | high-pressure chemical processing |
| | Friedrich Bergius | Germany | |
| **1932** | Irving Langmuir | United States | surface chemistry |
| **1933‡** | | | |
| **1934** | Harold Urey | United States | discovery of heavy hydrogen |
| **1935** | Frédéric Joliot | France | synthesis of new radioactive elements |
| | Irène Joliot-Curie | France | |
| **1936** | Peter Debye | Germany, b. Netherlands | dipole moments and X-ray diffraction |
| **1937** | Norman Haworth | United Kingdom | carbohydrates and vitamin C |
| | Paul Karrer | Switzerland | vitamins A and B12 |

| Year | Recipient(s) | Country | Work |
|---|---|---|---|
| 1938 | Richard Kuhn | Germany, b. Austria | carotenoids and vitamins |
| 1939 | Adolf F. J. Butenandt | Germany | sex hormones |
| | Leopold Ruzicka | Switzerland, b. Hungary | polymethylenes and terpenes |
| 1940‡ | | | |
| 1941‡ | | | |
| 1942‡ | | | |
| 1943 | George DeHevesy | Sweden, b. Hungary | isotope tracers |
| 1944 | Otto Hahn | Germany | fission of heavy nuclei |
| 1945 | Artturi I. Virtanen | Finland | agricultural and food chemistry and preservation of fodder |
| 1946 | James Sumner | United States | crystallization of enzymes |
| | John H. Northrop | United States | preparation of proteins and enzymes in pure form |
| | Wendell Stanley | United States | |
| 1947 | Robert Robinson | United Kingdom | alkaloids |
| 1948 | Arne W. K. Tiselius | Sweden | electrophoresis and serum proteins |
| 1949 | William F. Giauque | United States | low temperature thermodynamics |
| 1950 | Otto Diels | Germany | diene synthesis |
| | Kurt Alder | Germany | |
| 1951 | Edwin McMillan | United States | chemistry of transuranium elements |
| | Glenn Seaborg | United States | |
| 1952 | Archer J. P. Martin | United Kingdom | invention of partition chromatography |
| | Richard L. M. Synge | United Kingdom | |
| 1953 | Hermann Staudinger | Germany | macromolecular chemistry |
| 1954 | Linus Pauling | United States | chemical bonding and molecular structure of proteins |
| 1955 | Vincent Du Vigneaud | United States | sulfur compounds of biological importance; synthesis of polypeptide hormone |
| 1956 | Cyril Hinshelwood | United Kingdom | mechanisms of chemical reaction |
| | Nikolay Semenov | USSR | |
| 1957 | Alexander Todd | United Kingdom | nucleotides and their co-enzymes |
| 1958 | Frederick Sanger | United Kingdom | protein structure; insulin |
| 1959 | Jaroslav Heyrovsky | Czechoslovakia | polarographic methods of analysis |
| 1960 | Willard Libby | United States | carbon-14 dating |
| 1961 | Melvin Calvin | United States | $CO_2$ assimilation in plants |
| 1962 | Max Perutz | United Kingdom, b. Austria | structure of globular proteins |
| | John Kendrew | United Kingdom | |

*(Continued)*

| Year | Recipient(s) | Country | Work |
|---|---|---|---|
| 1963 | Giulio Natta | Italy | high polymers |
| | Karl Ziegler | Germany | |
| 1964 | Dorothy Crowfoot Hodgkin | United Kingdom | X-ray techniques of the structures of biochemical substances |
| 1965 | Robert Woodward | United States | organic synthesis |
| 1966 | Robert Mulliken | United States | chemical bonds and electronic structure of molecules |
| 1967 | Manfred Eigen | Germany | study of very fast chemical reactions |
| | Ronald Norrish | United Kingdom | |
| | George Porter | United Kingdom | |
| 1968 | Lars Onsager | United States, b. Norway | thermodynamic of irreversible processes |
| 1969 | Derek Barton | United Kingdom | conformation |
| | Odd Hassel | Norway | |
| 1970 | Luis Leloir | Argentina | sugar nucleotides and carbohydrate biosynthesis |
| 1971 | Gerhard Herzberg | Canada, b. Germany | structure and geometry of free radicals |
| 1972 | Christian Anfinsen | United States | ribonuclease, amino acid sequencing and biological activity |
| | Stanford Moore | United States | chemical structure and catalytic activity of ribonuclease |
| | William Stein | United States | |
| 1973 | Ernst Fischer | Germany | organometallic sandwich compounds |
| | Geoffrey Wilkinson | United Kingdom | |
| 1974 | Paul Flory | United States | macromolecules |
| 1975 | John Cornforth | United Kingdom | stereochemistry of enzyme-catalyzed reactions; |
| | Vladmir Prelog | Switzerland, b. Bosnia | stereochemistry of organic molecules |
| 1976 | William Lipscomb | United States | structure of borane and bonding |
| 1977 | Ilya Prigogine | Belgium, b. Russia | theory of dissipative structures |
| 1978 | Peter Mitchell | United Kingdom | chemiosmotic theory |
| 1979 | Herbert Brown | United States, b. United Kingdom | organic synthesis of boron and phosphorus compound |
| | George Wittig | Germany | |
| 1980 | Paul Berg | United States | recombinant DNA |
| | Walter Gilbert | United States | nucleic acid base sequences |
| | Frederick Sanger | United Kingdom | |

| Year | Recipient(s) | Country | Work |
|---|---|---|---|
| 1981 | Kenichi Fukui | Japan | chemical reactions and orbital theory |
| | Roald Hoffman | United States, b. Poland | |
| 1982 | Aaron Klug | United Kingdom | crystallographic electron microscopy applied to nucleic acids and proteins |
| 1983 | Henry Taube | United States, b. Canada | electron transfer in metal complexes |
| 1984 | Bruce Merrifield | United States | chemical synthesis |
| 1985 | Herbert Hauptman | United States | crystal structures |
| | Jerome Karle | United States | |
| 1986 | Dudley Herschbach | United States | chemical elementary processes |
| | Yuan Lee | United States, b. Taiwan | |
| | John Polanyi | Canada | |
| 1987 | Donald Cram | United States | development of molecules with |
| | Jean-Marie Lehn | France | highly selective structure |
| | Charles Pedersen | United States, b. Korea (Norwegian) | specific interactions |
| 1988 | Johann Deisenhofer | Germany and United States, b. Germany | photosynthesis |
| | Robert Huber | Germany | |
| | Hartmut Michel | Germany | |
| 1989 | Sidney Altman | United States, b. Canada | catalytic properties of RNA |
| | Thomas R. Cech | United States | |
| 1990 | Elias J. Corey | United States | organic synthesis |
| 1991 | Richard Ernst | Switzerland | nuclear resonance spectroscopy |
| 1992 | Rudolph Marcus | United States, b. Canada | electron transfer in chemical systems |
| 1993 | Kary Mullis | United States | invention of PCR method; |
| | Michael Smith | Canada, b. United Kingdom | mutagenesis and protein studies |
| 1994 | George Olah | United States, b. Hungary | carbocation chemistry |
| 1995 | Paul Crutzen | Germany, b. Netherlands | atmospheric chemistry |
| | Mario Molina | United States, b. Mexico | stratospheric ozone depletion |
| | F. Sherwood Rowland | United States | |
| 1996 | Robert Curl Jr. | United States | discovery of fullerenes |
| | Harold Kroto | United Kingdom | |
| | Richard Smalley | United States | |

*(Continued)*

| Year | Recipient(s) | Country | Work |
|---|---|---|---|
| 1997 | Paul Boyer | United States | enzyme mechanism of ATP |
| | John Walker | United Kingdom | discovery of ion transport enzyme NA+, K+-ATPase |
| | Jens Skou | Denmark | |
| 1998 | Walter Kohn; | United States | density function theory; |
| | John Pople | United Kingdom | computational methods in quantum chemistry |
| 1999 | Ahmed Zewail | United States, b. Egypt | transition states using femto spectroscopy |
| 2000 | Alan J. Heeger | United States | discovery of conductive polymers |
| | Alan MacDiarmid | United States | |
| | Hideki Shirakawa | Japan | |
| 2001 | William Knowles | United States | chirally catalyzed hydrogenation reactions |
| | Ryoji Noyori | Japan | |
| | K. Barry Sharpless | United States | |
| 2002 | John B. Fenn | United States | identification of biological macromolecules |
| | Koichi Tanaka | Japan | |
| | Kurt Wüthrich | Switzerland | NMR analysis of biological macromolecules |
| 2003 | Peter Agre | United States | discovery of water channels; |
| | Roderick MacKinnon | United States | structural and mechanistic studies of ion channels |
| 2004 | Aaron Ciechanover | Israel | discovery of ubiquitin-mediated protein degradation |
| | Avram Hershko | Israel, b. Hungary | |
| | Irwin Rose | United States | |
| 2005 | Yves Chauvin | France | development of the metathesis method in organic synthesis |
| | Robert H. Grubbs | United States | |
| | Richard R. Schrock | United States | |
| 2006 | Roger D. Kornberg | United States | molecular bases of eukaryotic transcription |
| 2007 | Gerhard Ertl | Germany | chemical processes on solid surfaces |

Source: Nobelprize.org.

* No prize awarded. Prize money allocated to the Special Fund of the Nobel Prize section for chemistry.

‡ No prize awarded. Prize money was allocated as follows: 1/3 to the Nobel Prize main fund, and 2/3 to the Special Fund of the Nobel Prize section for chemistry.

# APPENDIX C

# Nobel Laureates in Physics (1901–2007)

Nobel laureates are listed consecutively by year starting with 1901 and ending with 2007. The first country shown indicates where the physicist did major work. If the physicist's country of birth is different than where the major work was, or is, done, this is indicated after the "b." In some years, several physicists shared the Nobel Prize for work in similar areas of physics, either working collaboratively, or independently, and their names are listed as such. When the prize has been shared for work in an entirely different area, the name, country, and work are shown as a separate entry.

| Year | Recipient(s) | Country | Work |
|---|---|---|---|
| **1901** | Wilhelm Röentgen | Germany | discovery of X-rays |
| **1902** | Hendrik A. Lorentz | Netherlands | effect of magnetism on radiation |
| | Pieter Zeeman | Netherlands | |
| **1903** | Henri Becquerel | France | spontaneous radioactivity; |
| | Marie Curie | France, b. Poland | joint research on Becquerel's discovery of radiation phenomena |
| | Pierre Curie | France | |
| **1904** | John W. Strutt (aka Lord Rayleigh) | United Kingdom | discovery of argon; density of gases |
| **1905** | Philipp von Lenard | Germany, b. Austria-Hungary | cathode rays |
| **1906** | Sir J. J. Thomson | United Kingdom | electrical conductivity of gases |
| **1907** | Albert A. Michelson | United States, b. Germany | spectroscopic metrological investigations with optical precision instruments |

*(Continued)*

| Year | Recipient(s) | Country | Work |
|---|---|---|---|
| 1908 | Gabriel Lippmann | France, b. Luxembourg | photographic reproduction of color |
| 1909 | Karl F. Braun | Germany | wireless telegraphy |
| | Guglielmo Marconi | United Kingdom, b. Italy | |
| 1910 | Johannes D. Van der Waals | Netherlands | equation of state for gases and fluids |
| 1911 | Wilhelm Wien | Germany | laws on radiation of heat |
| 1912 | Nils Gustaf Dalén | Sweden | automatic regulators for gas |
| 1913 | Heike Kamerlingh-Onnes | Netherlands | matter at low temperature |
| 1914 | Max von Laue | Germany | X-ray diffraction with crystals |
| 1915 | Lawrence Bragg | United Kingdom, b. Australia | crystal structure using X-rays |
| | William Bragg | United Kingdom | |
| 1916* | | | |
| 1917 | Charles G. Barkla | United Kingdom | characteristic Röentgen radiation of elements |
| 1918 | Max Planck | Germany | energy quanta |
| 1919 | Johannes Stark | Germany | Doppler effect and splitting of spectral lines in electric field |
| 1920 | Charles Guillaume | Switzerland | anomalies in nickel and steel alloys |
| 1921 | Albert Einstein | Germany and Switzerland, b. Germany | photoelectric effect |
| 1922 | Niels Bohr | Denmark | atomic structure and radiation |
| 1923 | Robert Millikan | United State | elementary electric charge |
| 1924 | Karl M. G. Siegbahn | Sweden | X-ray spectroscopy |
| 1925 | James Franck | Germany | impact of electron on atom |
| | Gustav Hertz | Germany | |
| 1926 | Jean Perrin | France | discontinuous structure of matter; sedimentation equilibria |
| 1927 | Arthur H. Compton; | United States | Compton effect; |
| | Charles T. R. Wilson | United Kingdom | invention of cloud chamber |
| 1928 | Owen Richardson | United Kingdom | Richardson's law, electron emission of hot metals |
| 1929 | Louis de Broglie | France | wave nature of electrons |
| 1930 | C. V. Raman | India | Raman effect, light diffusion |
| 1931* | | | |
| 1932 | Werner Heisenberg | Germany | quantum mechanics |
| 1933 | Paul Dirac | United Kingdom | discovery of new productive forms of atomic theory |
| | Erwin Schrödinger | Germany, b. Austria | |

| Year | Recipient(s) | Country | Work |
|---|---|---|---|
| 1934‡ | | | |
| 1935 | James Chadwick | United Kingdom | discovery of neutron |
| 1936 | Carl D. Anderson | United States | discovery of positron; |
| | Victor F. Hess | Austria | discovery of cosmic rays |
| 1937 | Clinton Davisson | United States | crystal diffraction of electrons |
| | George Thomson | United Kingdom | |
| 1938 | Enrico Fermi | Italy | neutron irradiation and discovery of new elements |
| 1939 | E. O. Lawrence | United States | invention of cyclotron |
| 1940‡ | | | |
| 1941‡ | | | |
| 1942‡ | | | |
| 1943 | Otto Stern | United States, b. Germany | magnetic moment of the proton |
| 1944 | Isidor Rabi | United States, b. Austria-Hungary | magnetic resonance of atomic nuclei |
| 1945 | Wolfgang Pauli | United States and Switzerland, b. Austria | exclusion principle of electrons |
| 1946 | Percy Bridgman | United States | high-pressure physics |
| 1947 | Edward Appleton | United Kingdom | upper-atmosphere physics |
| 1948 | Patrick Blackett | United Kingdom | nuclear and cosmic physics with cloud chamber |
| 1949 | Hideki Yukawa | United States and Japan, b. Japan | prediction of mesons |
| 1950 | Cecil F. Powell | United Kingdom | photographic method for mesons |
| 1951 | John D. Cockcroft | United Kingdom | transmutation of atomic nuclei |
| | Ernest T. S. Walton | Ireland | |
| 1952 | Felix Bloch | United States, b. Switzerland | nuclear magnetic resonance methods |
| | Edward Purcell | United States | |
| 1953 | Frits Zernike | Netherlands | phase-contrast microscopy |
| 1954 | Max Born | United Kingdom, b. Germany | statistical interpretation of wave function; |
| | Walther Bothe | Germany | coincidence method |
| 1955 | Polykarp Kusch | United States, b. Germany | magnetic moment of the electron; |
| | Willis Lamb Jr. | United States | structure of hydrogen spectrum |
| 1956 | John Bardeen | United States | semiconductors and discovery of transistor effect |

*(Continued)*

| Year | Recipient(s) | Country | Work |
|---|---|---|---|
| | Walter Brattain | United States | semiconductors and discovery of transistor effect |
| | William Shockley | United States | |
| 1957 | Tsung-Dao Lee | United States, b. China | parity laws in their application to elementary particles |
| | Chen Ning Yang | United States, b. China | |
| 1958 | Pavel Cherenkov | USSR | discovery and interpretation of Cherenkov effect |
| | Ilya Frank | USSR | |
| | Igor Tamm | USSR | |
| 1959 | Owen Chamberlain | United States | discovery of antiproton |
| | Emilio Segrè | United States, b. Italy | |
| 1960 | Donald Glaser | United States | invention of bubble chamber |
| 1961 | Robert Hofstadter | United States | structure of nucleus; |
| | Rudolf Mössbauer | United States, b. Germany | absorbance and emission of photons |
| 1962 | Lev Landau | USSR | condensed matter and liquid helium |
| 1963 | J. Hans Jensen | Germany | nuclear shell structure; |
| | Maria Goeppert-Mayer | United States, b. Germany | |
| | Eugene Wigner | United States, b. Hungary | fundamental symmetry principles |
| 1964 | Nikolay Basov | USSR | maser-laser principles |
| | A M. Prokhorov | USSR | |
| | Charles Townes | United States | |
| 1965 | Richard Feynman | United States | quantum electrodynamics |
| | Julian Schwinger | United States | |
| | Sin-Itiro Tomonaga | Japan | |
| 1966 | Alfred Kastler | France | Hertzian resonance in atoms |
| 1967 | Hans Bethe | United States, b. Germany | energy production in stars |
| 1968 | Luis W. Alvarez | United States | elementary particles and resonance states |
| 1969 | Murray Gell-Mann | United States | quark model and elementary particles |
| 1970 | Hannes Alfvén | Sweden | magneto hydrodynamics; |
| | Louis-Eugene Néel | France | ferrimagnetism and anti-ferromagnetism |
| 1971 | Dennis Gabor | United Kingdom, b. Hungary | development of holography |

| Year | Recipient(s) | Country | Work |
|---|---|---|---|
| 1972 | John Bardeen | United States | theory of superconductivity, known as the BCS Theory |
| | Leon N. Cooper | United States | |
| | John Schrieffer | United States | |
| 1973 | Leo Esaki | United States, b. Japan | tunneling in semiconductors |
| | Ivar Giaever | United States, b. Norway | and superconductors, respectively; |
| | Brian Josephson | United Kingdom | supercurrents/Josephson effects |
| 1974 | Antony Hewish | United Kingdom | radioastrophysics and pulsars |
| | Sir Martin Ryle | United Kingdom | |
| 1975 | Aage N. Bohr | Denmark | atomic nucleus structure |
| | Ben Mottelson | Denmark, b. United States | |
| | James Rainwater | United States | |
| 1976 | Burton Richter | United States | discovery of J/psi particle |
| | Samuel C. Ting | United States | |
| 1977 | Philip Anderson | United States | electronic structure of magnetic disordered systems |
| | Sir Nevill F. Mott | United Kingdom | |
| | John Van Vleck | United States | |
| 1978 | Pyotr Kapitsa | USSR | helium liquefaction; |
| | Arno Penzias | United States, b. Germany | cosmic microwave background radiation |
| | Robert W. Wilson | United States | |
| 1979 | Sheldon Glashow | United States | unification of electromagnetic and weak interaction |
| | Abdus Salam | Pakistan | |
| | Steven Weinberg | United States | |
| 1980 | James Cronin | United States | violation of symmetry principles in the decay of neutral k-mesons |
| | Val Fitch | United States | |
| 1981 | Nicolaus Bloembergen | United States, b. Netherlands | laser spectroscopy; |
| | Arthur Schawlow | United States | |
| | Kai Siegbahn | Sweden | electron spectroscopy |
| 1982 | Kenneth G. Wilson | United States | phase transitions |
| 1983 | Subrahmanyan Chandrasekhar | United States, b. India; | stellar evolution; |
| | William A. Fowler | United States | element formation in universe |
| 1984 | Simon van der Meer | Switzerland, b. Netherlands | discovery of W and Z particles |
| | Carlo Rubbia | Switzerland, b. Italy | |
| 1985 | Klaus von Klitzing | Germany | quantized Hall effect |

*(Continued)*

| Year | Recipient(s) | Country | Work |
|---|---|---|---|
| 1986 | Gerd Binnig | Switzerland, b. Germany | scanning tunnel microscopy; |
| | Heinrich Rohrer | Switzerland | |
| | Ernst Ruska | Germany | electron microscopy |
| 1987 | J. Georg Bednorz | Switzerland | superconductivity in ceramics |
| | K. Alexander Müller | Switzerland | |
| 1988 | Leon Lederman | United States | discovery of muon neutrino |
| | Melvin Schwartz | United States | |
| | Jack Steinberger | United States and Switzerland, b. Germany | |
| 1989 | Hans Dehmelt | United States, b. Germany | ion trap technique |
| | Wolfgang Paul | Germany | |
| | Norman Ramsey | United States | atomic clocks |
| 1990 | Jerome Friedman | United States | deep inelastic scattering of electrons and quark discovery |
| | Henry Kendall | United States | |
| | Richard E. Taylor | Canada | |
| 1991 | Pierre-Gilles de Gennes | France | order transitions in liquid crystals |
| 1992 | Georges Charpak | France, b. Poland | multiwire proportional chamber detector |
| 1993 | Russell Hulse | United States | discovery of binary pulsar |
| | Joseph Taylor Jr. | United States | |
| 1994 | Bertram Brockhouse | Canada | neutron spectroscopy |
| | Clifford Shull | United States | neutron diffraction technique |
| 1995 | Martin Perl | United States | discovery of tau lepton; |
| | Frederick Reines | United States | detection of the neutrino |
| 1996 | David Lee | United States | superfluidity in heavy helium |
| | Douglas Osheroff | United States | |
| | Robert Richardson | United States | |
| 1997 | Steven Chu | United States | laser cooling and trapping of atoms |
| | Claude Cohen-Tannoudji | France, b. Algeria | |
| | William Phillips | United States | |
| 1998 | Robert Laughlin | United States | fractional quantum Hall effect |
| | Horst Störmer | United States, b. Germany | |
| | Daniel Tsui | United States, b. China | |
| 1999 | Gerardus 't Hooft | Netherlands | quantum structure of electroweak interactions in physics |
| | Martinus Veltman | Netherlands | |

| Year | Recipient(s) | Country | Work |
|---|---|---|---|
| 2000 | Zhores Alferov | Russia | semiconductor heterostructures in high-speed and opto-electronics |
| | Herbert Kroemer | United States, b. Germany; | semiconductor heterostructures in high-speed and opto-electronics; |
| | Jack St. Clair Kilby | United States | invention of integrated circuits |
| 2001 | Eric Cornell | United States | Bose-Einstein condensation of alkali atoms |
| | Wolfgang Ketterle | United States, b. Germany | |
| | Carl E. Wieman | United States | |
| 2002 | Raymond Davis Jr. | United States | detection of cosmic neutrinos; |
| | Masatoshi Koshiba | Japan | |
| | Riccardo Giacconi | United States, b. Italy | astrophysics and cosmic X-ray sources |
| 2003 | Alexei Abrikosov | United States, b. Russia | superconductivity and superfluids |
| | Vitaly Ginzburg | Russia | |
| | Anthony Leggett | United States, b. United Kingdom | |
| 2004 | David J. Gross | United States | asymptotic freedom and the theory of the strong interaction |
| | David H. Politzer | United States | |
| | Frank Wilczek | United States | |
| 2005 | Roy J. Glauber | United States | quantum theory of optical coherence; |
| | John L. Hall | United States | laser-based precision spectroscopy |
| | Theodor Hänsch | Germany | |
| 2006 | John C. Mather | United States | blackbody form and anistophy of cosmic microwave background radiation |
| | George F. Smoot | United States | |
| 2007 | Albert Fert | France | discovery of giant magneto-resistance resistance |
| | Peter Grunberg | Germany | |

Source: Nobelprize.org.

* No prize awarded. Prize money allocated to the Special Fund of the Nobel Prize section for physics.

‡ No prize awarded. Prize money was allocated as follows: 1/3 to the Nobel Prize main fund, and 2/3 to the Special Fund of the Nobel Prize section for physics.

# APPENDIX D

# Nobel Laureates in Physiology or Medicine (1901–2007)

Nobel laureates are listed consecutively by year starting with 1901 and ending with 2007. The country indicates where the physician and/or researcher did major work. If their country of birth is different than where the major work was or is done, this is indicated after the "b." In a number of years, several physicians and or researchers shared the Nobel Prize for work in similar areas, either working collaboratively or independently, and their names are listed as such. When the prize has been shared for work in entirely different areas, the name, country, and work are shown as a separate entry.

| Year | Recipient(s) | Country | Work |
|---|---|---|---|
| 1901 | Emil von Behring | Germany | serum therapy for diphtheria |
| 1902 | Ronald Ross | United Kingdom | discoveries related to malaria |
| 1903 | Niels R. Finsen | Denmark, b. Faroe Islands | treatment of diseases, mainly lupus, using light radiation |
| 1904 | Ivan Pavlov | Russia | work on physiology of digestion |
| 1905 | Robert Koch | Germany | discoveries related to tuberculosis |
| 1906 | Camillo Golgi | Italy | work on the structure of the nervous system |
| | Santiago Ramón y Cajal | Spain | |
| 1907 | Alphonse Laveran | France | disease-causing protozoa |
| 1908 | Ilya Mechnikov | Russia | work on immunity |
| | Paul Ehrlich | Germany | |
| 1909 | Theodor Kocher | Switzerland | physiology, pathology, and surgery of thyroid gland |
| 1910 | Albrecht Kossel | Germany | protein cell chemistry, including nucleic substances |

*(Continued)*

| Year | Recipient(s) | Country | Work |
|---|---|---|---|
| 1911 | Allvar Gullstrand | Sweden | dioptrics of the eye |
| 1912 | Alexis Carrel | France | vascular suture and transplantation of blood vessels and organs |
| 1913 | Charles Richet | France | work on anaphylaxis |
| 1914 | Robert Bárány | Austria | physiology and pathology of the vestibular apparatus |
| 1915* | | | |
| 1916* | | | |
| 1917* | | | |
| 1918* | | | |
| 1919 | Jules Bordet | Belgium | discoveries relating to immunity |
| 1920 | August Krogh | Denmark | capillary motor-regulating mechanism |
| 1921* | | | |
| 1922 | Archibald V. Hill | United Kingdom | production of heat in the muscle; |
| | Otto F. Meyerhof | Germany | fixed relationship between oxygen consumption and metabolism of lactic acid in muscle |
| 1923 | Frederick Banting | Canada | discovery of insulin |
| | John J. R. Macleod | Canada | |
| 1924 | Willem Einthoven | Netherlands, b. Dutch East Indies | discovery of electrocardiogram |
| 1925* | | | |
| 1926 | Johannes Fibiger | Denmark | discovery of Spiroptera carcinoma |
| 1927 | Julius Wagner-Jauregg | Austria | discovery of therapeutic value of malaria inoculation for dementia paralytica |
| 1928 | Charles Nicolle | France | work on typhus |
| 1929 | Christiaan Eijkman | Netherlands | discovery of antineuritic vitamin; |
| | Frederick Hopkins | United Kingdom | discovery of growth-stimulating vitamins |
| 1930 | Karl Landsteiner | United States and Netherlands b. Austria | discovery of human blood groups |
| 1931 | Otto Warburg | Germany | nature and mode of action of respiratory enzyme |
| 1932 | Charles Sherrington | United Kingdom | functions of neurons |
| | Edgar Adrian | United Kingdom | |
| 1933 | Thomas Morgan | United States | role of chromosomes in heredity |
| 1934 | George Whipple | United States | liver therapy in cases of anemia |
| | George Minot | United States | |
| | William Murphy | United States | |

| Year | Recipient(s) | Country | Work |
|---|---|---|---|
| 1935 | Hans Spemann | Germany | organizer effect in embryonic development |
| 1936 | Henry Dale | United Kingdom | chemical transmission of nerve impulses |
| | Otto Loewi | Austria | |
| 1937 | Albert Szent-Györgyi | Hungary | biological combustion processes, relative to vitamin C, and catalysis of fumaric acid |
| 1938 | Corneille Heymans | Belgium | sinus and aortic mechanisms in regulation of respiration |
| 1939 | Gerhard Domagk | Germany | discovery of antibacterial effects of prontosil (sulfa drug) |
| 1940‡ | | | |
| 1941‡ | | | |
| 1942‡ | | | |
| 1943 | Henrik Dam | Denmark | discovery of vitamin K; |
| | Edward Doisy | United States | discovery of chemical nature of vitamin K |
| 1944 | Joseph Erlanger | United States | differentiated functions of single nerve fibers |
| | Herbert Gasser | United States | |
| 1945 | Alexander Fleming | United Kingdom | discovery of penicillin |
| | Ernst Chain | United Kingdom, b. Germany | |
| | Howard Florey | United Kingdom | |
| 1946 | Hermann Muller | United States | production of mutations by means of X-ray irradiation |
| 1947 | Carl F. Cori | United States, b. Austria | catalytic conversion of glycogen; |
| | Gerty T. Cori | United States, b. Austria; | |
| | Bernardo Houssay | Argentina | role of pituitary lobe hormone in the metabolism of sugar |
| 1948 | Paul Müller | Switzerland | high efficiency of DDT as poison against certain arthropods |
| 1949 | Walter Hess | Switzerland | functional organization of inter-brain that coordinates activities of internal organs; |
| | Egas Moniz | Portugal | therapeutic value of leucotomy (lobotomy) in certain psychoses |
| 1950 | Edward Kendall | United States | structure and biological effects of hormones of adrenal cortex |
| | Tadeus Reichstein | Switzerland, b. Poland | |
| | Philip Hench | United States | |

*(Continued)*

| Year | Recipient(s) | Country | Work |
|---|---|---|---|
| **1951** | Max Theiler | United States, b. South Africa | discoveries concerning yellow fever and how to combat it |
| **1952** | Selman Waksman | United States, b. Russia | discovery of streptomycin |
| **1953** | Hans Krebs | United Kingdom, b. Germany; | discovery of citric acid cycle; |
| | Fritz Lipmann | United States, b. Germany | discovery of coenzyme A and importance for intermediary metabolism |
| **1954** | John Enders | United States | growth of poliomyelitis viruses in cultures of various types of tissues |
| | Thomas Weller | United States | |
| | Frederick Robbins | United States | |
| **1955** | Hugo Theorell | Sweden | nature and mode of action of oxidation enzymes |
| **1956** | André Cournand | United States, b. France | heart catheterization and pathological changes in circulatory system |
| | Werner Forssmann | Germany | |
| | Dickinson Richards | United States | |
| **1957** | Daniel Bovet | Italy, b. Switzerland | effects of synthetics on vascular system and skeletal muscles |
| **1958** | George Beadle | United States | genes that act by regulating definite chemical events genetic recombination and genetic material of bacteria |
| | Edward Tatum | United States | |
| | Joshua Lederberg | United States | |
| **1959** | Severo Ochoa | United States, b. Spain | biological synthesis of RNA and DNA |
| | Arthur Kornberg | United States | |
| **1960** | Frank M. Burnet | Australia | discovery of acquired immunological tolerance |
| | Peter Medawar | United Kingdom | |
| **1961** | Georg von Békésy | United States, b. Hungary | physical mechanism of stimulation within cochlea |
| **1962** | Francis Crick | United Kingdom | molecular structure of nucleic acids (DNA) |
| | James Watson | United States | |
| | Maurice Wilkins | United Kingdom | |
| **1963** | John Eccles | Australia | ionic mechanisms in peripheral and central portions of nerve cell membranes |
| | Alan Hodgkin | United Kingdom | |
| | Andrew Huxley | United Kingdom | |
| **1964** | Konrad Bloch | United States, b. Germany | mechanism and regulation of cholesterol and fatty acid metabolism |
| | Feodor Lynen | Germany | |
| **1965** | François Jacob | France | genetic control of enzyme and virus synthesis |
| | André Lwoff | France | |
| | Jacque Monod | France | |

| Year | Recipient(s) | Country | Work |
|---|---|---|---|
| 1966 | Peyton Rous | United States | discovery of tumor-inducing viruses |
| | Charles Huggins | United States | hormonal treatment of prostatic cancer |
| 1967 | Ragnar Granit | Sweden, b. Finland | primary physiological and chemical visual processes in the eye |
| | Haldan Hartline | United States | |
| | George Wald | United States | |
| 1968 | Robert Holley | United States | genetic code and function in protein synthesis |
| | HarGobind Khorana | United States, b. India | |
| | Marshall Nirenberg | United States | |
| 1969 | Max Delbrück | United States, b. Germany | replication mechanism and the genetic structure of viruses |
| | Alfred Hershey | United States | |
| | Salvador Luria | United States, b. Italy | |
| 1970 | Bernard Katz | United Kingdom | humoral transmitters in nerve terminals, and storage, release, and inactivation mechanisms |
| | Ulf von Euler | Sweden | |
| | Julius Axelrod | United States | |
| 1971 | Earl Sutherland, Jr. | United States | action of hormones |
| 1972 | Gerald Edelman | United States | chemical structure of antibodies |
| | Rodney Porter | United Kingdom | |
| 1973 | Karl von Frisch | Germany, b. Austria | organization and elicitation of individual and social behavior patterns |
| | Konrad Lorenz | Austria | |
| | Nikolaas Tinbergen | United Kingdom, b. Netherlands | |
| 1974 | Albert Claude | Belgium | structural and functional organization of the cell |
| | Christian de Duve | Belgium and United States | |
| | George Palade | United States, b. Romania | |
| 1975 | David Baltimore | United States | interaction between tumor viruses and the genetic material of the cell |
| | Renato Dulbecco | United Kingdom, b. Italy | |
| | Howard Temin | United States | |
| 1976 | Baruch Blumberg | United States | new mechanisms for the origin and dissemination of infectious diseases |
| | Carleton Gajdusek | United States | |
| 1977 | Roger Guillemin | United States, b. France | peptide hormone production of the brain; |
| | Andrew Schally; | United States, b. Poland; | |
| | Rosalyn Yalow | United States | radioimmunoassay of peptide hormones |

*(Continued)*

| Year | Recipient(s) | Country | Work |
|---|---|---|---|
| 1978 | Werner Arber | Switzerland | restriction enzymes and their application to problems of molecular genetics |
| | Daniel Nathans | United States | |
| | Hamilton Smith | United States | |
| 1979 | Allan M. Cormack | United States, b. South Africa | computer assisted tomography |
| | Godfrey Hounsfield | United Kingdom | |
| 1980 | Baruj Benacerraf | United States, b. Venezuela | genetically determined structures on cell surfaces that regulate immunological reactions |
| | Jean Dausset | France | |
| | George Snell | United States | |
| 1981 | Roger Sperry | United States | functional specialization of the cerebral hemispheres information processing in the visual system |
| | David Hubel | United States, b. Canada | |
| | Torsten Wiesel | United States, b. Sweden | |
| 1982 | Sune Bergström | Sweden | prostaglandins and related biologically active substances |
| | Bengt Samuelsson | Sweden | |
| | John R. Vane | United Kingdom | |
| 1983 | Barbara McClintock | United States | discovery of mobile genetic elements |
| 1984 | Niels Jerne | Switzerland, b. London (Danish citizen) | development and control of immune system and principle production of monoclonal antibodies |
| | Georges Köhler | Switzerland, b. Germany | |
| | César Milstein | United Kingdom, b. Argentina | |
| 1985 | Michael Brown | United States | regulation of cholesterol metabolism |
| | Joseph Goldstein | United States | |
| 1986 | Stanley Cohen | United States | discoveries of growth factors |
| | Rita Levi-Montalcini | Italy/United States, b. Italy | |
| 1987 | Susumu Tonegawa | United States, b. Japan | genetic principle for generation of antibody diversity |
| 1988 | James W. Black | United Kingdom | discoveries of important principles for drug treatment |
| | Gertrude Elion | United States | |
| | George Hitchings | United States | |
| 1989 | J. Michael Bishop | United States | cellular origin of retroviral oncogenes |
| | Harold Varmus | United States | |
| 1990 | Joseph Murray | United States | organ and cell transplantation in treatment of human disease |
| | E. Donnall Thomas | United States | |

| Year | Recipient(s) | Country | Work |
|---|---|---|---|
| 1991 | Erwin Neher | Germany | function of single ion channels in cells |
| | Bert Sakmann | Germany | |
| 1992 | Edmond Fischer | United States, b. China | reversible protein phosphorylation as a biological regulatory mechanism |
| | Edwin Krebs | United States | |
| 1993 | Richard Roberts | United States, b. United Kingdom | discoveries of split genes |
| | Phillip Sharp | United States | |
| 1994 | Alfred Gilman | United States | G-proteins and their role in signal transduction in cells |
| | Martin Rodbell | United States | |
| 1995 | Edward Lewis | United States | genetic control of early embryonic development |
| | Christiane Nüsslein-Volhard | Germany | |
| | Eric Wieschaus | United States | |
| 1996 | Peter Doherty | United States, b. Australia | specificity of the cell-mediated immune defense |
| | Rolf Zinkernagel | Switzerland | |
| 1997 | Stanley Prusiner | United States | discovery of Prions |
| 1998 | Robert Furchgott | United States | nitric oxide as a signaling molecule in the cardiovascular system |
| | Louis Ignarro | United States | |
| | Ferid Murad | United States | |
| 1999 | Günter Blobel | United States, b. Germany | intrinsic signals of proteins that govern their transport and localization in the cell |
| 2000 | Arvid Carlsson | Sweden | signal transduction in the nervous system |
| | Paul Greengard | United States | |
| | Eric Kandel | United States, b. Austria | |
| 2001 | Leland Hartwell | United States | key regulators of the cell cycle |
| | R. Timothy Hunt | United Kingdom | |
| | Paul M. Nurse | United Kingdom | |
| 2002 | Sydney Brenner | United Kingdom, b. South Africa | genetic regulation of organ development and programmed cell death |
| | H. Robert Horvitz | United States | |
| | John Sulston | United Kingdom | |
| 2003 | Paul Lauterbur | United States | magnetic resonance imaging |
| | Peter Mansfield | United Kingdom | |
| 2004 | Richard Axel | United States | odorant receptors and the organization of olfactory system |
| | Linda Buck | United States | |
| 2005 | Barry Marshall | Australia | discovery of *Heliobacter pylori* in gastric and peptic ulcer disease |
| | J. Robin Warren | Australia | |

*(Continued)*

| Year | Recipient(s) | Country | Work |
|---|---|---|---|
| 2006 | Andrew Fire | United States | RNA interference-gene silencing by double-stranded RNA |
| | Craig Mello | United States | |
| 2007 | Mario Capecchi | United States, b. Italy | gene modifications in mice by the use of embryonic stem cells |
| | Martin Evans | United Kingdom | |
| | Oliver Smithies | United States, b. United Kingdom | |

Source: Nobelprize.org.

* No prize awarded. Prize money allocated to the Special Fund of the Nobel Price section for physiology or medicine.

‡ No prize awarded. Prize money was allocated as follows: 1/3 to the Nobel Prize main fund, and 2/3 to the Special Fund of the Nobel Prize section for physiology or medicine.

# Selected Bibliography


Adair, Eleanor, R., ed. *Microwaves and Thermoregulation*. New York: Academic Press, 1983.

Adams, Fred and Greg Laughlin. *The Five Ages of the Universe*. New York: The Free Press, 1999.

Allaby, Michael and Derek Gjertsen. *The Makers of Science* (5 volumes). Cambridge, UK: Oxford University Press, 2002.

American Association for the Advancement of Science. *Science for All Americans: A Project 2061 Report on Literacy Goals in Science, Mathematics, and Technology*. Washington, DC: AASS Publication, 1989.

Angier, Natalie. *The Canon: A Whirligig Tour of the Beautiful Basics of Science*. New York: Houghton Mifflin Co., 2007.

Asimov, Isaac. *Asimov's Biographical Encyclopedia of Science and Technology*. New York: Doubleday, 1964.

———. *Asimov's Chronology of Science & Discovery*. New York: Harper & Row, 1989.

———. *Beginnings: The Story of Origins—Of Mankind, Life, the Earth, and the Universe*. New York: Berkeley Books, 1987.

———. *Isaac Asimov's Guide to Earth and Space*. New York: Fawcett Crest, 1991.

Bacon, Francis. *Novum Organum*. 1620. Reprinted as *Physical & Metaphysical Works of Lord Bacon, Including the Advancement of Learning & Novum Organum*. St. Clair Shores, MI: Scholarly Press, 1976.

Baeyer, Hans Christian von. *Maxwell's Demon: Why Warmth Disperses and Time Passes*. New York: Random House, 1998.

Barnes-Svarney, Patricia, ed. *The New York Public Library Science Desk Reference*. New York: Macmillan, 1995.

Barr, Stephen M. *Modern Physics and Ancient Faith*. Notre Dame, IN: University of Notre Dame Press, 2003.

Barrow, John D. *Theories of Everything: The Quest for Ultimate Explanation*. Oxford, UK: Oxford University Press, 1991.

Beckmann, Petr. *A History of PI (II)*. New York: St. Martin's Press, 1971.

Bendick, Jeanne. *Archimedes and the Door of Science*. Bathgate, ND: Bethlehem Books, 2000.

Bolles, Edmund Blair, ed. *Galileo's Commandment: An Anthology of Great Science Writing*. New York: W. H. Freeman, 1997.

Boorstin, Daniel. *The Discoverers: A History of Man's Search to Know His World and Himself*. New York: Vintage Books, 1985.

Bruno, Leonard C. *Landmarks of Science. From the Collection of the Library of Congress*. 1987. Reprint, New York: Facts on File, 1990.

Brian, Denis. *Einstein: A Life*. New York: John Wiley & Sons, Inc., 1996.

Bunch, Bryan. *Handbook of Current Science & Technology*. Detroit, MI: Gale, 1996.

Bryson, Bill. *A Short History of Nearly Everything*. New York: Broadway Books, 2003.

Campbell, Norman. *What is Science?* New York: Dover, 1953.

Campbell, Steve. *Statistics You Can't Trust: A Friendly Guide to Clear Thinking About Statistics in Everyday Life*. Parker, CO: Think Twice Publishing, 1999.

Carnap, Rudolph. *An Introduction to the Philosophy of Science*, ed. Martin Gardner. New York: Dover, 1995.

Carne, John. *Eyewitness to Science*. Cambridge, MA: Harvard University Press, 1995.

Cheney, Margaret. *Tesla, Man Out of Time*. New York: Touchstone, 2001.

Close, Frank E. and Philip R. Page. "Glueballs." *Scientific American* (November 1998).

*Concise Science Dictionary*, 3rd ed. New York: Oxford, 1996.

Connor, James A. *Kepler's Witch: An Astronomer's Discovery of Cosmic Order Amid Religious War, Political Intrigue, and the Heresy Trial of His Mother*. New York: HarperCollins, 2004.

Coutts, Timothy J. and Mark C. Fitzgerald. "Thermophotovoltaics." *Scientific American* (September 1998).

Crombie, A. C. *The History of Science: From Augustine to Galileo*. New York: Dover, 1995.

Cromer, Alan. *Uncommon Sense: The Heretical Nature of Science*. New York: Oxford University Press, 1993.

Crystal, David, ed. *The Cambridge Paperback Encyclopedia*. Avon, UK: Cambridge University Press, 1993.

Daintith, John, and Sarah Mitchell, Elizabeth Tootill, and Derek Gjertsen. *Biographical Encyclopedia of Scientists*, 2d ed. 2 vols. Bristol, UK and Philadelphia: Institute of Physics Publishing, 1981–94.

Darwin, Charles. *On the Origin of the Species*. New York: Mentor, 1958.

Davis, Paul. *About Time*. New York: Touchstone/Simon & Schuster, 1995.

Derry, T. K. and Trevor I. Williams. *A Short History of Technology: From Earliest Times to A.D. 1900*. New York: Dover, 1960.

Disney, Michael. "A New Look at Quasars." *Scientific American* (June 1998).

Dobzhansky, Theodosius. *Genetics and the Origin of Species*, 1937. Reprint, New York: Columbia University Press, 1982.

Ehrlich, Paul. *The Population Bomb*. New York: Sierra Club/Ballantine, 1968.

*Einstein Revealed*. A NOVA Production by Green Umbrella Ltd. for WGBH/Boston in association with BBC-TV and Sveriges Television. 1996. Videocassette.

*Encyclopedia Brittannica CD*. Chicago, 1999.

Euclid. *Elements of Geometry. The Thirteen Books of Euclid's Elements*, ed. Gail Kay Haines. New York: Dover, 1989.

Ferris, Timothy. *Coming of Age in the Milky Way*. New York: Doubleday, 1988.

Feynman, Richard P. *Six Easy Pieces: Essentials of Physics*. New York: Helix/Addison-Wesley, 1995.

———. *The Character of Physical Law*. Cambridge, MA: MIT Press, 22d Printing, 1995.

———. *The Pleasure of Finding Things Out*. Cambridge, MA: Perseus Books, 1999.

Feynman, Michelle, ed. *Perfectly Reasonable Deviations: From the Beaten Track: Letters of Richard P. Feynman*. New York: Basic Books, 2005.

Finkebeiner, Ann. *The Jasons: The Secret History of Science's Postwar Elite*. New York: Viking, 2006.

Forrester, Jay. *World Dynamics*, 2d ed. Cambridge, MA: Wright-Allen, 1973.
Freeman, Ira M. *Physics Made Simple* (revised by William J. Durden). New York: A Made Simple Book/Doubleday, 1990.
Gamow, George. Mr. *Tompkins in Paperback*. Reprint, Cambridge, UK: Cambridge University Press, Reissued 1994.
Gerstein, Mark and Michael Levitt. "Simulating Water and the Molecules of Life." *Scientific American* (November 1998).
Gleick, James. *Chaos, Making a New Science*. New York: Penguin, 1987.
Gonzalez, Frank I. "Tsunami!" *Scientific American* (May 1999).
Green, Brian. *The Elegant Universe: Superstrings, Hidden Dimensions, and the Quest for the Ultimate Universe*. New York: Vintage Books, 2000.
Gribbin, John. *The Search for Superstrings, Symmetry, and the Theory of Everything*. Boston: Little, Brown, 1998.
———. *Q is for Quantum, an Encyclopedia of Particle Physics*. New York: Free Press, 1998.
———. *In Search of Schrodinger's Cat*. New York: Free Press, 1996.
*Handy Science Answer Book (The)*. Compiled by the Science and Technology Department of the Carnegie Library of Pittsburgh. Detroit, MI: Visible Ink Press, 1994.
Hardin, Garrett. *Living Within Limits: Ecology, Economics and Population Taboos*. New York: Oxford University Press, 1993.
Hawking, Stephen. *A Brief History of Time*. New York: Bantam, 1988.
———. *The Universe in a Nutshell*. New York: Bantam, 2001.
Harris, Stephen L. *Agents of Chaos*. Missoula, MT: Mountain Publishing Co., 1990.
Henry, Patrick, Ulrich G. Briel, and Hans Bohringer. "The Evolution of Galaxy Clusters." *Scientific American* (December 1998).
Herzenberg, Caroline L. *Women Scientists From Antiquity to the Present: An Index*. West Cornwall, CT: Locust Hill Press, 1986.
Hogan, Craig J., Robert P. Kirshner, and Nicholas B. Suntzeff. "Survey Space-time with Supernovae." *Scientific American* (January 1999).
Holmyard, E. J. *Alchemy*. New York: Dover, 1990.
Horgan, John. *The End of Science*. New York: Addison-Wesley, 1996.
Howard, Philip K. *The Death of Common Sense*. New York: Random House, 1994.
Huber, Peter. *Hard Green: Saving the Environment from the Environmentalists*. New York: Basic Books, 1990.
Isaacs, Alan, ed. *A Dictionary of Physics*. New York: Oxford University Press, 1996.
Jaffe, Bernard. *Crucibles: The Story of Chemistry*. New York: Dover, 1976.
Jones, Judy and William Wilson. *An Incomplete Education*. New York: Ballantine, 1995.
Kaku, Michio. *Visions: How Science Will Revolutionize the 21st Century*. New York: Anchor Books, 1997.
Karukstis, Kerrry K. and Gerald R. Van Hecke. *Chemistry Connections, the Chemical Basis of Everyday Phenomena*, 2d.ed. New York: Academic Press, 2003.
Kevles, Daniel J. *The Baltimore Case: A Trial of Politics, Science, and Character*. New York: W.W. Norton & Co., 1998.
Kohn, Alexandre. *From the Closed World to the Infinite Universe*. Baltimore: Johns Hopkins University Press, 1957.
Krebs, Robert E. *The History and Use of Our Earth's Chemical Elements*, 2d ed. Westport, CT: Greenwood, 2006.
———. *The Basics of Earth Science*. Westport, CT: Greenwood, 2003.
———. *Scientific Development and Misconceptions Through the Ages*. Westport, CT: Greenwood, 1999.
———. *Scientific Laws, Principles, Laws and Theories*. Westport, CT: Greenwood, 2001.
———. *Groundbreaking Scientific Experiments, Inventions & Discoveries of the Middle Ages and the Renaissance*. Westport, CT: Greenwood, 2004.

Krebs, Robert E. and Carolyn A. Krebs. *Groundbreaking Scientific Experiments, Inventions & Discoveries of the Ancient World.* Westport, CT: Greenwood, 2003.

Krupp, E. C. *Beyond the Blue Horizon.* New York: Oxford University Press, 1991.

Landy, Stephen D. "Mapping the Universe." *Scientific American* (June 1999).

Lasota, Jean-Pierre. "Unmasking Black Holes." *Scientific American* (May 1999).

Lennox, James. G. *Aristotle's Philosophy of Biology.* Cambridge, UK: Cambridge University Press. 2001.

Lerner, Rita G., and George L. Trigg. *Encyclopedia of Physics,* 2d ed. New York: VCH Publishers, 1991.

Levi, Primo. *The Periodic Table.* New York: Schocken/Random House, 1984.

Lindley, David. *Boltzmann's Atom: The Great Debate that Launched a Revolution in Physics.* New York: The Free Press, 2001.

Livingston, James D. "100 Years of Magnetic Memories." *Scientific American* (November 1998).

Lovelock, James. *Gaia: A New Look at Life on Earth.* New York: Oxford University Press, 1995.

Luminet, Jean-Pierre, Glenn D. Starkman and Jeffrey R. Weeks. "Is Space Finite?" *Scientific American* (April 1999).

MacDonald, Ian R. "Natural Oil Spills." *Scientific American* (November 1998).

Macrone, Michael. *Eureka! What Archimedes Really Meant.* New York: Cader, 1994.

Malthus, Thomas R. *An Essay on the Principle of Population,* ed. Anthony G. Flew. New York: Penguin, 1985.

Margotta, Roberto. *The History of Medicine.* London: Reed International Books, 1996.

Margulis, Lynn. *Symbiotic Planet: A New Look at Evolution.* New York: Basic Books, 1998.

Margulis, Lynn and Dorion Sagan. *Slanted Truths: Essays on Gaia, Symbiosis, and Evolution.* New York: Copernicus, 1997.

Mayewski, Paul Andrew and Frank White. *The Ice Chronicles: The Quest to Understand Global Climate Changes.* Lebanon, NH: The University Press of New England, 2002.

*McGraw-Hill Concise Encyclopedia of Science and Technology,* 3d. ed., ed. in chief Sybil P. Parker. New York: McGraw-Hill, 1994.

*McGraw-Hill Dictionary of Scientific and Technical Terms, 5th ed.,* ed. in chief Sybil P. Parker. New York: McGraw-Hill, 1994.

Millar, David, Ian Millar, John Millar, and Margaret Millar. *The Cambridge Dictionary of Scientists.* Cambridge, UK: Cambridge University Press, 1996.

Moran, Jeffrey B. *How Do We Know the Laws of Thermondynamics.* New York: Rosen Publishing Group, Inc., 2001.

Motz, Lloyd. *The Story of Physics.* New York: Avon, 1989.

Motz, Lloyd and Jefferson Hane Weaver. *Conquering Mathematics: From Arithmetic to Calculus.* New York: Plenum, 1991.

Moyer, Albert E. *A Scientist's Voice in American Culture: Simon Newcomb and the Rhetoric of Scientific Method.* Berkeley and Los Angeles: University of California Press, 1992.

Myers, Rusty L. *The Basics of Physics.* Westport, CT: Greenwood, 2006.

Nasar, Sylvia. *A Beautiful Mind: A Biography of John Nash.* New York: Touchstone, 1998.

National Academy of Sciences. *Science and Creationism: A View from the National Academy of Sciences, 2nd ed.* Washington, DC: National Academy Press, 1999.

Neimark, Peninah and Peter Rhoades Mott, eds. *The Environmental Debate: A Documentary History.* Westport, CT: Greenwood Press, 1999.

Nesse, Randolph M. and George C. Williams. "Evolution and the Origins of Disease." *Scientific American* (November 1998).

Newton, Isaac. *Philosopiae Naturalis Principia Mathematics.* 1650. *The Mathematical Papers of Isaac Newton,* ed. D. T. Whiteside. Cambridge, UK: Cambridge University Press, 1967–81.

Newton, Roger G. *The Truth of Science: Physical Theories and Reality.* Cambridge, MA: Harvard University Press, 1997.

New York Academy of Sciences. *The Flight from Science and Reason*, eds. Paul R. Gross, Norman Levitt, and Martin W. Lewis. New York: New York Academy of Sciences, 1996.

North, John. *Astronomy and Cosmology*. New York: Norton, 1995.

Norris, Christopher. *Against Relativism: Philosophy of Science, Deconstruction and Critical Theory*. Oxford, UK and Malden, MA: Blackwell Publishers, 1997.

O'Neill, John J. *Prodigal Genius: The Life of Nikola Tesla*. New York: Cosimo, 2006.

Oxlade, Chris, Corinne Stockley, and Jane Wertheim. *The Usborne Illustrated Dictionary of Physics*. London: Usborne Publishing, 1986.

Parsons, Michael L. *Global Warming: The Truth Behind the Myth*. New York: Insight Books (Plenum Publishers), 1995.

Pannekoek. A. *A History of Astronomy*. New York: Dover, 1961.

Penrose, Roger. *A Road to Reality: A Complete Guide to the Laws of the Universe*. New York: Alfred A. Knopf, 2004.

Randall, Lisa. *Warped Passages: Unraveling the Mysteries of the Universe; Hidden Dimensions*. New York: HarperCollins, 2005.

Read, John. *From Alchemy to Chemistry*. Toronto, ON, Canada: General Publishing, 1995.

Ronan, Colin. D. *Lost Discoveries: The Forgotten Science of the Ancient World*. New York: McGraw-Hill, n.d.

Rothmann, Milton A. *Discovering the Natural Laws: The Experimental Basis of Physics*. New York: Dover Publications, 1989.

Rothmann, Tony. *Instant: From Aristotle to Einstein, and Beyond*. New York: Fawcett Columbine, 1995.

Rudin, Norah. *Dictionary of Modern Biology*. New York: Carron's Educational Series, 1997.

Sacks, Oliver, et al. *Hidden Histories of Science*. New York: New York Review Book, 1995.

Sagan, Carl. *The Demon Haunted World: Science as a Candle in the Dark*. New York: Random House. 1995.

Santos, Miguel A. *The Environmental Crisis*. Westport, CT: Greenwood, 1999.

Scerri, Eric R. "The Evolution of the Periodic System." *Scientific American* (September 1998).

Schectman, Jonathan. *Groundbreaking Scientific Experiments, Inventions & Discoveries of the 18th Century*. Westwood, CT: Greenwood, 2003.

Schneider, Herman and Leo Schneider. *The Harper Dictionary of Science in Commonplace Language*. New York: Harper & Row, 1988.

*Science and Technology Desk Reference* Library of U. *Pittsburgh*. Detroit, MI: Gale, 1996.

Shabecoff, Philip. *A Fierce Green Fire: The American Environmental Movement*. New York: Hill and Wang, 1993.

Shamos, Morris H., ed. *Great Experiments in Physics: Firsthand Accounts: From Galileo to Einstein*. New York: Dover Publications, 1959.

Silver, Brian L. *The Ascent of Science*. New York: Oxford University Press, 1998.

Silvers, Robert B., ed. *Hidden Histories of Science*. New York: The New York Review of Books, 1995.

Simonis, Doris, ed. *Lives & Legacies: An Encyclopedia of People who Changed the World: Scientists, Mathematicians, and Inventors*. New York: Oryx Press. 1999.

Singer, Charles. *A History of Scientific Ideas*. New York: Barnes & Noble, 1966.

Singh, Simon. *The Big Bang*. New York: HarperCollins, 2004.

Smolin, Lee. *The Trouble With Physics: The Rise of String Theory, The Fall of a Science, and What Comes Next*. New York: Houghton Mifflin Company, 2006.

Sobel, Dava. *The Planets*. New York: Viking, 1995.

Sokal, Alan and Jean Bricmont. *Fashionable Nonsense: Postmodern Intellectuals' Abuse of Science*. New York: Picador, 1998.

Spangenburg, Ray and Diane K. Moser. *The History of Science: From Ancient Greeks to the Scientific Revolution*. New York: Facts on File, 1993.

———. *The History of Science: In the Eighteenth Century*. New York: Facts on File, 1993.

———. *The History of Science: In the Nineteenth Century*. New York: Facts on File, 1994.

———. *The History of Science: From 1946 to the 1990s*. New York: Facts on File, 1994.

Stachel, John, ed., with the assistance of Trevor Lipscombe and others. *Einstein's Miraculous Year: Five Papers That Changed the Face of Physics*. Princeton, NJ: Princeton University Press, 1998.

Stenger. Victor J. *The Comprehensible Cosmos: Where do the Laws of Physics Come From?* Amherst, New York: Prometheus Books, 2006.

Stone, Carol Leth. *The Basics of Biology*. Westport, CT: Greenwood, 2004.

Strahler, Arthur N. *Understanding Science: An Introduction to Concepts and Issues*. Buffalo, NY: Prometheus, 1992.

Stroup, Richard L. and Roger E. Meiners. *Cutting Green Tape: Toxic Pollutants, Environmental Regulations and the Law*. New Brunswick, NJ: Transaction Publishers, 2000.

Struik, Dirk J. *A Concise History of Mathematics*. New York: Dover Publications, Inc., 1987.

Susskind, Leonard. *The Cosmic Landscape: String Theory and the Illusion of Intelligent Design*. New York: Little, Brown & Co., 2006.

Swartz, Tim. *The Lost Journals of Nikola Tesla: HAARP—Chemicals and the Secret of Alternative 4*. New Brunswick, NJ: Global Communications, 2000.

Tesla, Nikola. *My Inventions: The Autobiography of Nikola Tesla*. Filiquarian Publishing, LLC, 2006. E-book.

Thain, M. and M. Hickman. *The Penguin Dictionary of Biology*. Reprint, New York: Penguin, 1995.

Trefil, James. *1001 Things Everyone Should Know About Science*. New York: Doubleday, 1992.

Tudge, Colin. *The Time Before History: 5 Million Years of Human Impact*. New York: Touchstone, 1996.

Verma, Surendra. *The Little Book of Scientific Principles, Theories, & Things*. Sydney, Australia: New Holland Publishers Pty Ltd., 2005.

Von Baeyer, Hans Christian. *Maxwell's Demon: Why Warmth Disperses and Time Passes*. New York: Random House, 1998.

Waldrop, Mitchell, M. *Complexity*. New York: Touchstone, 1992.

Weaver, Warren. *Scene of Change: A Lifetime in American Science*. Reprint, New York: Charles Scribner's Sons, n.d.

Webster, Charles. *From Paracelsus to Newton: Magic and the Making of Modern Science*. New York: Barnes & Noble, 1982.

*Webster's II New Riverside University Dictionary*. New York: Riverside Publishing, 1994.

Weinberg, Steven. *The First Three Minutes: A Modern View of the Origin of the Universe* (1977), Updated ed. New York: Basic Books, 1993.

———. *Dreams of a Final Theory: The Scientist's Search for the Ultimate Laws of Nature*. New York: Vintage Books, 1993.

———. *Facing Up: Science and Its Cultural Adversaries*. Cambridge, MA: Harvard University Press, 2001.

Weissman, Paul R. "The Oort Cloud." *Scientific American* (September 1998).

Whitfield, Philip. *From So Simple a Beginning: The Book of Evolution*. New York: Macmillan, 1993.

Wilson, David Sloan. *Darwin's Cathedral: Evolution, Religion, and the Nature of Society*. Chicago: University of Chicago Press. 2002.

Wilson, Edward O. *Naturalist*. Washington, DC: Island Press, 1994.

———. *Consilience: The Unity of Knowledge*. New York: Knopf, 1998.

———. *The Future of Life*. New York: Alfred A. Knopf, 2002.

Windelspecht, Michael. *Groundbreaking Scientific Experiments, Inventions & Discoveries of the 17th Century*. Westport, CT: Greenwood, 2002.

Wynn, Charles M. and Arthur W. Wiggins. *Quantum Leaps: in the Wrong Direction: Where Real Science Ends and Pseudoscience Begins*. Washington, DC: Joseph Henry Press, 2001.

Yonas, Gerold. "Fusion and the Z Pinch." *Scientific American* (August 1998).

## INTERNET REFERENCES

http://astro.berkeley.edu
http://astroboligy.nrc.nasa.gov
http://books/nap.edu
http://cfpa.berkeley.edu
http://chem.ch.huji.ac.il
http://chemweb.ussie/news
http://content.cdlib.or
http://csep10.phys.ytj.edu
http://dbhs.wvusd.k12.ca.us
http://dept.physics.upenn.edu
http://ed-thelen.org
http://ei.cs.vt.edu
http://enviroment.newscientist.com
http://euler.ciens.ucv.veEnglish/mathem
http://geography.about.com
http://geology.about.com
http://gsahist.org
http://history.math.csusb.edu
http://home.san.rr.com
http://homepages.ihug.co.nz
http://humanists.net
http://hyperphysics.phy.astr.gsu.edu
http://institute-of-brilland-failures.com
http://inventors.about.com
http://library.thinkquest.org
http://love.geology.yale.edu
http://mathdl.maa.org
http://micro.magnet.fsu.edu
http://nature.com
http://neic.usgs.gov/neis/seismology
http://news-service.stanford.edu
http://newton.nap.edu
http://nobelprize.org
http://particle.physics.ucdavis.edu
http://people.ccmr.cornell.edu
http://pslc.ws/macog.nonlin
http://pub.acs.org
http://scienceworld.wolfram.com
http://solidstate.physics.sunny
http://spaceboy.nasda.go
http://step.sdsc.edu/project95/chem
http://stephenschneider.stanford.edu
http://tqd.advanced.org
http://trucha.ens.uabc.mx/bcymii/notas/historia
http://turnbull.des.st-amd.ac.uk/~history/Biographies
http://weather.cod.edu
http://web.mit.edu
http://www.accessexcellence.org
http://www.agnescott.edu
http://www.aim25.ac.uk

http://www.aip.org/history/climate
http://www.archimedes-lab.org
http://www.astro.virginia.edu
http://www.bellspalsy.org.uk
http://www.bergen.org/AAST
http://www.cartage.org.lb/en/bioraphies
http://www.cartage.org
http://www.chemheritage.org
http://www.chemicalelements.com
http://www.corrosion-doctors.org
http://www.damtp.cam.ac.uk
http://www.depauw.edu
http://www.dieselduck.ca/library
http://www.edge.org
http://www.elgi.hu/museum/elete
http://www.enchantedlearning.com
http://www.exploratorium.edu
http://www.external.ameslab.gov
http://www.genomenewsnetwork.org/resources
http://www.groups.scs.st-and.ac.uk
http://www.hms.harvard.edu
http://www.holophile.con/history
http://www.ideesdefrance.fr/agenda/age
http://www.invectis.co.uk
http://www.invent.org/book
http://www.ippt.gov.pl
http://www.islandnet.com
http://www.ithaca.edu/
http://www.itl.nist.gov/div898/handbook
http://www.iupac.org
http://www.jbc.org
http://www.jcbmac.chem.brown.edu
http://www.jlab.org
http://www.kfki.hu/eotos/stepcikk
http://www.lapp.in2p3.fr/neutrinos
http://www.ma.iup.edu
http://www.minsocam.org
http://www.muskingum.edu
http://www.nahste.ac.uk
http://www.nap.edu
http://www.ncbi.nlm.nih.gov
http://www.netsrq.com
http://www.newsandevents.utoronto.com
http://www.nnbd.com/people
http://www.northwester.edu
http://www.nytimes.com
http://www.ornl.gov.sci/techresources/hyman
http://www.osti.gov
http://www.photolib.noaa.gov
http://www.physics.capcollege
http://www.polybiblio.com
http://www.psu.edu/nasa

http://www.sciam.com
http://www.science.uwaterloo.ca
http://www.scribridg.sds.edu
http://www.seds.org
http://www.sentex.net/-tcc/gtl-trends.html
http://www.soacelab.net
http://www.spartacus.schoonet.co.uk
http://www.students.wfu.edu
http://www.theharbinger.org
http://www.thetech.org/revolutionaries
http://www.tnr.com/doc
http://www.todayinsci.com
http://www.traderslog.com/biography
http://www.vigyanprasar.gov.in/scientists
http://www.wadham.ox.ac.uk
http://www.weburbia.com/physics
http://www.woonsockethigh.org/scientists
http://www.wramc.amedd.army.mil
http://www.writingup.com
http://www.aaas.org
http://www/wppdrpw/org
http://www-history.mcs.st-andrews.ac.uk
http://wwwxbi.ee/baer/laws.htm

# Index

Note: Page numbers for entries in text are **bold.**


Abbe, Ernst, 1
Abbe's Theory for Correcting Lens Distortions, **1**
Abegg, Richard, 2, 504
Abegg's Rule and Valance Theory, 2
Abel, Sir Frederick, 132
Abel, Niels Henrik, 2–3, 325
Abelian groups. *See* Abel's Theory of Groups
Abel Prize, 3
Abel's Theory of Groups, **2**
Abiogenesis. *See* Life
Absolute temperature. *See* Kelvin's Concepts of Energy
Ackerman, T. P., 487
Adams, John Quincy, 326
Adams, Roger, 3
Adams, Walter, 148
Adams' Concept of Hydrogenation, **3**
Adhemar, Joseph Alphonse, 4
Adhemar's Ice Age Theory, **4**
Aether (Ether), 22, 187, 335, 387–88, 446
Agassiz, Jean Louis R., 4–6, 90
Agassiz's Geological Theories, **4–6**
Agricola, Georgius, 6–7
Agricola's Theories of Earthquakes and Volcanoes, **6–7**
AIDS (Acquired Immune Deficiency Syndrome), 369; theories, 41, 214–15, 393–94
Air chemical composition, 492; dephlogisticated, 447–48; flammable, 81; liquid, 92, 302; masses, 57–58; pressure, 428, 536–37
Airy, George Biddell, 7, 158
Airy's Concepts of Geologic Equilibrium, **7**
Al-Battani, Abu Abdullah, 7–8
Al-Battani's Theories, **7–8**
Albedo, 46
Alchemy, 425
"All things are numbers," 453
Alpha particles, 481–82
Altimeter, 74, 428
Alvarez, Luis Walter, 8–9, 527
Alvarez, Walter, 9
Alvarez's Hypotheses of Subatomic Collisions, **8–9**
Ambartsumian, Viktor Amazaspovich, 9–10
Ambartsumian's Theory of Stellar Associations, **9–10**
Amdahl, Gene Myron, 10–11
Amdahl's Law, **10–11**
Amontons, Guillaume, 283
Ampère, André-Marie, 11–12, 15, 177
Ampère's Theories of Electrodynamics, **11–12**
Anatomy, theories: ancient physiology, 164–65; autonomic nervous system, 332–33; circulatory system, 208, 256–57; female reproduction, 176; germ-layer, 126; kidneys, 209; nervous system, 208–9; spinal nerve roots, 48; structure of animals, 363–64; Vesalius', 551–52
Anaximander of Miletus, 12–13

Anaximander's Concepts and Ideas, **12–13**
Anderson, Carl, 13, 88, 136, 271, 272, 581
Anderson, Philip Warren, 13–14, 550
Anderson-Hamiltonian theory, 14
Anderson's Positron Theory, **13**
Anderson's Theories and Model, **13–14**
Andrews, Thomas, 73
Ångström, Anders Jonas, 14–15
Ångström's Principle of Spectrum Analysis and Related Theories, **14–15**
Animal classification, 109. *See also* Taxonomy
Anthropology, 295–96, 338–40, 568–69
Antibiotics, 139–41, 188, 191
Antibodies, 294–95, 444–45, 535–36
Antimatter, 13, 136
Antisepsis/antiseptics, 354, 430
Antitoxins, 47
"Anything that can go wrong will go wrong.," 395
Apollo 15, 211
Arago, Dominique Francois J., 15
Arago's Wave Theory of Light, **15**
Arber, Werner, 15, 401
Arber's Concept of the Structure of DNA, **15**
Archer, Frederick Scott, 112
Archimedes of Syracuse, 16–19, 352
Archimedes' Theories, **16–19**
Arecibo Observatory, 143
Aristarchus of Samos, 96
Aristotle of Macedonia, 19–22, 26, 68, 96, 180, 209, 409, 486, 564, 565
Aristotle's Theories, **19–22**
Arrhenius, Svante August, 22–25, 277, 470, 519
Arrhenius' Theories, Principles, and Concepts, **22–25**
Arsphenamine, 140
Artificial intelligence (AI), 391–92, 538–39
Ascorbic acid, 468
Aspirin, 227
Asteroid gap theory, 316–17
Aston, Francis William, 25, 256
Aston's Whole Number Rule, **25**, 256
Astronomical Units (AU), 59, 80–81, 322, 513–14
Astronomy: ancient theories, 284–85; atmospheric pressure, 537; comets, 252, 316, 421, 564–65; cosmic neutrinos, 38–39; dark matter, 30, 39, 479, 520, 572; Doppler principle, 141–42; Drake equation, 142–43; Earth's ecliptic, 8; Earth's latitude, 313; Galileo's, 212–13; formation of chemical elements, 70; gravitational microlensing, 30; Hubble's law and constant, 278–79; Interstellar gases, 507; interstellar matter, 519–20; Kepler's three laws of planetary motion, 308–9; Kirkwood's asteroid gap, 316–17; Kuiper belt, 322; motions of the moon, 8; Olber's paradox, 418–19; Oort's galaxy and comet cloud, 421; perfect cosmological principle, 236; planetary life, 142–43; planetary orbits, 59, 220; pulsars, 236, 272–73, 525–26; radio, 290, 357–58; Roche limit, 475; spectroscopic, 280–81; thermal ionization, 489; x-ray, 290. *See also* Big Bang; Black holes; Galaxies; Planets; Solar System; Stars; Universe
Atom smasher. *See* Cyclotron
Atomic: bomb, 381, 437, 489, 507, 526, 545–46; pile, 423, 513, 567; project, 422–23, 513, 522, 543, 585; reactor, 585; structure, 60–61, 334, 480–82, 531–32; theories, 219–20, 583–84; transmutation, 480–81. *See also* Nuclear, fission, Quantum theory
Atomism Theories, 26–27, 219–20
Atropine, 333
Auger, Pierre Victor, 27
Auger Effect, **27**
Aurora Borealis, 15, 56, 220
Australopithecus afarensis, 295–96
Australopithecus africanus, 295
Australopithecus anamenis, 340
Autopoiesis. *See* Life
Avery, Oswald, 103
Avogadro, Lorenzo Romano Amedeo, 27–28, 52, 115, 222
Avogadro's constant, 76
Avogadro's Law, Hypotheses, and Number, **27–28**, 76, 222
Avogadro's number, 28, 52
Axiom, 168, 434

Baade, Wilhelm Heinrich W., 29–31
Baade's Theories of Stellar Phenomena, **29–31**
Babbage, Charles, 32
Babbage's Theory of Computing, **32**
Babinet, Jacques, 33, 37
Babinet's Principle, **33**
Babo, Lambert Heinrich C. von, 33–34
Babo's Law, **33–34**
Bacon, Sir Francis, 34, 559

Bacon's Concept of Inductive Reasoning, **34–35**
Bacteriology: antisepsis, 354; bactericide hypothesis, 188; bacteriolytic, 132–33; destruction of bacteria, 442; dyes as antibiotics, 139–41; germ-disease postulate, 318–19; infectious disease, 94; microscopic life, 345; mucus secretions, 191; transmission of yellow fever, 465–66. *See also* Germs
Bacteriophage, 15, 129–30, 133, 343, 401, 413, 491
Baekeland, Leo Hendrik, 35–36, 38
Baekeland's Concept of Synthetic Polymerization, **35–36**
Baer, Karl von, 36–37, 521
Baer-Babinet law of current flow, 37
Baer's Laws of Embryonic Development, **36–37**
Baeyer, Johann Friedrich Adolph von, 36, 37–38
Baeyer's Strain Theory for Compound Stability, **37–38**
"Baghdad boils," 347
Bahcall, John Noris, 38–39
Bahcall's Theory for the Solar Neutrino Model, **38–39**
Bakelite, 36, 38
Bakker, Robert, 39–40
Bakker's Dinosaur Theory, **39–40**
Balloons, hot air, 222, 271–72
Balmer, Johann Jakob, 40, 482
Balmer Series, **40–41**, 482
Baltimore, David, 41, 144
Baltimore's Hypothesis for the Reverse Transfer of RNA to DNA, **41**
Banach, Stefan, 41–42
Banach-Tarski paradox, 42
Banach's Theory of Topological Vector Spaces, **41–42**
Banting, Sir Frederick Grant, 42–43
Banting's Theory for Isolating Pancreatic Insulin, **42–43**
Barbiturates, 38
Bardeen, John, 43, 107, 297, 414, 503
Bardeen's Theory of Superconductivity, **43–44**
Barnard, Edward Emerson, 303
Barnard's star, 303
Barometers, 119, 428, 536–37
Barringer, Daniel Moreau, 44–45
Barringer's Impact Theory of Craters, **44–45**
Baryons, 203, 225–26
Basov, Nikolai G., 362, 363, 537
Bassi, Agostino, 198
Bateson, William, 119
Bauer, Georg. *See* Agricola
Bayly, Helen Maria, 254
BCS (Bardeen Cooper Schrieffer) theory, 43–44, 297
Beadle, George Wells, 343, 525
Beagle, H.M.S., 122
Beaumont, Jean Baptiste, Elie de, 45
Beaumont's Theory for the Origin of Mountains, **45**
Becher, Johann, 514
Becquerel, Antoine Henri, 46, 107, 437
Becquerel's Hypothesis of X-Ray Fluorescence, **46**
Beer, August, 46–47, 328
Beer-Lambert-Bouguer Law, 46, 328–29
Beer's Law, **46–47**
Behavioral theories, 433–34, 569–70, 576
Behring, Emil Adolph von, 47–48
Behring's Theory of Immunology, **47–48**
Bell, Alexander Graham, 182
Bell, Sir Charles, 48
Bell's Law (Bell-Magendie Law), **48**
Bell's palsy/spasm, 48
Bell-Magendie Law. *See* Bell's Law
Berg, Otto, 411
Berg, Paul, 230, 491
Bergeron, Tor Harold P., 49
Bergeron-Findeisen theory, 49
Bergeron's Theory of Cloud Processes, **49**
Bernal, John Desmond (J.D.), 275
Bernoulli, Daniel, 50–51, 54, 101
Bernoulli, Jakob (Jacques), 49–50
Bernoulli, Johann, 50, 51
Bernoulli, Nicolaus, 50
Bernoulli's Law of Large Numbers, **49–50**
Bernoulli's Principle, **50–51**
Berson, Solomon, 577–78
Berthollet, Claude-Louis, 450
Berzelius, Jöns Jakob, 52–53, 227, 352, 392–93
Berzelius' Chemical Theories, **52–53**
Bessel, Friedrich, 51, 53–54
Bessel's Astronomical Theories, **53–54**
Bessel functions, 51, 54
Bessemer process, 506
Best, Charles, 42–43
Beta decay. *See* Nuclear
Betatron, 310–11
Bethe, Hans Albrecht, 54–55, 527, 561
Bethe's Theory of Thermonuclear Energy, **54–55**
Bevatron. *See* Particle accelerators
$B_2FH$ Theory. *See* Burbidge-Burbidge-Fowler-Hoyle Theory
Big Bang: foundations of, 149, 217–18, 290; modifications of, 196, 224, 570–71; nucleosynthesis, 70;

static universe theory and, 276–77, 347–48, 490–91, 494; support for, 258, 392, 508; theories, 133–34, 276–77, 439–40, 483. *See also* Universe
Big Crunch, 260
Binoculars, 1
Biochemistry theories: adrenal steroids, 307–8, 468; alcohol breakdown, 530–31; cyclol theory, 573–74; gamma globulin, 444–45; metabolism of cholesterol, 237–38; peptide synthesis, 383–84; projection formulas, 186; proteinoid microspheres, 196–97; separating/identifying sugars, 186
Biogeography, 556
Biology. *See* Bacteriology; Cells; DNA; Embryology; Evolution; Genetics; Immunology; Life; Medicine; Reproduction; RNA; Species; Taxonomy;
Biopoiesis. *See* Life
Biot, Jean Baptiste, 15, 55–56, 222
Biot-Savart Law, **55–56**
Birkeland, Kristian Olaf B., 56
Birkeland-Eyde process, 56
Birkeland's Theory of the Aurora Borealis, **56**
Bjerknes, Carl, 57
Bjerknes, Jacob, 57
Bjerknes, Vilhelm Friman K., 57–58
Bjerknes' Theory of Air Masses, **57–58**
Blackbody radiation, 315, 441, 515, 566
Black, James Whyte, 160–61
Black, Joseph, 58–59, 337
Black holes, 236, 259, 437–38, 494, 497–98. *See also* Stars
Black's Theories of Heat, **58–59**
Blackett, Patrick, 13
Blagden, Charles, 34
Bloch, Felix, 166, 367, 452, 453
Blood: groups, 331–32; human gamma globulin, 444–45; molecular structure of hemoglobin, 441–42; rate of evolution, 585; serum proteins, 533
Bode, Johann Elert, 59
Bode's Law for Planetary Orbits, **59**
Bohm, David Joseph, 59–60
Bohm's Interpretation of the Uncertainty Theory for Electrons, **59–60**
Bohr, Niels Hendrik David, 60–61, 129, 169, 190, 262, 331, 334, 380, 432, 455, 496
Bohr's Quantum Theory of Atomic Structure, **60–61**, 396, 504
Bok, Bart Jon, 61–62
Bok, Priscilla, 62
Bok's Globules Theory of Star Formation, **61–62**
Bolides, 9
Boltzmann, Ludwig Edward, 62–63, 86, 374–75, 380, 515
Boltzmann's Laws, Hypotheses, and Constant, **62–63**
Bonnet, Charles, 63
Bonnet's Theories of Parthenogenesis and Catastrophism, **63**
Boole, George, 63
Boole's Theory of Symbolic Logic, **63–64**
Boolean algebra and logic, 64
Born, Max, 64–65, 262, 380, 422, 468
Born-Haber Theory of Cycle Reactions, **64–65**
Born-Oppenheimer approximation, 422
Bose, Satyendra Nath, 330, 478
Bose-Einstein qualities, 330, 478
Bosons, 478–79, 548
Bothe, Walter, 64, 83
Bouguer, Pierre, 46, 47, 328
Boyle, Robert, 34, 35, 51, 65–66, 115, 220, 283
Boyle's Law, **65–66**, 283–84
Bradley, James, 66–67
Bradley's Theory of a Moving Earth, **66–67**
Brahe, Tycho, 67–68, 80, 84, 252–53, 308–9
Brahe's Theory of the Changing Heavens, **67–68**
Branly, Edouard Eugene D., 368
Brattain, Walter, 43, 44, 414, 503
Braun, Karl Ferdinand, 368
"Breeder reactor," 585
British Antarctic Society (BAS), 106
Brown, Alexander Crum, 305
Brown, Michael, 237
Brown, Robert, 151, 440
Brownian motion, 151–52, 221, 440–41
Bruno, Giordana, 197
Bubble chamber, 9, 88, 232–33
Buckminsterfullerene (Buckyballs), 108–9, 321–22
Buffon, George Louis Leclerc de, 68–69
Buffon's Theories of Nature, **68–69**
Bullen, Keith, 293
Bunsen, Robert, 65, 69–70, 202, 280, 316
Bunsen burner, 69
Bunsen's Theory of the Spectrochemistry of Elements, **69–70**
Burbidge, Eleanor Margaret Peachey, 70–71
Burbidge, Geoffrey, 70–71
Burbidge-Burbidge-Fowler-Hoyle Theory, **70–71**
Burnell, Jocelyn Bell, 236, 272–73

Butterfly effect, 356–57
Byerly, Perry, 293

Cagniard De La Tour, Charles, 73
Cagniard De La Tour's Concept of "Critical State," **73**
Cailletet, Louis, 73, 74
Cailletet's Concept for Liquefying Gases, **74**
Calculus. *See* Mathematics
Calendars, 284
Caloric, 479–80, 506
Calorimeter, 479–80
Calvin, Melvin, 74–75
Calvin's Carbon Cycle, **74–75**
Camera obscura, 111–12
Cancer, 144
Candolle, Augustin Pyrame de, 75–76
Candolle's Concept of Plant Classification, **75–76**
Cannizzaro, Stanislao, 28, 76, 386
Cannizzaro's Theory of Atomic and Molecular Weights, **76**
Cantor, Georg Ferdinand Ludwig P., 77–78
Cantor's Mathematical Theories, **77–78**
Capacitation factor, 85
Capacitors, 178
Carbon atoms: asymmetric, 342; benzene/hexagonal ring, 178, 304–5, 575; carbohydrate ring, 261; fullerenes, 321–22; inorganic/organic silicons, 314–15; molecular structure, 556–57; structure of, 37–38, 549. *See also* Chemistry, organic
Carbon cycle, 74–75
Carbon-nitrogen cycle, 54–55, 562
Carbonation, 337, 448
Cardano, Gerolamo, 78, 524–25
Cardano's Cubic Equation, **78**
Carnot, Nicholas Leonard Sadi, 78–79, 306, 374
Carnot's Theories of Thermodynamics, **78–79**
Carroll, James, 466
Carson, Rachel, 410, 466
Cartesian coordinate system, 130
Casimir, Hendrick Brugt Gerhard, 79
Casimir Force (Effect), **79–80**
Caspersson, Torbjorn, 80
Caspersson's Theory of Protein Synthesis, **80**
Cassini, Giovanni, 80–81, 476
Cassini's Hypothesis for the Size of the Solar System, **80–81**
Catalytic Converter, 138, 334
Catastrophism. *See* Evolution
Cavendish, Henry, 81, 87, 101, 461
Cavendish's Theories and Hypothesis, **81**
Celluoid, 428
Cells: animal, 497; cancer cell transformation, 144; cell enzyme synthesis, 426–27; cell pathology, 552; chromosomes, 555–56; cytology laws, 519; differences, 160–61; endosymbiotic, 368–69; growth, 350; human origins, 418; living cell, 218–19; mitosis/meiosis, 119; neuron theory, 276, 460–61, 555–56; nuclein theory, 388–89; "patch clamp," 403–4; plant, 493–94; protoplasm, 94, 560–61; somatic mutation, 295; sickle cell, 286–87; staining, 150–51; vesicles, 418
Cells (batteries). *See* Electricity
Celsius, Anders, 82
Celsius Temperature Scale, **82**, 306
Celsus, 425
Centigrade, 306
CERN (Conseil Européen pour la Recherche Nucléaire), 89, 459, 548
Cetus Corporation, 397
Chadwick, James, 82–83, 380, 419–20, 433, 510
Chadwick's Neutron Hypothesis, **82–83**
Chain, Ernst, 188, 191
Chain reaction. See Nuclear, fission
Chamberlain, Owen, 500
Chamberlain, Thomas, 293
Chambers, Robert, 83–84
Chambers, William. 84
Chambers' Theory for the Origin of Life, **83–84**
Chandler, Seth Carlo, Jr., 313
Chandrasekhar, Subrahmanyan, 84, 195
Chandrasekhar Limit, **84**
Chang, Min Chueh, 85
Chang Heng, 84–85
Chang's Theories and Concepts, **84–85**
Chang's Theory of Capacitation, **85**
Chaos theory, 449–50. *See also* Mathematics, complex systems
Chapman, Sydney, 86, 106
Chapman-Enskog Kinetic Theory of Gases, **86**
Chargaff, Erwin, 86, 103
Chargaff's Hypothesis for the Composition of DNA, **86**
Charles II of England, 71
Charles, Jacques Alexandre Cesar, 86–87, 283
Charles, Robert, 87, 115
Charles' Law, **86–87**, 283–84
Charney, Jule Gregory, 87–88
Charney's Theoretical Meteorology, **87–88**
Charpak, Georges, 88–89

Charpak's Concept of Tracking Particles, **88–89**
Charpentier, Jean de, 89–90
Charpentier's Glacier Theory, **89–90**
Chemical equivalents and types: *See* Periodic Table of the Chemical Elements
Chemistry: acid-base pairs, 23; adiabatic demagnetization, 228; addition hydrogenation, 567; adsorption, 334; Babo's law, 33–34; "buckyballs," 108–9, 321–22; carbon compounds, 37–38; carbon atoms, 108–9, 321–22; carbon cycle, 74–75; catalysts, 138–39, 424, 567; catalytic hydrogenation, 287–88; chemical bonding, 64–65, 175–76, 333–34, 350–52, 396, 431–32; chemical equivalents, 335–36; chemical proportions, 52; chemical thermodynamics, 228–29, 229–30; citric acid (tricarboxylic) cycle, 320–21; colloidal, 240; combustion, 336, 514; compounds, 115; coordinate bonds, 504–5; coordination compounds, 562–63; covalent bonds, 350–52; definite/constant composition, 273–74; definite proportions, 450–51; diffusion, 239; dissociation, 22–23; dyes, 139–41; effusion, 240; electrochemical, 52, 244; electrolysis, 123–25; electrolytes, 127; equilibrium, 342–43; fatty acids, 90; glycogen conversion, 100–101; Haber process, 243–44; heat, 58–59; iatro-chemistry, 425–26; ionic, 22–23, 151, 510–11; isomers, 342, 352; isomorphism, 392–93; isoprenes, 556–57; isotopes, 255–56; law of dilution, 424; mesomerism, 285–86; microchemistry, 426; molecular orbital theory, 279; nanostoichiometric, 450; octet theory, 405–6; orbital symmetry, 275; organic, 3–4, 186, 227–28, 261, 285–86, 304–5, 314–15, 336, 342, 352, 412–13, 556–57, 51, 572–73; Parkes process, 427–28; pH scale, 510–11; photochemical, 412–13; polymers, 35–36, 191–92, 287, 401–2, 584; projection formulas, 186; quantitative, 76; radicals, 52, 145, 352; rate law, 24; respiration, 336–37; retrosynthetic analysis, 98–99; reversible chemical reactions, 567–68; separating lanthanides, 512; solutions, 22; spectrochemistry, 69–70; stereochemistry, 430–31, 549–50; synthesizing hydrocarbons, 575; thermite reaction, 512–13; valence, 2, 199, 416; whole number rule, 25, 255–56; Williamson's synthesis, 568. *See also* Biochemistry; Elements; Gases
Chemotherapy, 151
Cherenkov, Pavel, 523
Cherenkov effect, 523–24
Chéseaux, Philippe Loys de, 501
Chevreul, Michel Eugene, 90
Chevreul's Theory of Fatty Acids, **90**
Chicxulub crater, 9, 160
Chiral bag model. *See* Nucleons
Chi-square test of statistical significance, 436
Chlorofluorocarbons, 106–7, 477–78
Cholesterol, 237–38
Christian IV of Denmark, 68
Christian of Lyons, 82
Chromatography, 75, 369–72, 512
Chromosphere, 147, 290
Chu, Paul Ching-wu, 90–91
Chu's Hypothesis for "High Temperature" Superconductivity, **90–91**
Churchill, Winston, 140
Circulatory system, 208
Citric acid cycle. *See* Krebs cycle
Clarke, Cyril, 91
Clarke's Supergene Theory, **91–92**
Classification. *See* Taxonomy
Claude, Georges, 92
Claude's Concept for Producing Liquid Air, **92**
Clausius, Rudolf Julius Emmanuel, 92–93, 306
Clausius' Laws and Theory of Thermodynamics, **92–93**
Climate. *See* Dendrochronology; Meteorology; Solar
Cloud chamber, 13, 88–89, 232–33, 390
Cloud condensation nuclei (CCN), 49
Cockcroft, Sir John Douglas, 93–94, 310, 337
Cockcroft-Walton Artificial Nuclear Reaction, **93–94**
Codons, 311–12
Cohen, Stanley, 350
Cohn, Ferdinand, 94
Cohn's Bacteria and Cell Theories, **94**
Cold fusion theory. *See* Nuclear
Collapsed star phenomena. *See* Black holes
Collip, James, 42–43
Columbus, Christopher, 245, 253, 446, 467
Comets, 252, 564–65
Compass, 417
Compton, Arthur Holly, 94–95
Compton's Wave/Particle Hypothesis, **94–95**

Computer-Aided Tomography (CT or CAT) scan, 453
Computer (human calculators), 32, 341, 542
Computer science theories: artificial automata, 553–54, 571; artificial intelligence, 391–92, 538–39; complex systems, 571–72; difference engine, 32; ENIAC, 87–88; Metcalfe's law, 385–86; Moore's law, 394–95; parallel computing, 10–11; symbolic logic, 63; very-large scale integration, 331
Conservation of mass, 337
Conservation of matter and energy, 263–64, 479–80
Conservation of momentum, 77, 130, 406–7, 412
Constellations. *See* Stars
Continental drift. *See* Geology
Contraception, 137
Conway, John Horton, 95–96
Conway's Game of Life Theory, **95–96**
Cooper, Leon, 43, 44, 297
Cooper pairs. *See* Electrons
Copenhagen Interpretation, 169–70. *See also* Bohr
Copernicus, Nicolaus, 8, 22, 96–98, 197, 231, 452, 471
Copernicus' Cosmology Theories, **96–98**
Corey, Elias James, 98–99
Corey's Theory of Retrosynthetic Analysis, **98–99**
Cori, Carl, 100–101
Cori, Gerty Radnitz, 100–101
Cori Theory of Catalytic Conversion of Glycogen, **100–101**
Coriolis, Gustave-Gaspard, 99–100
Coriolis effect, 247
Coriolis' Theory of Forces Acting on Rotating Surfaces, **99–100**
Correlation coefficient, 436
Cortisone, 307–8, 468
Cosmic radiation theory, 272
Cosmology. *See* Universe
Coster, Dirk, 380
Coulomb, Charles de, 101–2
Coulomb's Laws, **101–2**
Couper, Archibald Scott, 102–3, 575
Couper's Theory for the Structure of Carbon Compounds, **102–3**
Covalent reaction, 504
Cowan, Clyde, 180, 469
Craters, 9, 44–45
Crelle, August, 2, 3
Crick, Francis, 80, 86, 103–5, 160, 199–200, 218, 230, 349, 384, 395, 432, 579
Crick-Watson Theory of DNA, **103–5**
Critical state, 73
Crookes, William, 105, 461
Crookes dark space, 25
Crookes' Radiation Theories, **105**
Crutzen, Paul, 106–7
Crutzen's Theory of Ozone Depletion, **106–7**, 477
Cryogenics, 301–2
Crystallography, 257–58, 303–4, 441–42
Crystals, 438
Curie, Jacques, 107
Curie, Marie Sklodowska, 46, 101, 107–8, 249, 256, 297, 332
Curie, Pierre, 46, 101, 107–8, 249, 297, 332
Curies' Radiation Theories and Hypotheses, **107–8**
Curl, Robert, 108–9, 321
Curl's Hypothesis for a New Form of Carbon, **108–9**
Curtis, Heber D., 501
Cuvier, Georges Leopold Chrétien, 5, 76, 109–10
Cuvier's Theories of Anatomy and Taxonomy, **109–10**
Cyclols, 573–74
Cyclotron, 310–11, 337–38, 379. *See also* Particle accelerators
Cytogenetics. *See* Genetics
Cytology. *See* Cells

Daguerre, Louis-Jacques-Mandé, 111–12
Daguerre's Concept of How to "Freeze" Images Made by the Camera Obscura, **111–12**
Daguerreotypes, 112, 143–44
Dale, Henry, 113
Dale's Theory of Vagus Nerve Stimuli, **113**
D'Alembert, Jean le Rond, 113–14
D'Alembert's Principle of Fluid Dynamics, **113–14**
Dalton, John, 52, 81, 87, 114–15, 124, 222, 273–74, 450
Dalton's Laws and Theories, **114–15**
Dana, James Dwight, 116–17
Dana's Theory of Geosyncline, **116–17**
Daniell, John Frederic, 117–19
Daniell's Concept of the Electro-Chemical Cell, **117–19**
Dark matter, 30, 39, 479, 520, 572
Darlington, Cyril Dean, 119–20
Darlington's Theory of Cell Nuclear Divisions, **119–20**
Dart, Raymond, 295
Darwin, Charles, 27, 63, 76, 84, 197, 421: and genetics 215, 383; natural selection, 120–22, 248; organic evolution, 238, 359; predecessors of, 76, 328, 359
Darwin's Theory of Evolution by Natural Selection, **120–22**. *See also* Evolution

Davisson, Clinton Joseph, 122–23, 127
Davisson's Theory of Diffraction of Electrons, **122–23**
Davy, Humphry, 123–25, 177
Davy's Concept that Electric Current Can be Used to Separate Elements, **123–25**
Dawkins, Richard, 125–26
Dawkins' Theory of Evolution, **125–26**
DDT (dichlorodiphenyltrichloroethane), 410, 466
Dead water, 7, 158
De Beer, Gavin Rylands, 126
De Beer's Germ-Layer Theory, **126**
De Broglie, Louis, 123, 127, 496
De Broglie's Wave theory of Matter, **127**
Debye, Peter, 127–28, 228, 279
Debye-Hückel Theory of Electrolytes, **127–28**
Deduction, 168
De Forest, Lee, 167, 190
Dehmelt, Hans George, 128, 462
Dehmelt's Electron Trap, **128**
Delbrück, Max, 129–30, 343
Delbrück's and Luria's Phage Theory, **129–30**
Del Ferro, Scipione, 524–25
Delisle, Joseph-Nicolas, 82
Democritus of Abdera, 26, 114, 130, 164, 584
Democritus' Atomic Theory of Matter, 26, 65, **130**, 219
Demon Paradox, 375–76. *See also* Maxwell, James Clerk
Dendrochronology, 142
Density, 18–19
Department of Defense (United States), 527
Desaga, Peter, 69
Descartes, Rene du Perron, 104, 130–31, 429
Descartes' Theories and Philosophy, **130–31**
Desertification, 88
De Sitter, Willem, 149
Determinism, 334
Deuterons, 420
De Vries, Hugo, 131
De Vries' "Pangenes" Theory of Evolution, **131**
Dewar, Sir James, 131–32, 301
Dewar's Concept of Liquefying Gases, **131–32**
D'Herelle, Felix, 132–33
D'Herelle's Bacteriolytic Theory, **132–33**
Diabetes, 43
Dicke, Robert Henry, 133–34
Dicke's Theory of the Big Bang, **133–34**
Diesel, Rudolf Christian Carl, 134
Diesel's Concept of an Internal Combustion, Engine, **134–35**
Diffusion, 185–86
Difference engine. *See* Babbage
Dinosaurs, 39–40, 139, 463
Diodes, 182–83, 190
Diphtheria, 47–48
Dirac, Paul, 13, 135–37, 457, 496, 498, 500
Dirac's Relativistic Theories, **135–37**
Disease. *See* Germ theory; Medicine
Djerassi, Carl, 137
Djerassi's Theory of Synthetic Oral Contraception, **137**
DNA (deoxyribonucleic acid); beginning of life, 218; cell enzyme synthesis, 426–27; chimpanzee/human ratio, 353; colinearity of protein, 579; composition, 86; double helix, 103–4, 579; endosymbiotic cell theory, 368–69; evolutionary, 544–43, 568–69; fingerprinting, 292; genes with genes, 491; genetic profiling, 291–92; helix structure, 199–200; human/ape divergence, 568–69; molecular clock, 492; molecular weight, 80; nuclein theory, 388–89; nucleotide/nucleosides, 533–34; PCR method, 396–97; phage theory, 129–30; regulator (operon) genes, 289; replicators/replication, 125–26, 384, 396–97; restriction enzymes, 401; reverse transfer, 41, 528; sequencing, 230–31; splicing, 501–2; split genes, 474–75; tetra-nucleotide, 349
Döbereiner, Johann, 138–39
Döbereiner's Law of Triads, **138–39**
Dobzhansky, Theodosius, 139
Dobzhansky's Theory of Genetic Diversity, **139**
Domagk, Gerhard, 139–40
Domagk's Concept of Dyes as an Antibiotic, **139–40**
Donovan, Charles, 346
Doppler, Christian Johann, 141–42, 280
Doppler effect, 141, 148, 187, 514
Doppler-Fizeau shift, 141, 187
Doppler's Principle, **141–42**
Dorn, Friedrich Ernst, 461
Douglass, Andrew Ellicott, 142
Douglass' Theory of Dendrochronology, **142**
Drake, Frank Donald, 142–43
Drake Equation, **142–43**
Draper, John William, 143–44

Draper's Ray Theory, **143–44**
Drift chamber, 88
Drugs. *See* Pharmacology
DuFay, Charles, 201, 552
Dulbecco, Renato, 41, 144
Dulbecco's Cancer Cell Theory, **144**
Dumas, Jean Baptiste André, 145, 227
Dumas' Substitution Theory, **145**
Duppa, B.F., 199
Dynamo, 177–78, 188–89, 266, 528–29, 546
Dyson, Freeman John, 145–46, 487, 498
Dyson's Theory of Quantum Electrodynamics, **145–46**. *See also* Quantum theory

Earth: age, 69; aurora borealis, 15, 56, 220; axial movement, 98; calculating the shape, 54; circumference, 165, 445–46; companion star, 463; core, 293; Coriolis effect, 99–100; cyclic temperature, 218; declination/inclination of magnetic field, 164; density, 446; distance from sun, 80–81, 97, 165; ecliptic, 8, 165; formula for rotation, 192–94; global warming, 4, 24–25, 313, 373, 469–71; gravitational constant, 446; interior structure, 393; latitude, 313; life on, 236–37; magnetism and rotation, 231–32; magnetosphere, 546; mass, 81; meridian lines, 284; motions of Earth's moon, 8; nuclear winter, 486–87; orbit around sun, 66–67; precession, 4, 67, 313; radius, 136; temperature, 515; tidal effect, 86
Earthquakes. *See* Geology
Eastman, George, 35
Ecology, 68, 161–62, 248
Eddington, Sir Arthur Stanley, 55, 147–49
Eddington-Adams confirmation of special relativity, 148
Eddington's Theories and Concepts, **147–49**
Edelman, Gerald Maurice, 444, 533
Edison, Thomas Alva, 149–50, 166–67, 182, 528
Edison effect, 149–50, 167, 190
Edison's Theory of Thermionic Effect, **149–50**
Ehrlich, Paul, 48, 140, 150–51
Ehrlich's "Designer" Drug Hypothesis, **150–51**
Eigen, Manfred, 151, 412
Eigen's Theory of Fast Ionic Reactions, **151**
Einstein, Albert, 30, 136, 169, 244, 380, 412, 490, 537, 585: atom bomb, 181; Brownian motion, 151–52, 441; curved space, 148, 204–5; GUT, 156, 263; gravity 154–55, 525, 565; Isaac Newton, 152–53, 156; letter to FDR, 423, 522; light, 123, 148, 152, 155, 157, 456, 580; mass/motion, 152–53; Maxwell and relativity, 377; relativity, foundations of, 157; relativity, general, 155, 235, 259, 390, 437, 498, 570–71; relativity, special, 148, 153–54, 187, 356; relativity, Stephen Hawking, 258; saddle-shape universe, 204–5, 217; string theory, 156; TOE, 156; unified field theory, 155, 56, 490, 560
Einstein's Theories, Hypotheses, and Concepts, **151–57**
Einthoven, Willem, 157–58
Einthoven's Theory that the Heart Generates an Electric Current, **157–58**
Eisenhower, General Dwight D., 394
Ekman, Vagn Walfrid, 7, 158–59
Ekman's Hypothesis of the Coriolis Effect on Ocean Currents, **158–59**
Elasticity, 275–76
Eldredge, Niles, 110, 159–60, 238–39, 463
Eldredge-Gould Theory of Punctuated Evolution, **159–60**
Electricity: Ampère's law, 11–12; animal tissue, 216–17; Biot-Savart law,, 55–56; birefringence, 309–10; Casimir force, 79–80; cathode ray, 105, 531; cells (batteries), 117–19; circuits, 266; collision ionization, 538; conduction, 319; conductivity of ions, 319–20; Coulomb's law, 101–2; current and voltage laws, 315; dielectrics, 178; dynamos, 177–78, 188–89, 266, 506; Edison effect, 149–50, 167; electric/magnetic flux, 221; electric motor, 265–66, 178, 506; electrodynamics, 11–12, 145–46; electron theory, 518–19, 531–32; electrostatic generator, 546–47; fluid theory, 200–201, 557; force, 447; fundamental law, 101–2; galvanization, 216; Hall effect, 253–54, 317–18; high voltage ac, 528–29; induction, 177–78; Josephson effect/junction, 297–98; Joule's law, 298–99; Kerr effect, 309–10; lightning, 201; magnetic induction, 15; n-p-n junction, 504; Ohm's law, 417–18; piezoelectricity,

107, 257, 332; positive/negative force, 201; pyroelectricity, 257; rectifying AC to DC, 190; right/left-hand rule, 188–89; semiconductors, 14, 166–67, 297–98, 321, 331; Stark effect, 515; superconductivity, 14, 372; thermionic effect, 149–50; thermoelectricity, 499–500; transverse waves, 306–7; voltaic pile, 552–53. *See also* Electromagnetism; Electrons

Electrocardiogram/electrocardiograph (ECG/EKG), 158

Electrodynamics, 145–46, 183–84, 329–30, 498

Electrolysis, 177, 279–80

Electrolytes, 319

Electromagnetic force (field), 235

Electromagnetic spectrum, 40, 202, 315–16, 377, 533

Electromagnetism, 15: amber effect, 231; amplifying electromagnetic waves, 537–38; Babinet's theorem, 33; blackbody radiation, 515–16; cathode ray, 105, 441; Compton effect, 94–95; contraction, 186–87; Coulomb's theory, 101–2; displacement law, 566; electroweak interactions, 233–35; fields, 306–7; geon theory, 564; Hall effect, 317–18; Hertzian waves, 269–70; induction, 177–78, 265; least-time principle, 180; Lenz's law, 348–49; Lorentz force, 355; maser/laser, 362–63; Mössbauer effect, 155; Maxwell's theory, 376–77; Moseley's law, 395; Oersted's theory, 416–17; penning trap, 128; photons, 148, 281, 537–38, 558–59; Poynting vectors, 446; radiation, 152, 281; rays, 143–44; right/left hand rules, 188–89; Stark effect, 515; telegraph, 266; unified field theory, 155–56; unifying with weak force, 559–60; Zeeman effect, 541, 583. *See also* Electrons; Light; Weak Force

Electrometer, 107

Electromotive force, 348–49

Electronegativity, 396, 431–32. *See also* Chemistry, chemical bonding

Electrons: anti-electrons (positrons), 500; antimatter, 136; Auger effect, 27; charge, 390; complementary principle, 61; Cooper pairs, 79; coordinate bonds, 504–5; correspondence principle, 61; diffraction, 122–23; discovery, 456; discrete absorption, 198–99; Doppler effect, 514–15; embedded electrons, 532; emission, 348; ionization, 538; kinetic energy, 472; Larmor precession, 335; layered structure, 333–34; Lorentz' theory, 355; magnetic moment, 322–23; measuring outer energy, 539; metals undergoing transition, 14; negative energy, 136; Pauli exclusion principle, 432–33; penning trap, 128; scattering from partons, 203–4; Stoney's theory, 518–19; thermionic diode, 182–83; thermionic emission, 472; Thomson's theory, 531–32; tunnel diodes, 166–67; two-electron theory, 79; uncertainty principle, 262; wave characteristics, 123, 127; wave function, 496–97. *See also* Superconductivity

Electrophoresis, 286, 292, 371

Electroscopes, 271

*Elements* (Euclid), 168, 185, 525

Elements: ancient symbols, 116; artificial, 296–97, 410–11; atomic number, 395; atomic theory, 114–15; atomic transformation, 419–20; atomic transmutation, 411; atomic weights, 145; "eka," 138–39, 382; formation, 70, 218, 277; functional group, 145; gaseous diffusion, 542; in sunspots, 250–51; isotopes, 255–56, 510; law of triads, 138–39; naming, 52; octet theory, 405–6; paramagnetic materials, 550–51; Parkes process, 427–28; periodicity, 382–83, 386–87, 405–6, 482; purifying uranium, 512–13; radioactive decay, 108; separating lanthanides, 512; solar helium, 290–91, 355; theory of types, 145; transmutation, 249; transmutation of heavy into light elements, 297; transuranium, 499; valance, 199; whole number rule, 25. *See also* Periodic Table of the Chemical Elements

Elion, Gertrude Belle, 160–61

Elion's Theory for Cell Differences, **160–61**

Elizabeth I of England, 34

Elton, Charles, 161–62

Elton's Theory of Animal Ecology, **161–62**

Elvius the Elder, 82

Embryology theories, 36–37, 126, 173–74, 247–48, 374, 382–83, 521

Empedocles, 22

Encke, Johann, 80

Enders, John, 162–63, 474

Enders' Theory for Cultivation of Viruses, **162–63**
Energy: cold fusion, 187–88; conservation of energy laws, 50–51, 263–64, 479–80; definitions, 263; negative, 136; quantum theory, 443; regenerating heat, 505–6; thermonuclear, 54–55. *See also* Entropy; Nuclear; Thermodynamics
Engineering, 134–35
Enigma code, 394
Enskog, David, 86
Enskog theory. *See* Chapman-Enskog Kinetic Theory of Gases
Entropy, 62, 92, 263, 331, 375–76, 507. *See also* Thermodynamics
Environment, 106–7
Enzymes, 413, 415–16, 525, 530
Eötvös, Baron Roland von, 163–64
Eötvös' Rule, **163–64**
Epicurus of Samos, 26, 219
Epicurus' theory of the atom, 26
Epigenesis. *See* Evolution
Equinoxes, 8
Erasistratus of Chios, 164–65, 208
Erasistratus' Theory of Anatomy and Physiology, **164–5**
Eratosthenes of Cyrene, 54, 165, 446
Eratosthenes' Mathematical Concepts, **165**
Ergotism, 113
Ernst, Richard Robert, 165–66
Ernst's Theory of the Magnetic Moment of Atomic Nuclei, **165–66**
Esaki, Leo, **166–67, 297**
Esaki's Theory of Tunnel Diodes, **166–67**
Esperanto, 435
Ether. *See* Aether
Euclid, 167–68, 467
Euclid's Paradigm for All Bodies of Knowledge, **167–68**
Eudoxus, 168
Eudoxus' Theory of Planetary Motion, **168**
Eugenics. *See* Genetics
Euler, Leonhard, 50, 169, 313, 374, 474
Euler's Contributions in Mathematics, **169**
Eve hypothesis, 569
Event horizon, 259, 437–38, 498
Everett, Hugh, 169–70
Everett's Multiple-Universe Theory of Reality, **169–70**
Evolution: acquired characteristics, 327–28; biparental heredity, 374; catastrophism, 5, 45, 63, 110, 159–60; clandestine, 126; common function of parts, 109–10; cyclic extinction, 462–63; disclaimer for reptile/bird, 277–78; endosymbiotic, 369–70; epigenesis, 36–37; ESS, 377–78; evidence for, 121–22; fossils and evolution, 126, 339–40; general uniformatarianism, 122; genetic drift, 573; gerontomorphosis, 126; group selection, 576; hierarchical reductionalism, 125–26; human, 295–96, 339–40; human/ape divergence, 568–69; inorganic origin, 248; mutations, 312; natural selection, 120–22, 248, 250, 555, 556; organic, 248; "Out of Africa," 568–69; packing, 161–62; pangenes, 131; pedomorphosis, 126; population catastrophe, 364–65; preformation, 63, 374, 521; punctuated equilibrium, 159–60, 238–39; rate, 585; recapitulation, 37, 239; Roman Catholic church, 197; social Darwinism, 125, 248; statistical theories, 435–36; theist, 40; Wallace line, 556; Weismann barrier, 561. *See also* Embryology; Genetics; Species
Ewing, William Maurice, 170
Ewing's Hypothesis for Undersea Mountain Ridges, **170**
Exner, Franz, 380
Explorer satellites, 545
Eyde, Samuel, 56
Eyring, Henry, 170–71
Eyring's Quantum Theory of Chemical Reaction Rates, **170–71**

Fabricius' Theory of Embryology, **173–74**
Fabrizio, Girolamo, 173–74
Fahrenheit, Daniel, 82, 174
Fahrenheit's Concept of a Thermometer, **174**
Fairbank, William, 174–75
Fairbank's Quark Theory, **174–75**
Fajans, Kasimir, 175–76
Fajans' Rules for Chemical Bonding, **175–76**
Fallopius, Gabriel, 176
Fallopius' Theories of Anatomy, **176**
Faraday, Michael, 12, 34, 69, 70, 124, 177–79, 265, 376, 518
Faraday's Laws and Principles, **177–79**
Farman, Joe, 106
Ferdinand II of Tuscany, 174
Fermat, Pierre de, 179–80, 374, 430
Fermat's Principles and Theories, **179–80**
Fermentation, 430
Fermi, Enrico, 136, 180–82, 344, 380, 411, 516, 548, 574, 585
Fermi-Dirac statistics, 478

Fermi's Nuclear Theories, **180–82**
Ferrari, Ludovico, 324–25
Ferrel cell, 247
Fertilizer, 56, 65, 243
Fessenden, Reginald Aubrey, 182–83
Fessenden's Concept of the Thermionic Diode, **182–83**
Feynman, Richard Phillips, 145, 183–84, 457, 487, 498, 534, 564
Feynman's Theory of Quantum Electrodynamics (QED), **183–84**
Fibonacci, Leonardo, 184–85
Fibonacci's Numbering System, **184–85**
Fick, Adolf Eugen, 185–86
Fick's Laws of Diffusion, **185–86**
Findeisen, Walter, 49
Finlay, Carlos, 465
*Finnegans Wake*, 226
Fior, Antoniomaria, 524
Fischer, Emil Hermann, 186
Fischer, Ernst, 567
Fischer's Projection Formulas, **186**
Fisher, Sir Ronald A., 573
Fission. *See* Nuclear
Fitch, Val, 527
Fittig, Wilhelm Rudolph, 575
Fitzgerald, George Francis, 186–87, 355–56
Fitzgerald's Concept of Electromagnetic Contraction, **186–87**
Fizeau, Armand Hippolyte Louis, 15, 141, 187, 192, 280
Fizeau's Theory of the Nature of Light as a Wave, **187**
Flamsteed, John, 252
Flavell, Richard, 291–92
Fleischmann, Martin, 187–88
Fleischmann's Theory for Cold Fusion, **187–88**
Fleming, Alexander, 188, 191
Fleming, John Ambrose, 167, 182, 188–90
Fleming's Bactericide Hypothesis, **188**
Fleming's Rule for Determining Direction of Vectors, **188–90**
Flerov, Georgii Nikolaevich, 190–91
Flerov's Theory of Spontaneous Fission, **190–91**
Florey, Howard Walter (Baron of Florey of Adelaide), 188, 191
Florey's Theory of Mucus Secretions, **191**
Flory, Paul John, 191–92
Flory's Theory of Nonlinear Polymers, **191–92**
Fluorescence, 518
Food chain, 162
"Form follows function," 109, 209
Fossil theories, 5–6, 110, 339–40, 358–59, 462–63, 464, 492, 516–17
Foucault, Jean Bernard Leon, 15, 187, 192–94
Foucault's Theories of Light and Earth's Rotation, **192–94**
Fourier, Jean-Baptiste-Joseph, 195, 417
Fourier's Theories of Heat Conduction and Harmonic Wave Motion, **195**
Fowler, William, 70, 195–96
Fowler's Theory of Stellar Nucleosynthesis, **195–96**
Fox, Sidney Walter, 196–98
Fox's Theory of Proteinoid Microspheres, **196–98**
Fracastoro, Girolamo, 198
Fracastoro's Theory of Disease, **198**
Fractals, 438–39, 571–72
Franck, James, 198–99
Franck's Theory of Discrete Absorption of Electrons, **198–99**
Frank, Ilya, 523
Frankland, Sir Edward, 199, 416
Frankland's Theory of Valence, **199**
Franklin, Benjamin, 200–201, 229, 326, 447, 531, 552
Franklin, Rosalind, 103–5, 199–200, 349, 395
Franklin's Concept of DNA Structure, **199–200**
Franklin's Theories of Electricity, **200–201**
Fraunhofer, Josef von, 201–2, 316
Fraunhofer's Theory of White Light, **201–2**
Frederick II of Denmark, 68
Frege, Gottlob, 64
Fresnel, Augustin Jean, 15, 202–3
Fresnel's Theory for Multiple Prisms, **202–3**
Friedman, Jerome Isaac, 203–4
Friedman's Theory of the Quark Structure of Nucleons, **203–4**
Friedmann, Alexsandr Alexandrovich, 204–5
Friedmann's Theory of an Expanding Universe, **204–5**
Frisch, Otto Robert, 61, 205, 380–81, 411, 437, 521
Frisch's Theory of a Chain Reaction, **205**
Fukui, Kenichi, 275
Fuller, R. Buckminster, 322
Fundamental forces (fields) of the universe, 235, 560
Furnaces, 506
Fusion. *See* Thermonuclear energy

Gabor, Dennis, 207–8
Gabor's Theory of Reproducing Three-Dimensional Images, **207–8**
Galaxies: Andromeda, 341, 507–8; classification, 278; dark matter, 30, 39, 479,

520, 572; evolution, 29–31, 585–86; interstellar gases, 507; Milky Way, 266, 303, 341, 421, 500–501, 508, 572; NGC205 (galaxy), 267; non-uniform distribution, 224; quasars, 482–83, 490; rotation, 302–3; Sagittarius, 290; structure, 266. *See also* Stars
Galen, 208–9, 256, 425, 551–52
Galen's Theories of Anatomy and Physiology, **208–9**
Galilean transformation, 153
Galileo, Galilei, 77, 197, 220, 309, 407: Aristotle, 21; gravity, 209–11; meteorology, 283; and planetary motion, 309, 212–14; pendulums, 193, 211–12, 471; thermoscope, 174; timekeeping, 211–12
Galileo's Theories, **209–14**
Galle, Johann Gottfried, 53
Gallo, Robert, 214–15, 394
Gallo's HIV-AIDS Theory, **214–15**
Galton, Sir Francis, 215–16
Galton's Theory of Eugenics, **215–16**
Galvani, Luigi, 118, 157, 216–17, 552–53
Galvani's Theories of Galvanization and Animal Tissue Electricity, **216–17**
Galvanization, 216
Galvanometer, 158, 177, 216–17
Game theory, 377–78
Gamow, George, 70, 196, 217–19, 224, 240, 276, 561
Gamow's Theories of the Universe and DNA, **217–19**
Gardiner, Bryan, 106
Gardner, Martin, 95–96
Garrod, Sir Archibald Edward, 219
Garrod's Theory of Congenital Metabolic Disorders, **219**
Gases: adsorption, 334; Avogadro's hypothesis, law, number, 27–28, 76; Boltzmann's laws, hypotheses, constant, 62–63; Boyle's law, 65–66; catalytic converters, 334; Charles' law, 86–87; collision ionization, 538; combining volumes, 222; combustion, 336; critical state, 73; Dalton's law, 114; equation for gas molecules, 548–49; fluorescence, 518; gaseous diffusion, 185–86, 436–37, 542–43; generalized gas law, 283–84; Ideal gas law, 283–84, 548–49; inert, 461; ionization, 271–72; Joule-Thomson effect, 299; kinetic theory, 86, 209, 305–6, 374–5; liquefying, 74, 92, 131–32, 302, 379; measuring energy level of molecules, 539; Meissner effect, 379; noble, 461; partial pressure, 114; perfect gases, 283–84; spectrum analysis, 14–15; two-fluid model for helium, 330–31; Tyndall effect, 540; weight, 264–65. *See also* Elements
Gassendi, Pierre, 219–20
Gassendi's Theories, **219–20**
Gates, Frederick, 442
Gauss, Karl Friedrich, 77, 169, 220–21, 245
Gauss' Mathematics and Electromagnetism Theorems, **220–21**
Gay-Lussac, Joseph-Louis, 56, 87, 90, 115, 222, 283, 352
Gay-Lussac's Law of Combing Volumes, **222**, 283–84
Geiger, Hans (Johannes) Wilhelm, 222–24
Geiger counter, 223–24, 546
Geiger-Nutter Law (Rule) for Decay of Radioactive Isotopes, **222–24**
Geissler tube, 531
Geller, Margaret Joan, 224, 236
Geller's Theory of a Nonhomogenous Universe, **224**
Gell-Mann, Murray, 175, 224–27, 527
Gell-Mann's Theories for Subatomic Particles, **224–27**
General theory of relativity. *See* Einstein
Generators. *See* Electricity, high voltage ac
Genetics: acquired characteristics; 359; artificial genes, 311–12; assimilation, 555; biogenetics, 37; cloning, 16; cytogenetics, 378; dating human/ape divergence, 492; diversity, 139; dynamic equilibrium, 569–70; engineering, 16, 343, 491; eugenics, 215–16; gene-controlling enzymes, 525; genetic code, 311–12; genetic drift, 573; genome, 129; inherited characteristics, 555; jumping genes, 295, 378; mapping, 401; Mendelian, 139, 382–83; mitosis/meiosis, 119–20; mutations, 312, 395–96; operons, 289; phages, 129–30; population, 249–50, 255; sickle cell, 286–87; split genes, 474–75; supergene theory, 91; *See also* DNA; Evolution; RNA; Species
Geology: craters, 9, 44–45; earthquakes, 85, 472–73; continental drift, 35, 520–21; crystallization, 257–58; geodesy, 509; geologic equilibrium, 7; geosyncline, 116–17; glaciers, 4–5, 89–90; interior structure

of earth, 293, 393; mineralogy, 6–7; Neptunian theory, 563; origin of mountains, 45; plate tectonics, 37, 45, 270–71; Richter scale, 472–73; seismological tables, 293; spreading of ocean floor, 270–71, 470; stratification, 5–6; submarine canyons, 502–3; subterranean gases, 6–7; undersea mountains, 170; uniformitarianism, 358–59; volcanoes, 6–7, 45. *See also* Fossils
Geometrodynamics. *See* Geon theory
Geometry. *See* Mathematics
Geon theory, 564
George III of England, 219, 266–67, 448
Gerhardt, Charles Frédéric, 227–28, 568
Gerhardt's Type Theory for Classifying Organic Compounds, **227–28**
Gerlach, Walter, 459, 517
Germer, Lester, 123
Germs: antisepsis, 354; disease postulate, 198, 318–19; fermentation, 430–31; germ layer theory, 126, 198; germ plasm, 560–61; parasites, 346–47; typhus, 409–10. *See also* Bacteriology; Medicine
Ghiorso, Albert, 190
Giaevar, Ivar 166, 297
Giauque, William Francis, 228–29
Giauque's Theory of Adiabatic Demagnetization, **228–29**
Gibbs, Josiah Willard, 229–30
Gibbs' Theory of Chemical Thermodynamics, **229–30**
Gilbert, Walter, 230–31, 491
Gilbert, William, 214, 231–32
Gilbert's Theory for DNA Sequencing, **230–31**
Gilbert's Theory of Magnetism, **231–32**
Gill, David, 303
Gillespie, Elizabeth, 286
Glaciers. *See* Geology
Glaser, Donald Arthur, 88, 232–33
Glaser, Otto Charles, 574
Glaser's Concept for a Bubble Chamber for Detecting Subnuclear Particles, **232–33**
Glashow, Sheldon, 233–35, 345, 489, 559
Glashow's Unifying Theory of the Weak Forces, **233–35**
Global warming, 4, 24–25, 313, 373, 469–71, 495–96
Gmelin, Leopold, 138
Godel, Kurt, 235
Godel's Incompleteness Theorem, **235**
Goeppert-Meyer, Maria, 566
Göhring, Otto, 176
Gold, 244
Gold, Thomas, 236–37, 273
Gold rush (California), 117
Gold's Cosmological Theories, **236–37**
Goldstein, Eugen, 56, 531
Goldstein, Joseph Leonard, 237–38
Goldstein's Theory for the Metabolism of Cholesterol, Fats, and Lipids, **237–38**
Golgi, Camillo, 460–61
Gondwanaland, 520–21, 559
Goodsir, John, 552
Gorgas, William Crawford, 466
Gorter, Cornelius, 79
Goudsmit, Samuel, 541
Gould, Stephen Jay, 159–60, 238–39, 463
Gould's Hypothesis of "Punctuated Equilibrium," **238–39**
Graham, Thomas, 239
Graham's Laws of Diffusion and Effusion, **239–40**
Grand unification theories (GUT), 156, 241, 263, 559–60
Graviton, 154, 282
Gravity: accelerating force, 209–10; compensator, 163; dark matter, 30, 39, 479, 520, 572; definition, 560; Eötvös effect, 164; falling bodies, 209–11, 220, 471–72; geon theory, 564; gravitational constant, 137, 472; gravitational force, 155–56, 235; gravitational waves, 525–26, 558–59; harmonics of planetary motion, 276; laws, 154–55, 407–8; mechanistic nature, 282–82; pendulums, 193–94, 211–12, 276, 282, 471–72; principles, 154–55; quantum, 260; specific, 18–19. *See also* Weak force
Great Atlantic rift, 470
Great global rift, 170, 270
Green Bank equation, 143
Greenhouse effect, 24–25, 469–71, 540. *See also* Global warming
Griffith, Frederick, 103
Groves, General Leslie, 423
Guillemin, Roger, 577
Gunpowder, 132
Gutenberg, Beno, 293
Guth, Alan, 240–41
Guth's Theory of an Inflationary Universe, **240–41**

Haber, Fritz, 65, 243–44
Haber process, 56, 65, 243
Haber's Theories, **243–44**
Hadamard, Jacques Salomon, 244–45
Hadamard's Theory of Prime Numbers, **244–45**
Hadley, George, 245–46
Hadley's Hypothesis for the Cause of the Trade Winds, **245–47**
Hadrons, 440
Haeckel, Ernst, 247–48

Haeckel's Biological Theories, **247–48**
Hahn, E. Vernon, 286
Hahn, Otto, 61, 190, 248–49, 297, 380–81, 521
Hahn's Theories of Nuclear Transmutations, **248–49**
Haldane, J.B.S. (John Burdon Sanderson), 197, 249–50
Haldane's Theories of Genetics, Evolution, and Origins of Life, **249–50**
Hale, George, 250–51
Hale's Solar Theories, **250–51**
Hall, Edwin Herbert, 253–54
Hall Effect of Electrical Flow, **253–54**, 317–18
Halley, Edmond, 252–53, 501
Halley's comet, 53, 252, 421, 565
Halley's Theories for Comets and Stars, **252–53**
Hamilton, William Rowan, 254, 325
Hamilton's Mathematical Theories, **254**
Hanford, WA, 423
Hardy, Godfrey Harold, 255
Hardy's Mathematical Theories, **255**
Harkins, William Draper, 255–56
Harkins' Nuclear Theories, **255–56**
Harmonic analysis, 195
Harsanyi, John, 400
Harvey, William, 173–74, 256–57, 332, 364, 465, 511, 552
Harvey's Theory for the Circulation of the Blood, **256–57**
Hauptman, Herbert, 303–4
Haüy, René Just, 257–58
Haüy's Geometric Law of Crystallization, **257–58**
Hawking, Stephen William, 258–61, 437, 570–71
Hawking's Theories of the Cosmos, **258–61**
Haworth, Walter, 261
Haworth's Formula, **261**
Heat: absolute entropy, 404; caloric, 124; Carnot cycle, 78–79; chemical thermodynamics, 228–29, 229–30; conduction, 195; critical state, 73; definition, 58, 263; demon paradox, 375–76; exchange of heat radiation, 447; heat sink, 228; heat/work equivalency, 263–64; latent heat of fusion, 58; latent heat of vaporization, 58; measurements, 82, 174; mechanical equivalent, 299; pyroelectricity, 257; regenerating, 505–6; resistance, 417; specific heat, 58; work to heat theory, 479. *See also* Thermodynamics
Heezen, Bruce Charles, 170
Heisenberg, Werner Karl, 65, 254, 262–63, 456, 496, 498
Heisenberg's Uncertainty Principle and Theory of Nucleons, **262–63**
Helical pump, 17, 19
Helmholtz, Herman Ludwig Ferdinand von, 32, 55, 263–64
Helmholtz's Theories and Concepts, **263–64**
Helmont, Jan Baptista van, 264–65, 486, 536
Helmont's Theory of Matter and Growth, **264–65**
Hench, Philip, 307, 468
Henderson, Thomas, 53
Henry, Joseph, 229, 265–66
Henry's Principles of Electromagnetism, **265–66**
Henseleit, Kurt, 320
Heraclitus of Ephesus, 530
Herrick, James B., 286, 432
Herschel, Caroline Lucretia, 266–67
Herschel, William, 59, 266–67, 303
Herschels' Stellar Theories and Discoveries, **266–67**
Hershey, Alfred, 129
Hertz, Gustav, 198
Hertz, Heinrich Rudolf, 269–70, 348
Hertz's Theory for Electromagnetic Waves, **269–70**
Hertzsprung, Ejnar, 267–69, 341
Hertzsprung–Russell diagram, 268–69
Hertzsprung's Theory of Star Luminosity, **267–69**
Hess, Harry Hammond, 270–71
Hess, Victor Francis, 13
Hess' Sea–Floor Spreading Hypothesis, **270–71**
Hess' Theory for the Ionization of Gases, **271–72**
Hewish, Antony, 236, 272–73, 482
Hewish's Theory of Pulsars, **272–73**
Hidalgo, 31
Hiero of Greece, 18
Higgins, William, 147, 273–74
Higgins' Law of Definite Composition, **273–74**
Higgs, Peter Ware, 274
Higgs' Field and Boson Theories, **274**
Hipparchus of Nicaea, 452
Hippocrates, 165, 209
Hisinger, Wilhelm, 52, 392
Hitchings, GeorgeHerbert, 160–61
HIV (human immunodeficiency virus), 41, 214–15, 369, 393–94
Hobbits, 296
Hockfield, Susan, 535
Hodgkin, Dorothy Crowfoot, 274–75
Hodgkin's Theory of Organic Molecular Structure, **274–75**
Hoffman, Felix, 227
Hoffman-LaRoche, 397
Hoffmann, Roald, 275

Hoffmann's Theory of Orbital Symmetry, **275**
Hofman, Wilhelm, 2
Hohenheim, Phillippus Aureolus Theophrastus Bombastus von. *See* Paracelsus
Holley, Robert W., 311–12
Hologram/holograph, 207–8
Hominids/hominoids, 492
Homo erectus, 296
Homo sapiens, 295–96, 569
Hooke, Robert, 35, 66, 195, 257, 263, 275–76, 418, 493
Hooke's Laws, Theories, and Ideas, **275–76**
Hoover, Herbert, 6
Hoover, Lou Henry, 6
Hopkins, B. Smith, 512
Hoppe-Seyler, Felix, 389
Horse latitudes, 246–47
Hounsfield, Godfrey, 367
Houssay, Bernardo, 100
Hoyle, Fred, 70, 196–97, 217, 276–78, 483, 562
Hoyle's Theories of the Universe, **276–78**
Hubble, Edwin, 30, 260, 278–79, 508
Hubble effect. *See* Red shift
Hubble's Law and Constant, **278–79**
Hubble Space Telescope, 419
Huchra, John, 224
Hückel, Erich Armand Arthur J., 127, 279–80
Hückel's MO Theory or Rule and the Debye-Hückel Theory, **279–80**
Huggins, Lady (Margaret Lindsay Murray), 281
Huggins, Sir William, 280–81
Huggins' Theory of Spectroscopic Astronomy, **280–81**
Hulse, Russell, 525
Hulst, Hendrik van de, 62
Human genome project, 16
Hund, Friedrich, 396
Hurricanes, 99–100
Hutton, James, 358, 563
Huygens, Christiaan, 194, 257, 281–82
Huygens' Theories of Light and Gravity, **281–82**
Hyatt, John Wesley, 427–28
Hybrids, 249
Hydraulics. *See* Hydrostatics
Hydrocracking, 3
Hydrodynamics, 518
Hydrogen: discovery, 81; H-bomb, 55, 420, 525–28, 542–43; heavy, 419–20, 542–43; helium energy reaction, 256; nuclear fusion, 489; quantum states of hydrogen atom, 329–30. *See also* Thermonuclear energy
Hydrogenation, 3–4, 567
Hydrostatics, 429
Hygrometer, 119, 329
Hypnotism, 385

Ice age, 4, 5
Ice floes, 158
Ideal Gas Law, 222, **283–84**, 548–49
Ihle, Abraham, 501
I-Hsing, 284–85
I-Hsing's Concepts of Astronomy, **284–85**
Immunology: antibodies, 535–36; antitoxins, 47–48; cell differences, 160–61; cell enzyme synthesis, 426–27; "designer drugs," 150–51; gamma globulin, 444–45; inoculation, 162–63, 293–94, 431; parasites, 409–10; polio vaccine, 162–63, 474, 485–86; radioimmunoassay, 577–78. *See also* Medicine
Inclined planes, 210
Indeterminacy, 262–63, 456. *See also* Uncertainty principle
Indigo dye, 38
Inductive reasoning. *See* Philosophy
Inertia, 406. *See also* Laws; Motion
Infinity, 21–22, 77. *See also* Mathematics
Influenza, 442
Ingenhousz, Jan, 285
Ingenhousz's Theory of Photosynthesis, **285**
Ingold, Sir Christopher Kelk, 285–86
Ingold's Theory for the Structure of Organic Molecules, **285–86**
Ingram, Vernon Martin, 286
Ingram's Sickle Cell Theory, **286–87**
Innocent III, Pope, 364
Inoculation. *See* Immunology
Insecticides, 410
Insulin, 42–43
Intel Corporation, 394, 414
Interconnected networks, 385–86
Interferometer, 388
Interferon, 288
Internal combustion engine, 134–35
International metric system. *See* Metric system
International Union of Pure and Applied Chemistry (IUPAC), 191
"Invisible College, 66
In vitro fertilization. *See* Reproduction
Ipatieff, Vladimir Nikolayevitch, 287–88
Ipatieff's Theory of High Pressure Catalytic Reactions, **287–88**
Isaacs, Alick, 288
Isaacs' Theory of Proteins Attacking Viruses, **288**
"I think, therefore I am," 131

Jacob, François, 289, 426
Jacob-Monod Theory of Regulator Genes, **289**
James I of England, 34
Jansky, Karl Guthe, 290, 357
Jansky's Theory of Stellar Radio Interference, **290**

Janssen, Pierre-Jules-Cesar, 290–91, 354–55, 461
Janssen, Zacharias, 212–13
Janssen's Theory of Spectral Lines of Sunlight, **290–91**
The Jasons, 527
Javan, Ali, 363
JB tables, 293
Jeans, James Hopwood, 291, 293
Jeans' Tidal Hypothesis for the Origin of the Planets, **291**
Jefferson, Thomas, 362
Jeffreys, Alec john, 291–92
Jeffreys, Sir Harold, 293
Jeffreys Seismological Theories, **293**
Jeffreys' Theory of Genetic (DNA) Profiling, **291–92**
Jenner, Edward, 293–94
Jenner's Inoculation Hypothesis, **293–94**
Jensen, J. Hans, 566
Jerne, Niels Kaj, 294–95
Jerne's Theory of Clonal Selection of Antibodies, **294–95**
Johanson, Donald Carl, 295–96, 339
Johanson's Theory for the Evolution of Humans, **295–96**
John Paul II, Pope, 197
Johnston, Herrick L., 229
Joint Institute for Nuclear Research (JINR), 190
Joliot-Curie, Frederic, 191, 296–97
Joliot-Curie, Irène, 296–97
Joliot-Curie's Theory of Artificial Radioactivity, **296–97**
Jordan, Ernst Pascual, 65
Josephson, Brian David, 166, 297–98
Josephson's Theory of Semiconductors, **297–98**
Joule, James, 73, 263, 298–99, 306
Joule's Law and Theories, **298–99**, 499
Joule-Lenz law, 349
Joule-Thomson effect, 74, 301
Joyce, James, 226

Kala-Azar, 346
Kamerlingh-Onnes, Heike, 301–2, 372
Kamerlingh-Onnes Theory of Matter at Low Temperature, **301–2**
Kapitsa, Pyotr, 302, 439
Kapitsa's Theory of Superfluid Flow, **302**
Kapteyn, Jacobus Cornelius, 253, 302–3
Kapteyn's Theory of Galactic Rotation, **302–3**
Karle, Isabella Lugoski, 303–4
Karle, Jerome, 303–4
Karle's Theory for Determining Molecular Structure, **303–4**
Karrer, Paul, 261
Kater, Henry, 194
Kattwinkel, Wilhelm, 339
Kekule von Stradonitz, Friedrich, 103, 178, 549, 575
Kekule's Theory of Carbon Compounds, **304–5**
Kellogg, Paul, 545
Kelvin, Lord William Thomson, 82, 263, 299, 305–7, 500
Kelvin's Concepts of Energy, **305–7**
Kendall, Edward, 307–8, 468
Kendall, Henry, 203
Kendall's Theory for Isolating Adrenal Steroids, **307–8**
Kendrew, John, 441
Kenyanthropus plotyops, 340
Kepler, Johannes, 68, 80, 84, 97, 308–9, 479
Kepler's Three Laws of Planetary Motion, **308–9**
Kerr, John, 309–10
Kerr's Theory of Quadratic Electro-Optic Effect (Kerr Effect), **309–10**
Kerst, Donald William, 310–11
Kerst's Theory for Accelerating Nuclear Particles, **310–11**
Khorana, Har Gobind, 311–12
Khorana's Theory of Artificial Genes, **311–12**
Kilby, Jack S., 414
Kimura, Hisashi, 313
Kimura, Motoo, 312
Kimura's Neo-Darwinian Theory for Mutations, **312**
Kimura's Theory for Variations in Earth's Latitudes, **313**
Kipping, Frederick Stanley, 314–15
Kipping's Theory of Inorganic-Organic Chemistry, **314–15**
Kirch, Gottfried, 501
Kirchhoff, Gustav, 69, 70, 202, 280, 315–16, 443
Kirchhoff's Laws and Theories, **315–16**
Kirkwood, Daniel, 316–17
Kirkwood's Asteroid Gap Theory, **316–17**
Kitasato, Shibasaburo, 47
Klaproth, Martin Heinrich, 52, 108
Klitzing, Klaus von, 317–18
Klitzing's Theory for the Quantization of the Hall Effect, **317–18**
Koch, Heinrich Hermann Robert, 47, 94, 318–19, 442
Koch's Germ-Disease Postulate, **318–19**
Kohler, Georges, 294
Kohlrausch, Friedrich Wilhelm Georg, 319–20
Kohlrausch's Law for the Independent Migration of Ions, **319–20**
Kornberg, Arthur, 415
Krebs, Hans Adolf, 320–21, 416

Krebs Cycle, **320–21**, 415–16
Kroto, Harold, 108, 321–22
Kroto's "Buckyballs," **321–22**
Kuiper, Gerard Peter, 322
Kuiper's Theory for the Origin of the Planets, **322**
Kusch, Polykarp, 322–23, 329
Kusch's Theory for the Magnetic Moment of the Electron, **322–23**

Lacunary space, 244
Ladder of Life, 353. *See also* Taxonomy
Lagrange, Joseph-Louis, 169, 325, 327, 447
Lagrange's Mathematical Theorems, **325–27**
Lamarck, Jean Baptiste de Monet, Chevalier de, 76, 327–28, 359, 555, 561
Lamarck's Theories of Evolution, **327–28**
Lamb, Willis Eugene, 322, 329–30
Lamb's Theory for the Quantum States of the Hydrogen Atom, **329–30**
Lambert, Johann Heinrich, 46, 47, 328–29
Lambert's Theories, **328–29**
Landau, Lev Davidovich, 330–31
Landau's "Two-Fluid Model" for Helium, **330–31**
Landauer, Rolf, 331
Landauer's Principle for Very-Large-Scale Integration, **331**
Landsteiner, Karl, 331–32
Landsteiner's Theories of Blood Groups, **331–32**
Langevin, Paul, 332
Langevin's Concept for Use of Ultrasound, **332**
Langley, John Newport, 332
Langley's Theories of the Nervous System, **332–33**
Langmuir, Irving, 333–34, 352, 504
Langmuir's Theories of Chemical Bonding and Adsorption Surface Chemistry, **333–34**
Laplace, Pierre Simon de, 15, 84, 291, 334–35, 429
Laplace's Theories and Nebular Hypothesis, **334–35**
Large Electron Positron (LEP) collider, 274
Large Hadron Collider (LHC), 274, 548
Larmor, Joseph, 335, 452
Larmor's Theories of Matter, **335**
Laser, 208, 362–63, 538
Laser Interferometer Gravitational-Wave Observatory (LIGO), 155
Latitude, 313
Laurent, Auguste, 145, 335–36, 568
Laurent's Theories for Chemical "Equivalents" and "Types," **335–36**
Lauterbur, Paul C., 367
Lavoisier, Antoine, 81, 327, 336–37, 449, 514
Lavoisier's Theories of Combustion, Respiration, and Conservation of Mass, **336–37**
Lawrence, Ernest Orlando, 310, 311, 337–38, 379
Lawrence Livermore Laboratory, 10
Lawrence's Theory for the Acceleration of Charged Particles, **337–38**
Laws: Amdahl's, 10–11; Ampere's, 11–12; Avogadro's, 28; Babo's 33–34; Baer's 36–37; Beer-Lambert-Bouguer, 328–29; Bell's 48; Bernoulli's large numbers, 49–50; Biot-Savart, 55–56; Boyle's, 65–66; Charles', 86–87; chemical bonding, 175–76; chemical dynamics, 549–50; combining volumes, 222; conservation of energy, 92, 263–64, 412; conservation of mass, 264, 337; conservation of momentum, 77, 412; circuital, 11–12; crystallization, 257; cytology, 519; definite or constant composition, 273–74; definite proportions, 115, 450–51; diffusion, 185–86, 239; dilution, 424; displacement, 566; dominance, 383; Eddington's astronomy, 148; effusion, 249; elasticity, 275–76; electrical charge, 102; electric current, 315; electricity, 101–2; electrolysis, 177; embryology, 247; Eötvös, 163–64; equilibrium, 343; equipartition, 62; exchange of radiation, 447; fluorescence, 518; Gauss', 221; generalized gas, 283–84; gravity, 101–2, 154–55, 210, 407–8; Hardy-Weinberg, 255; Hubble's, 278; hydraulics, 429; hydrodynamics, 578; ideal gas law, 283–84; independent migration of ions, 319–20; inertia, 406; inheritance, 382–83; inverse square law, 276; isomorphism, 392–93; Joule's, 298–99; Lambert-Beer, 328–29; Lenz's, 348–49; light scattering, 463–64; magnetic flux, 221; mechanical equivalent of heat, 299; Metcalfe's, 385–86; Moore's, 321, 394–95; Moseley's, 395; motion, 21, 152–53, 406–8; multiple proportions, 115; Newton's three laws, 406–8; octaves, 405–6; Ohm's, 417–18; partial pressure, 114; perfect gases, 283–84; planetary

motion, 308–9; planetary orbits, 59; polarization, 365–66; radiation, 315; radioactive decay of isotopes, 222; radioactive displacement, 510; Raoult's, 462; recapitulation, 126, 149; rate, 24; refraction, 509–10; Snell's, 130, 374, 509–10; Stefan-Boltzmann, 515; thermionic emission, 472; thermodynamics, 306, 506–7; triads, 138–39; voltage, 315
Lazear, Jesse William, 466
Leakey, Deborah, 340
Leakey, Jonathan, 340
Leakey, Louis Seymour, 296, 338–39
Leakey, Louise, 339, 340
Leakey, Maeve Epps, 338, 340
Leakey, Mary Douglas Nicol, 338–39
Leakey, Philip, 340
Leakey, Richard Erskine Frere, 296, 338–40
Leakeys' Anthropological Theories, **338–40**
Leavitt, Henrietta Swan, 30, 340–42
Leavitt'sTheory for the Periodicity/Luminosity Cycle of Cepheid Variable Stars, **340–42**
Le Bel, Joseph Achille, 342
Le Bel's Theory of Isomers, **342**
Le Chatelier, Henri-Louis de, 342–43
Le Chatelier's Principle, **342–43**
Lederberg, Joshua, 343, 525
Lederberg's Hypothesis for Genetic Engineering, **343**
Lederman, Leon Max, 344, 516
Lederman's Two-Neutrino Hypothesis, **344**, 440
Lee, Tsung-Dao, 344–45, 578
Lee's Theories of Weak Nuclear Interaction, **344–45**
Leeuwenhoek, Anton van, 197, 345
Leeuwenhoek's Theory of Microscopic Life, **345**
Leibniz, Gottfried, 50, 77, 169, 345–56, 409
Leibniz's Theory for "The Calculus," **345–46**
Leishman, William Boog, 346–47
Leishman's Hypothesis for Parasitic Diseases, **346–47**
Leith, Emmett, 208
Lemaître, Georges Edouard, 196, 224, 347–48
Lemaître's Theory for the Origin of the Universe, **347–48**
Lenard, Philipp Eduard Anton, 152, 348
Lenard's Theory for Electron Emission, **348**
Lenses: contact, 186; prisms, 202–3
Lenz, Heinrich Friedrich Emil, 348–49
Lenz's Law of Electromagnetics, **348–49**
Leonardo da Vinci, 551
Lepton theory, 440
Leucippus of Miletus, 26, 455
Levene, Phoebus Aaron, 103, 349
Levene's Tetra-Nucleotide Hypothesis, **349**
Leverrier, Urbain, 15, 53
Levers, 17–18
Levi-Montalcini, Rita, 350
Levi-Montalcini Cell Growth Theory, **350**
Lewis, Gilbert Newton, 334, 350–52, 456, 504, 512
Lewis' Theory of Covalent Bonds, **350–52**
Leyden jar, 201, 557
Liebig, Justus von, 352
Liebig's Theory of Isomers and Organic Compound Radicals, **352**
Life: abiogenesis, 196; autopoiesis, 197, 422; biopoiesis, 187; chemical evolution, 389–90; contraception, 187; cyclic extinction, 562–63; Earth and Mars, 236–37, 358, 493; extraterrestrial life, 143, 487–88; origin, 20, 24, 83–84, 218, 250, 418, 421–22, 444; panspermia, 24, 197; self-organizing molecules, 449–50; spontaneous generation, 20, 196, 422, 430, 464–65, 511–12. *See also* Evolution
Light: aether, 387–88; antimatter, 136; Beer's law, 46–47; Beer-Lambert law, 328–29; birefringence, 309–10; blackbody radiation, 315; corpuscular, 408–9; Doppler principle, 141–42; Einstein's theory, 153, 155; electromagnetic contraction, 186–87; electromagnetic rays, 143–44; electromagnetic waves, 33, 148; Faraday effect, 178; fluorescence, 518; holograms/holograph, 207–8; least time principle, 180, 373–74; light spectrum in chemical analysis, 315–16; Lorentz-Fitzgerald contraction, 355–56; magnetic effect, 583; Newton's theories, 408–9; Olber's paradox, 418–19; particle/wave duality, 94, 136, 152, 155, 276; photoelectric effect, 348; polarization, 33, 365–66; Raman effect, 460; red shift, 148; speed, 67, 153–54, 187, 192, 405–4, 476, 523; quantum theories, 145–46, 443, 523–24; reflection, 130; rotation effect, 178–79; scattering, 123, 460, 463–64, 540; Snell's law, 130, 374,

509–10; transmission through gases, 540; transverse waves, 202–3; wave theories, 180, 187, 192, 281–82, 579–80; white light, 201–2, 315–16. *See also* Electrons; Electromagnetism
Lincoln, Abraham, 292
Lindemann, Carl Louis Ferdinand von, 352
Lindemann's Theory of Pi, **352**
Linnaeus, Carolus, 76, 82, 121, 353–54, 464
Linnaeus' Theories for the Classification of Plants and Animals, **353–54**
Lipman, Fritz Albert, 320
Lippershey, Hans, 212
Lippmann, Gabriel, 208
Lister, Joseph, 354
Lister's Hypothesis for Antisepsis, **354**
Lockheed Martin Corporation, 10
Lockyer, Sir Joseph Norman, 290–91, 354–55
Lockyer's Solar Atmosphere Theories, **354–55**
Lodge, Oliver Joseph, 368
Loewi, Otto, 113
Logic: action-at-a-distance, 565–66; Aristotelian, 22; logical positivism, 363, 467–68
London, Fritz, 379
London, Heinz, 379
Longitude, 467
Lorentz, Hendrik Antoon, 186–87, 355–56, 583
Lorentz-Fitzgerald contraction, 187, 355–50
Lorentz's Physical Theories of Matter, **355–56**
Lorenz, Edward Norton, 356–57
Lorenz's Theory for Complex/Chaotic Systems, **356–57**
Los Alamos National Laboratory, 10, 423
Lothar Meyer's curves, 387
Love, William T., 529
Love Canal, 529
Lovell, Sir Alfred Charles Bernard, 357–58
Lovell's Theory of Radio Astronomy, **357–58**
Lowell, Percival, 358
Lowell's Theory of Life on Mars, **358**
Loxodromes, 509
Lucretius (Titus Lucretius Carus), 27, 219, 220
Luria, Salvador, 129
Lwoff, André, 289
Lyell, Charles, 90, 122, 358
Lyell's Theory of Uniformitarianism, **358–59**
Lysenko, Trofim Denisovich, 359, 561
Lysenko's Theory of the Inheritance of Acquired Characteristics, **359**

MacArthur, Robert, 570
Mach, Ernst, 361–62
Mach's Number, **361–62**
Macleod, John, 42–43
Macromolecules, 191–92
Magellan cloud, 253
Magendie, François, 48
"Magic bullet," 150–51
Magnetic Moment: atomic nuclei, 165–66, 367, 452–53, 462; electron, 322–23, 459–60; proton, 517–18
Magnetic Resonance Imaging (MRI), 12, 517–18: theories, 165–66, 322–23, 367, 452–53, 459–60, 462
Magnetism: Ampère's laws, 11–12; adiabatic demagnetization, 228; animal, 384–85; antiferromagnetism, 402–3; Earth's, 231–32; electric and magnetic forces, 102, 177–78, 221, 231; electrodynamics, 11–12; electromotive force (EMF), 348–49; extraterrestrial magnetic fields, 251; Faraday effect, 178; ferromagnetism, 402–3; Hall effect, 253–54; lodestones, 231; magnetic strength, 417; mesmerism, 384–85; paramagnetism, 550–51; right/left hand rules, 188–89; semiconductors, 14; space quantization, 517–18; superconductivity, 14; Van Allen radiation belt, 545–46. *See also* Electromagnetism
Magnetometer, 221
Magnetosphere, 546
Maiman, Theodore Harold, 362–63
Maiman's Theory for Converting the Maser to the Laser, **362–63**
Malpighi, Marcello, 257, 363–64
Malpighi's Theory for the Detailed Structure of Animals and Plants, **363–64**
Malthus, Thomas Roberts, 122, 364–65
Malthusian Population Catastrophe Theory, **364–65**
Malus, Étienne Louis, 365–66
Malus' Law for the Polarization of Light, **365–66**
Manhattan Project, 9, 10, 181–82, 183–84, 423, 498, 507, 513, 527, 585
Manometer, 74
Mansfield, Sir Peter, 367
Mansfield's Theory of Magnetic Resonance, **367**
Maraldi, Jean-Dominique, 501
Marconi, Guglielmo Marchese, 368
Marconi's Theory of Radio Telegraphy, **368**
Margulis, Lynn, 197, 368–69
Margulis' Endosymbiotic Cell Theory, **368–69**
Mariner, Ruth, 488
Mariotte, Edme, 66
Mariotte's law, 66

Marius, Simon, 508
Marsden, Ernest, 223
Martin, Archer John Porter, 369–72
Martin's Theory of Chromatography, **369–72**
Maser, 362–63, 462, 538
Mass, 152–53, 479–80
Mass point, 169
Mass spectrometer, 25
Mathematics: Abelian groups, 2–3; algebraic equations, 325; aggregates, 221; automata, 553–54; Balmer series, 40–41; Boolean logic, 63–64; calculus, 50, 130, 345–46, 409; Cartesian coordinate system, 130; chaotic systems, 356–57; complex systems, 356–57, 449–50, 571–72; computers, 10–11, 32, 553–54; continuous curves, 434–35; cotangents, 8; cubic equation, 78, 524–25; diameter of circle, 530; embedding theorems, 400; equilibrium equation, 255; errors, 221; Euclidean, 167–68; Fermat's principles and theories, 179–80; Fibonacci sequence, 184–85; fluid dynamics, 113–14; fractals/tiles, 438–39; game of life, 95–96; game theory, 377, 400; geometry, 365, 473–74; golden ratio, 185; Hamiltonian functions, 254; Hardy-Weinberg law, 255; inclined plane, 17–18; incompleteness theorem, 235; infinity, 77; large number coincidences, 49–50, 136; least squares, 220–21; levers, 17; Lorenz attractor, 356–57; metric system, 326–27; notations, 169; Moore's law, 394–95; Noether current, 411–12; numbering system, 184–85; Pascal's theorem, 428, 429–30; Peano's axioms, 434; pi, 16, 85, 352; prime number theory, 165, 244–45; probability theories, 49–50, 179, 335, 429–30, 467–68, 542; Pythagoras' theorems, 453; quaternions, 254; Riemann's theory, 473–74; set theory, 77–78; sines, 8; statistical theory, 435–36; symmetry, 411–12; three–body problem, 169, 325; Torricellis theorem, 537; transfinite numbers, 77; trigonometry, 8, 466–67; Twistor theory, 438; volume of spheres, 17; wave analysis, 195
Matter: aether, 22; atomism, 26–27, 219–20; black holes, 236, 259, 437–39; Brownian motion, 151–52, 221, 440–41; condensed, 330–31; constants, 148; dark, 30, 39, 479, 520, 572; differential equation, 435; electron theories, 335; fluids, 325, 384–85, 537; growth, 264–65; hydrodynamics, 518; kinetic theory, 441; Lorentz invariant, 356; low temperatures, 301–2; molecular energy/motion, 440–41, 539; negative energy, 136; Raoult's law, 462; relativistic terms, 258; three states, 58; Van der Waals force, 549; wave theory, 127; Wigner's theory, 566–67. *See also* Universe
Matthews, Thomas, 494
Matthias, Bernd, 372
Matthias' Theory of Superconductivity, **372–73**
Matrix mechanics, 65, 262
Maunder, Edward Walter, 373
Maunder's Theory for Sunspots' Effects on Weather, **373**
Maupertuis, Pierre-Louis Moreau de, 373–74
Maupertuis' Principle of Least Action, **373–74**
Maxam, Allan M., 230
Maximillian of Habsburg, Emperor, 551
Maxwell, James Clerk, 86, 152, 176, 187, 188, 269, 355, 374–77, 380
Maxwell-Boltzmann distribution equation, 86, 375
Maxwell's Theories, **374–77**
Mayer, Julius von, 263, 299
Maynard Smith, John, 377–78
Maynard-Smith's Theory of Evolution, **377–78**
McCarthy, John, 391
McClintock, Barbara, 378
McClintock's Theory of Cytogenetics, **378**
McMillan, Edwin Mattison, 378–79, 499
McMillan's Concept of "Phase Stability," **378–79**
Mead, Carver, 394
Mean, 436
Median, 436
Medicine: antibiotics, 188, 191; antibodies, 535–36; antisepsis, 354; bacteriolysis, 442–43; Bell's law, 48; blood groups, 331–32; cancer cell transformation, 144; cell pathology, 552; cholesterol/fats/lipids, 237–38; disease theory, 198; dyes as antibiotics, 139–41; EKG/ECG, 157–58; humoral theory, 208–9; iatro-chemistry, 425–26; immunology, 150–51, 160–61, 162–63, 293–94, 294–95; insulin, 42–43; Krebs cycle, 320–21; metabolic disorder, 219; nerve growth factor, 350; parasitic diseases, 346–47; radioimmunoassay (endocrinology), 577–78; sickle cell, 286–87

Meiosis, 119
Meissner, Walther, 79, 379–80
Meissner Effect, **379**
Meitner, Lise, 61, 248–49, 380–81, 411, 437, 521
Meitner's Theory of Nuclear Fission, **380–81**
Memes. *See* Evolution
Mendel, Gregor, 121, 378, 382–83, 561
Mendel's Law of Inheritance, **382–83**
Mendeleev, Dmitri, 138, 336, 381–82, 393
Mendeleev's Theory for the Periodicity of the Elements, **381–82**
Mercalli, Giuseppe, 472
Merrifield, Bruce, 383–84
Merrifield's Theory of Solid-Phase Peptide Synthesis, **383–84**
Meselson, Matthew Stanley, 384
Meselson-Stahl Theory of DNA Replication, **384**
Mesmer, Franz Anton, 384–85
Mesmer's Theory of Animal Magnetism, **384–85**
Messier, Charles, 267, 501
Metabolic theories, 219, 320–21
Metallurgy, 427–28
Metcalfe, Robert Melancton, 385–86
Metcalfe's Law, **385–86**
Meter (standard), 33, 327
Meteorology: biological systems/climate change, 495–96; cloud processes, 49; complex/chaotic systems, 356–57; Coriolis effect, 99–100, 158–59; dendrochronology, 142; desertification, 88; sunspots' effects, 373; theoretical, 87–88; thermodynamics of regional air masses, 57–58; trade winds, 245–47
Metric system, 82, 326–27
Meyer, Julius Lothar, 382, 386–87
Meyer's Theory for the Periodicity of the Elements, **386–87**
Michelson, Albert, 153, 187, 192, 376, 387–88
Michelson's Theory for the "Ether," **387–88**
Michurin, Ivan, 359
Microbiology. *See* DNA; Enzymes; Immunology; Viruses
Microscopes, 207–8, 345
Miescher, Johann Friedrich, 388–89
Miescher's Nuclein Theory, **388–89**
Milky Way galaxy, 31, 62, 213
Miller, Stanley Lloyd, 197, 250, 389–90, 547
Miller's Theory for the Origin of Life, **389–90**
Millikan, Robert Andres, 272, 390
Millikan oil drop experiment, 175, 390
Millikan's Theory for the Charge of Electrons, **390**
Milstein, Cesar, 294
Mimicry, 91
Mineralogy, 6–7, 116–17, 257–58, 563
Minkowski, Hermann, 154, 390–91
Minkowski's Space-Time Theory, **390–91**
Minsky, Marvin Lee, 391–92
Minsky's Theory of Artificial Intelligence (AI), **391–92**
Misner, Charles William, 392
Misner's Theory for the Origin of the Universe, **392**
Mitosis, 119
Mitscherlich, Eilhard, 392–93
Mitscherlich's Law of Isomorphism, **392–93**
Mittag-Leffler, Gösta, 342
Mode, 436
Mohorovicic, Andrija, 393
Mohorovicic's Theory of the Earth's Interior Structure, **393**
Molecular biology, 129–30, 441–42
Molecular clock, 492, 569
Molecular motion (equipartition), 62
Molecular photoelectron spectroscopy, 539
Molina, Mario, 106, 477
Momentum, 77, 130, 406. *See also* Motion,
Monod, Jacques, 289, 426
Montagnier, Luc, 214–15, 393–94
Montagnier's Theory for the HIV Virus,
"Monte Carlo" system, 542
Moon (Earth's), 8, 86
Moore, Gordon Earl, 321, 394–95
Moore's Law, 321, **394–95**
Morgan, William, 62
Morley, Edward, 153, 187, 388
Moseley, Henry Gwyn, 395, 482
Moseley's Law, **395**
Mössbauer effect, 155
Motion: Aristotle's three laws, 21; Brownian motion, 151–52, 185; Descartes, 130; Eötvös effect, 164; fluid dynamics, 113–14; harmonic/wave analysis, 195; inertial/gravitational, 163; Lagrangian point, 325; Mach's principle, 362–63; molecular, 440–41; momentum, 412; Newton's three laws, 152, 169, 406–8; thermodynamics, 306, 506–7; three-body problem, 16
Mott, Nevill, 14, 550
Moulton, Forest, R., 293
Mountains, 45
MRI. *See* Magnetic Resonance Imaging

Muller, Hermann Joseph, 395–96
Müller, Johannes P., 48
Müller, Karl Alexander, 90
Müller, Walther, 223
Müller von Konigsberg, Johannes. *See* Regiomontanus
Muller's Theory of Mutation, **395–96**
Mulliken, Robert Sanderson, 396
Mulliken's Theory of Chemical Bonding, **396**
Mullis, Kary Banks, 396–97
Mullis' Theory for Enzymatic Replication of DNA, **396–97**
Mu-meson. *See* Muon
Muon, 13, 440, 581
Murphy's law, 395
Musschenbrock, Pieter van, 557

Nambu, Yoichiro, 399–400
Nambu's Theory for the "Standard Model," **399–400**
Nanotechnology, 79, 321–22
Nansen, Fridtjof, 158
Napoleon Bonaparte, 11, 257, 326
Nasar, Sylvia, 400
Nash, Alicia de Lardé, 400
Nash, John Charles Martin, 400
Nash, John F., Jr., 400
Nash's Embedding Theorems, **400**
Nathans, Daniel, 15, 401
Nathans' Theory for Restriction Enzymes, **401**
Natta, Giulio, 401–2, 584
Natta's Theory for High Polymers, **401–2**
Needham, John, 511–12
Néel, Louis Eugène Felix, 402–3
Néel's Theories of Ferrimagnetism and Antiferromagnetism, **402–3**
Neher, Erwin, 403–4
Neher's "Patch Clamp" to Record Small Ionic Currents, **403–4**
Nelmes, Sarah, 294
Nemesis theory, 463
Neon lighting, 92
Nephelometry, 540
Neptune (planet), 15
Nernst, Walther Hermann, 404, 506–7
Nernst's Heat Theorem, **404**
Nervous system, 113, 208–9, 332–33, 350, 460–61, 555–56
Neuron theory, 276, 460–61, 555–56
Neutrinos, 83, 180–81: theories, 38–39, 344, 469, 516
Neutrons, 180–81, 323, 432–33, 510, 521–22
Newcomb, Simon, 404–5
Newcomb's Theory for the Speed of Light, **404–5**
Newlands, John, 405–6
Newlands' Law of Octaves, **405–6**
Newton, Frank, 119
Newton, Isaac, 21, 35, 54, 69, 77, 82, 101, 152, 220, 257; calculus, 50, 346, 409; gravity, 210, 252, 276, 308; laws and principles, 325, 406–9; light, 192, 201, 281, 315, 408–9; motion, 169, 210–11, 252, 308, 334, 406–8
Newton's Law and Principles, **406–9**
Nicholas of Cusa, 409
Nicholas of Russia, Czar, 287
Nicholas' Theory of an Incomplete Universe, **409**
Nicholson, John William, 573
Nicolle, Charles Jules Henri, 409–10
Nicolle's Theory for the Cause of Typhus, **409–10**
Nitrogen, 336
Nirenberg, Marshall Warren, 311–12
Noddack, Ida Tacke, 410–11
Noddack, Walter, 411
Noddack's Hypothesis for Producing Artificial Elements, **410–11**
Noether, Amalie Emmy, 411–12
Noether's Theorem, **411–12**
Norrish, Ronald, 151, 412–13
Norrish's Theory of Very Fast Reactions, **412–13**
Northern lights. *See* Aurora Borealis
Northrop, John Howard, 413
Northrop's Hypothesis for the Protein Nature of Enzymes, **413**
*Novum organum, 34*
Noyce, Robert Norton, 412–14
Noyce's Concept for the Integrated Circuit, **413–14**
Nuclear: accelerating particles, 310–11; artificial reaction, 93–94; beta decay, 181, 423, 574–75, 580; chain reaction, 181–82, 297, 521–22; cold fusion, 187; fission, 205, 249, 380–81, 437; individual nuclei, 541; isomerism, 248–49; nuclear winter, 486–87; parity/symmetry, 566–67; slow neutrons, 180–81; spontaneous fission, 190–91; stellar nucleosynthesis, 195–96; strong force, 235, 560, 578, 580–81; weak force, 233–35, 344–45;. *See also* Thermonuclear
Nuclear Magnetic Resonance (NMR). *See* Magnetic Moment; Magnetic Resonance Imaging
Nucleon theory, 79–80, 262–63
Nucleotides/nucleosides, 533–35
Nucleus, 61

Oak Ridge National Laboratories, 423
Occhialini, Giuseppe, 13
Oceanography, 158–59
Ochoa de Albornoz, Severo, 415–16
Ochoa's Theory for the Synthesis of RNA, **415–16**
Ochsenfeld, Robert, 379
Octet theory, 2
Odling, William, 416
Odling's Valence Theory, **416**
Olitsky, Peter, 442
Oersted, Hans Christian, 12, 177, 416–17
Oersted's Theory of Electromagnetism, **416–17**
Ohm, Georg Simon, 417–18
Ohm's Law, **417–18**
Oken, Lorenz, 418
Oken's Cell Theory, **418**
Olbers, Heinrich, 418–19
Olbers' Paradox, 217–18, **418–19**
Oldham, Richard D., 293
Olduvai Gorge, 339
Oliphant, Marcus, 419–20
Oliphant's Concept of Isotopes for Light Elements, **419–20**
*On the Origin of Species by Means of Natural Selection*, 121
Ontogeny, 239, 247
Oort cloud, 322, 421
Oort, Jan, 290, 421
Oort's Galaxy and Comet Cloud Theories, **421**
Oparin, Alexsandr Ivanovich, 197, 421–22
Oparin's Theory for the Origin of Life, **421–22**
Ophthalmoscope, 32
Oppenheimer, Julius Robert, 422–23, 498, 528
Oppenheimer's Contributions to Theoretical Physics, **422–23**
Optics, 1, 7, 33, 309–10, 329
Ostwald, Friedrich Wilhelm, 424
Ostwald's Theories and Concept of Chemistry, **424**
Oxygen (mass figure), 229
Ozone layer, 106–7, 477–78

Packing fraction, 256
Paine, Thomas, 448
PaJaMo experiment, 426
Paleobiology, 397
Paleospecies, 160
Panama Canal, 465–66
Pangaea, 520, 559
Pangenes. *See* Evolution
Panspermia. *See also* Life
Paracelsus, 151, 425–26
Paracelsus' Concept of Medicine, **425–26**
Paradigm, definition of, 168
Parasites. *See* Medicine
Pardee, Arthur, 289, 426–27
Pardee's Theory for Cell Enzyme Synthesis, **426–27**
Parkes, Alexander, 427–28
Parkes' Theory for Separating Metals from Ores, **427–28**
Parthenogenesis, 63
Particle accelerators, 12, 83, 310–11, 337–38, 379, 456–57, 479, 500, 546–47, 547–48
Particle physics: acceleration of charged particles, 310–11, 337–38; antiprotons, 500; beta decay, 180–81, 574–75, 580–81; Cherenkov effect, 523–24; cloud chamber, 568; confirming weak force, 547–48; cosmic rays, 95; detecting subnuclear particles, 232–33; Doppler effect, 141, 514–15; electromagnetic spectrum, 40, 202, 315–16, 377, 533; exclusionary principle, 432–33; four forces, 490; Higgs boson, 274; indeterminacy, 262–63, 456; intermediate vector bosons, 478–79; j/psi particle, 532–33; lepton theory, 440; neutron bombardment, 82–83; nucleon theory, 79–80, 262–63; phase stability, 378–79; positrons, 500; properties of elementary particles, 489–90; quantum mechanics, 135–36; quarks, 174–75, 203–4, 224–27; standard model, 234, 399–400; Stark effect, 515; strangeness, 226–27; Stoney's electron, 518–19; subatomic particles, 225–26, 532–33; symmetry, 227; tracking, 88–89; transmuting atomic particles, 557; two-neutrinos, 516; uncertainty theory, 262, 456; wave/particle hypotheses, 496–97. *See also* Neutrinos; Nuclear; Radiation; Thermonuclear; Weak Force
Pascal, Blaise, 179, 428–30, 525
Pascal's Concepts, Laws, and Theorems, **428–30**
Pasteur, Louis, 24, 94, 197, 354, 422, 430–31, 535
Pasteur's Germ and Vaccination Theories, **430–31**
Paul, Wolfgang, 128, 462
Paulescu, Nicolae, 43
Pauli, Wolfgang, 262, 380, 432–33, 469, 498, 516, 574
Pauli Exclusion Principle, **432–33**
Pauling, Linus Carl, 104, 107, 285, 431–32
Pauling's Theory of Chemical Bonding, **431–32**
Pavlov, Ivan Petrovich, 433–34
Pavlov's Theory of Associative Learning by Respondent Conditioning, **433–34**
Peano, Giuseppe, 434–35
Peano's Axioms and Curve Theorem, **434–35**
Pearman, J. Peter, 143

Pearson, Karl, 435–36
Pearson's Statistical Theories, **435–36**
Peierls, Rudolph Ernst, 205, 436–37
Peierls' Concept for Separating U-235 from U-238, **436–37**
Pendulums, 193–94, 211–12, 276, 282, 471–72
Penicillin, 140–41, 188
Penning trap. *See* Electron trap
Penrose, Roger, 260, 437–39
Penrose's Theories for the Black hole, "Twistors," and "Tiling," **437–39**
Penzias, Arno Allan, 149, 218, 302, 439–40
Penzias' Theory for the Big Bang, **439–40**
Perfect exhaustion, theory of, 16–17
Periodic Table of the Chemical Elements, 115–16, 138–39, 249, 336, 372, 381–82, 387, 395, 405–6, 482
Perkin, Sir William Henry, 150
Perl, Martin Lewis, 440, 469
Perl's Theory for a New Lepton, **440**
Perrin, Jean Baptiste, 440–41
Perrin's Theory of Molecular Motion, **440–41**
Perutz, Max Ferdinand, 441–42
Perutz's Theory of Molecular Structure of Hemoglobin, **441–42**
Petrochemicals, 287
PET (Positron Emission Tomography), 12
Peurbach, Georg von, 466–67
Pfeiffer, Richard Friedrich, 442–43
Pfeiffer's Phenomenon: The Theory of Bacteriolysis, **442–43**
pH scale, 510–11
Pharmacology theories, 150–51, 160–61, 288, 425–26, 556–57
Philolaus, 96
Philosophers' stone, 372, 425, 481
Philosophy: action-at-a-distance, 565–66; ancient, 12–13; Aristotle's four causes, 19–20; Aristotle's three classes of living things, 20; Descartes, 131; Galen's 209; inductive reasoning, 34–35; Laplace's demon (determinism), 334; logical positivism, 467–68; water as basis of lie, 529–30; Zeno's paradox, 583–84
Phipps, James, 294
Phlogiston theories, 81, 336–37, 514
Photoelectric effect, 348
Photography, 111–12, 143–44, 208
Photometer, 329
Photosphere, 147
Photosynthesis, 74–75, 94, 285, 486
Phylogeny, 239, 247
Physiology, 433–34
Pi, 16–17, 85, 352, 509
Piazzi, Giuseppe, 59
Pierce, John Robinson, 414
Piezoelectricity. *See* Electricity
Pilocaine, 333
Pilot wave theory, 59–60
"Pinch effect," 12
Pincus, Gregory, 85
Pines, Herman, 287
Pion, 581
Pisano, Leonardo. *See* Fibonacci
Pitchblende, 108
Pitchfork, Colin, 292
Planck, Max, 152, 315, 380, 443, 456, 468
Planck's constant, 262, 443
Planck's Formula and Quantum Theory, **443**
Planets: brightness, 97; Cassini division, 317; classification, 267; discovery, 15, 31, 59, 266–67; distance from sun, 97; epicycle motion, 98; formation, 277, 562; Galileo's theories, 212–13; Gliese 581 c, 488; irregular motion, 168; Kuiper belt, 322; laws of motion, 308–9; life on, 236–37, 358, 487–88; Martial channels, 493; NASA expedition, 488; orbital laws, 59; origins, 69, 291, 322; positions, 96–97; Ptolemaic system, 452; Roche limit, 475; Saturn's rings, 493;
Plants: classification, 75–76, 353–54, 464, 530; cytology, 519; growth, 264–65; Mendel's law, 382–83; photosynthesis, 285, 486; structure, 363–64, 493–94. *See also* Taxonomy
Plastics, 401, 427–28, 584
Plate tectonics. *See* Geology
Plato, 120, 309
Playfair, John, 358
Plücker, Julius, 47
Pogson, Norman, 443–44
Pogson's Theory for Star Brightness, **443–44**
Polariscope, 33
Polarization. *See* Light
Polio virus, 162–63
Pollack, J.B., 487
Ponnamperuma, Cyril Andrew, 444, 488
Ponnamperuma's Chemical Theory for the Origin of Life, **444**
Pons, Stanley, 188
Popov, Alexsandr, 368
Population catastrophe theory, 364–65
Population genetics. *See* Genetics
Port Royal Society, 429
Porter, George, 151, 412

Porter, Rodney Robert, 444–45
Porter's Theory for the Structure of Human Gamma Globulin, **444–45**
Poseidonius of Apamea, 445–46
Poseidonius' Concept of the Earth's Circumference, **445–46**
Positron, 13, 500
Powell, Cecil, 581
Poynting, John Henry, 446
Poynting's Theories, **446**
Precession, 67, 313
Preformation. *See* Evolution
"The present is the key to the past.," 258
Prévost, Pierre, 447
Prévost's Theory for the Exchange of Heat Radiation, **447**
Priestley, Joseph, 101, 285, 336–37, 447–49
Priestley, Mary Wilkinson, 448
Priestley's Theories of Electrical Force and Dephlogisticated Air, **447–49**
Prigogine, Ilya, 449–50
Prigogine's Theories of Dissipative Structure and Complex Systems, **449–50**
Prisms, 202–3
Probability theories. *See* Mathematics
Progesterone, 137
Prokhorov, Alexsandr M., 362, 363, 537
Proposition, definition of, 168
Proteins theories, 80, 491, 533, 573–74
Proust, Joseph-Louis, 450–51
Proust's Law of Definite Proportions, **450–51**
Ptolemy of Alexandria, 7, 8, 96, 451–52
Ptolemy's Theory of a Geocentric Universe, **451–52**
Pulsars, 236, 272–73, 525–26
Punctuated equilibrium. *See* Evolution
Purcell, Edward Mills, 166, 367, 452–53
Purcell's Theory of Nuclear Magnetic Resonance (NMR), **452–53**
Pyrometer, 119
Pythagoras of Samos, 54, 179, 195, 453
Pythagoras' Theorems, 179, **453**

Quantum disorder, 14
Quantum electrodynamics (QED), 145–46, 183–84, 329–30, 498, 534–35
Quantum gravity, 260
Quantum leap, 60, 128–29, 198, 455
Quantum Theories: From 1900 to 2008, **455–57**
Quantum theories: action theory, 443; atomic structure, 60–61; correspondence principle, 61; energy, 198–99; exclusionary principle, 432–33, 478; Hall effect, 318; hydrogen atom, 329–30; intermediate vector bosons, 478–79; light, 443, 523–24; mechanics, 135–36; relativity, 135, 570–71; standard model, 399–400; tunnel diodes, 166–67; uncertainty principle, 262; unifying weak forces, 233–35; wave function, 169–70, 496–97
Quarks (theories), 174–75, 203–4, 224–27, 399–400, 344, 440
Quasars, 482–83, 490, 494

Rabi, Isidor Isaac, 166, 452, 459–60
Rabi's Theory of Magnetic Moment of Particles, **459–60**
Rabies vaccine, 431
Radar, 141, 557–58
Radiation: alpha/beta/gamma, 480; blackbody radiation, 315, 441, 515–16; cathode ray tube, 105, 475–76; Cherenkov effect, 523–24; cosmic, 476–77; cloud chamber, 568; Compton effect, 94–95; Crookes, 105; Curies' hypotheses, 107–8; diminishing sunlight, 329; Doppler effect, 540; electromagnetic, 376–77; exchange of heat, 447; fluorescence, 518; genetic mutation, 395–96; infrared, 266; ionization of gases, 271–72; Kirchhoff's law, 315; microwaves, 218, 240, 362–63; Mössbauer effect, 155; solar, 446; Van Allen radiation belt, 545–46; Wien's displacement law, 566. *See also* Electromagnetism; Radioacitivity
Radicals, 550–51
Radio: astronomy, 357–58, 482–83; broadcast, 182; radar, 141, 557–58; radioactive series, 510; stellar interference, 290; waves, 463
Radioactivity: artificial, 296–97, 410–11; atomic transmutation, 411, 480–81; beta decay, 180–81, 574–75, 580–81; cathode ray, 105; Curies' hypotheses, 108; decay of isotopes, 222–24; fallout, 272; isomerism, 248–49; packing fraction, 256; radioactive displacement law, 510; radioisotopes, 83; separating U-235 from U-238, 420, 423, 436–37, 543; x-ray fluorescence, 475–76. *See also* Radiation
Radiometer, 105
Raman, Sir Chandrasekhara Venkata, 460

Raman's Theory of Light Scattering, **460**
Ramanujan, Srinivasa, 255
Ramón y Cajal, Santiago, 460–61
Ramón y Cajal's Neuron Theory, **460–61**
Ramsay, Sir William, 291, 355, 461
Ramsay's Hypothesis for Inert Gases, **461**
Ramsey, Norman Foster, 128, 462
Ramsey's Chemical Shift Theory for Improved MRI, **462**
Rankine, William, 82
Raoult, François-Marie, 34, 462
Raoult's Law, **462**
Raup, David Malcolm, 462–63
Raup's Theory of Cyclic Extinction of Animals, **462–63**
Ray, John, 464
Ray's Theories of Fossils and Plant Classification, **464**
Rayleigh, Third Baron John, 461, 463–64
Rayleigh's Light Scattering Law, **463–64**
Reamur, Rene-Antoine, 82
Reber, Grote, 357
Recapitulation theory, 37, 239. *See also* Evolution
Redi, Francesco, 24, 196–97, 354, 422, 430, 464–65, 511
Redi's Theory of Spontaneous Generation, **464–65**
Red shift (redshift), 141, 148, 187, 278, 280, 508, 514. *See also* Doppler effect; Hubble's Law
Reed, Walter, 465–66
Reed's Theory of the Transmission of Yellow Fever, **465–66**
Regiomontanus, 466–67
Regiomantanus' Theory for Trigonometry, **466–67**
Regression analysis, 436
Reichenbach, Hans, 467–68
Reichenbach's Theory of Probability Based on Logical Empiricism (aka Logical Positivism), **467–68**
Reichstein, Tadeus, 261, 307, 468
Reichstein's Theory of the Chemical Role of the Adrenal Gland, **468**
Reines, Frederick, 180, 440, 469
Reines' Theory of Natural Neutrinos, **469**
Relatvity. *See* Einstein
Reproduction, 20–21, 85, 511–12. *See also* Life
Restriction enzymes, 16
Retrosynthetic analysis, 98–99
Retrovirus, 41, 214–15, 393–94
Revelle, Roger Randall, 469–71
Revelle's Theory of Global Warming, **469–71**
RH antibodies, 91
Ricciolo, Giovanni Battista, 471–72
Ricciolo's Theory of Falling Bodies, **471–72**
Richardson, Owen Williams, 472
Richardson's Law of Thermionic Emission, **472**
Richter, Burton, 532–33
Richter, Charles Francis, 472–73
Richter, Jean, 54
Richter scale, 472
Richter's Theory of Earthquake Magnitude, **472–73**
Riemann, Georg Friedrich B., 245, 473–74
Riemann's Theory for Differential Geometry, **473–74**
"Right hand rule," 12
RNA (ribonucleic acid): bacterial enzyme synthesis, 415–16; cell enzyme synthesis, 426–27; colinearity of DNA/proteins, 579; genetic information, 528; nucleotides/nucleosides, 534–35; protein nature of enzymes, 43; protein synthesis, 80; reverse transfer of RNA to DNA, 41, 528; splicing of DNA, 501–2; split genes, 474–75; synthesizing oligonucleotides, 311–12
Robbins, Frederick Chapman, 162–63, 474
Robbins' Theory for the Polio Virus, 162–63, **474**
Roberts, Richard John, 474–75, 501–2
Roberts' Theory of Split Genes, **474–75**
Robertson, Howard, 446
Roche, Edouard Albert, 475
Roche's "Limit" Theory, **475**
Röentgen, Wilhelm Conrad, 46, 475–76
Röentgen's Theory of X-Rays, **475–76**
Römer, Olaus (Olé), 82, 174, 476
Römer's Theory for the Speed of Light, **476**
Roosevelt, Eleanor, 137
Roosevelt, Franklin D., 140, 423, 522
Rossi, Bruno Benedetti, 476–77
Rossi's Theory for Cosmic Radiation, **476–77**
Rowland, F. Sherwood, 106, 477–78
Rowland, Henry Augustus, 229
Rowland's Theory of Chlorofluorocarbons' Effects on the Ozone, **477–78**
Royal Observatory, Greenwich, 71
Royal Society, 66
Rubbia, Carlo, 89, 478–79, 547–48
Rubbia's Theory of Intermediate Vector Bosons, **478–79**

Rubin, Vera, 479
Rubin's Theory of Dark Matter, **479**
Rumford, Benjamin Thomson, 299, 479–80
Rumford's Theory of Relating Work to Heat, **479–80**
Russell, Bertrand, 64, 565, 573
Russell, Henry Norris, 268, 480
Russell's Theory of Stellar Evolution, **480**
Rutherford, Ernest, 60, 223, 249, 419, 455, 478, 480–82, 510, 568
Rutherford's Theories of Radioactivity/Transmutation and Atomic Structure, **480–82**
Rydberg, Johannes Robert, 40, 482
Rydberg's Theory of Periodicity for Atomic Structure, **482**
Ryle, Sir Martin, 272, 482–83
Ryle's Theory of Using Radio Astronomy for Observing Distant Galaxies, **482–83**

Sabatier, Paul, 287
Sabin, Albert Bruce, 485–86
Sabin's Theory for Attenuated Live Polio Vaccine, **485–86**
Sachs, Julius von, 486
Sachs' Theory of Photosynthesis, **486**
Sagan, Carl Edward, 143, 486–88
Sagan's Theories of Nuclear Winter and the Cosmos, **486–88**
Saha, Meghnad N., 489
Saha's Theory of Thermal Ionization, **489**
Sakharov, Andrei Dmitriyevich, 489
Sakharov's Nuclear Fusion Theory, **489**
Sakman, Bert, 403
Salam, Abdus, 233, 345, 489–90, 559–60
Salam's Theory for the Properties of Elementary Particles, **489**
Salk, Jonas, 485
Salvarsan, 140
Sandage, Allan Rex, 490–91, 494
Sandage's Theories of Quasars and the Age of the Universe, **490–91**
Sandia National Laboratory, 11–12
Sanger, Frederick, 230, 491
Sanger's Theories of the Structure of Proteins and Gene Splitting, **491**
Sarich, Vincent, 492
Sarich's Theory of Utilizing Protein to Genetically Date Man/Ape Divergence, **492**
Sauveur, Joseph, 56
Savart, Felix, 55–56
Schally, Andrew, 577
Scheele, Karl Wilhelm, 492
Scheele's Theory of the Chemical Composition of Air, **492**
Schiaparelli, Giovanni Virginio, 251, 36–58, 492–93
Schiaparelli's Theory of Regularity in the Solar System, **492–93**
Schleiden, Matthias Jakob, 493–94, 497
Schleiden's Cell Theory for Plants, **493–94**
Schmidt, Maarten, 494
Schmidt's Theory of the Evolution and Distribution of Quasars, **494**
Schneider, Stephen Henry, 495–96
Schneider's Theory of Biological Systems and Climate Change, **495–96**
Schott, Otto, 1
Schrieffer, John, 43, 44, 297
Schrödinger, Erwin, 127, 136, 262, 496–97, 580
Schrödinger's cat, 496
Schrödinger's Theory of Wave Mechanics, **496–97**
Schwann, Theodor, 493–94, 497
Schwann's Theory of Animal Cells, **497**
Schwartz, Melvin, 344, 516
Schwarzschild, Karl, 497–98
Schwarzschild's "Black Hole" Theory, **497–98**
Schwinger, Julian Seymour, 145, 183, 498, 534
Schwinger's Theory for Renormalization, **498**
Scott, David, 211
Seaborg, Glenn Theodor, 378, 499, 585
Seaborg's Hypothesis for Transuranium Elements, **499**
Seebeck, Thomas Johann, 343, 499–500
Seebeck's Theory of Thermoelectricity, **499–500**
Segrè, Emilio Gino, 500
Segrè's Hypothesis for the Antiproton, **500**
Seismograph, 85
Seismological tables, 293
Selten, Reinhard, 400
Semiconductors, 166–67, 297–98, 321, 331, 413–14, 503–4
Sepkoski, J. John, 462
Serber, Robert, 423
Set theory. *See* Mathematics
SETI (Search for Extraterrestrial Intelligence), 143
Shankin, Jonathan, 106
Shapley, Harlow, 290, 341–42, 500–501
Shapley's Theory of Globular Clusters, **500–501**
Sharp, Phillip Allen, 474, 501–2
Sharp's Theory for the "Splicing" of DNA, **501–2**

Sharpey-Schafer, Sir Edward A., 43
Shepard, Francis Parker, 502–3
Shepard's Theory of Submarine Canyon Formation, **502–3**
Sherman, Irving J., 286
Shockley, William Bradford, 43, 44, 414, 503–4
Shockley's Theory of Semiconductors, **503–4**
Shoemaker-Levy comet, 421
Sickle cell theory, 286–87, 432
Sidereal period, 97
Siderophiles, 9
Sidgwick, Nevil Vincent, 504–5
Sidgwick's Theory of Coordinate Bonds, **504–5**
Siemens, Carl Wilhelm, 505–6
Siemens, Ernst Werner, 505–6
Siemens' Theory for Regenerating Heat, **505–6**
Sieve of Eratosthenes, 165
*Silent Spring*, 410, 466
Simon, Sir Francis (Franz), 506–7
Simon's Third Law of Thermodynamics, **506–7**
Simpson, O.J., 292, 397
Singer, S. Fred, 487, 545
Singularities. *See* Space/time theories
Sixtus IV, Pope, 467
Skinner, B. F. (Burrhus Fredeic), 433
Slipher, Vesto Melvin, 507–8
Slipher's Theory of Interstellar Gases and Andromeda, **507–8**
Smalley, Richard E., 108, 321
Smallpox, 293–94
Smith, Hamilton O., 15, 401
Smith, Michael, 396
Smith, William, 5
Smoot, George Fitzgerald, 508
Smoot's Theory of a Nonuniform Universe, **508**
SNARC (Stochastic Neural-Analog Reinforcement Computer), 391
Snell, Willebrord van Roijen, 509–10
Snell's Law, 374, **509–10**
Sociobiology, 433–34, 576
Soddy, Frederick, 176, 437, 481, 510
Soddy's Displacement Law for Radioactive Decay and Theory of Isotopes, **510**
Soddy-Fajans method, 176
Solar (sun): atmosphere, 354–55; core pressure, 147; core temperature, 147; diameter of sun's core, 147; distance from Earth, 80–81; helium, 290–91; layers, 147–48; magnitude, 444; Milky Way, 421; nebula, 421; neutrino model, 38–39; parallax, 513–14; perigree, 8; radiation, 446; sunspots, 142, 250–51, 373; temperature, 515; temperature cycles, 218; wind, 56; year 8
Solar System: asteroid belt, 316–17; dissipative structures, 449; heliocentric, 213; Kuiper belt, 322; movement, 266; nebula/planetary hypothesis, 334–35, 562; nemesis theory, 463; Oort's comet cloud, 421; origin, 277; regularity, 493–94; size, 80; sunspots, 142. *See also* Galaxies; Planets
Solution, definition of, 319
Somatic mutation, 295
Sommerfeld, Arnold, 432, 468
SONAR, 332
Sorensen, Soren Peter Lauritz, 510–11
Sorensen's Negative Logarithms Representing Hydrogen Ion Concentration, **510–11**
Sound: barrier, 361; Doppler principle, 141; Radar, 557–58; SONAR, 332; sonic boom, 361; speed, 220, 362; theories, 276, 361–62; ultrasound, 332
Southern cross, 253
Space/time theories: big bang, 217–18; Hawking's, 259–61; Lorentz invariant, 356; Minkowski's, 390–91; singularities, 437–38; special relativity153–54; superstrings, 570–71; twisters, 438
Spallanzani, Lazarro, 197, 511–12
Spallanzani's Theory Refuting Spontaneous Generation, **511–12**
Special relativity theory, 153–54
Species, 18, 109, 139, 161–62, 569–70. *See also* Evolution; Genetics; Taxonomy
Spectrograph, 25
Spectroheliograph, 251
Spectroscopy, 69–70, 280–81, 316, 355, 541
Spedding, Frank Harold, 512–13
Spedding's Theories, **512–13**
Spencer-Jones, Sir Harold, 513–14
Spencer-Jones' Concept for Measuring Solar Parallax, **513–14**
Spill, Daniel, 427–28
Spontaneous generation. See Life
Stadia, 446
Stahl, Franklin, 384
Stahl, Georg Ernst, 514
Stahl's Phlogiston Theory, **514**
Stalin, Joseph, 133, 190, 359, 561
Standard deviation, 436

Standard model for physics particles, 234, 345, 399–400, 478, 490
Stanford Linear Accelerator Center (SLAC), 203, 533
Stanley, Wendell, 413
Stark, Johannes, 514–15
Stark's Theories, **514–15**
Stars: black holes, 236, 259–60, 437–38, 494, 497–98; blue star objects (BSO), 490; brown dwarfs, 40–41; brightness, 148, 443–44; Cepheid variables, 340–41; chemical composition, 280–81; chemical elements, 70; classes, 268–69, 341; constellations, 303; Drake equation, 142–43; expansion/contraction, 341; equilibrium, 147; evolution, 480; formation, 61–62, 561–62; galactic rotation, 302–3; giants, 269; globular clusters, 500–501; groups, 9–10; Hertzsprung-Russell Diagram, 268–69, 480; interstellar gases, 507; interstellar matter, 519–20; mass/luminosity, 29–30, 148, 267–68; Mizar (double star), 471; motion, 252–53; nebulae, 267, 562; neutron, 236, 585–86; nova, 148; nucleosynthesis, 195–96; parallax, 53, 66–67, 253, 480; periodicity/luminosity, 340–41; populations, 29; pulsars, 236, 272–73, 525–26; quasars, 482–83, 490, 494; red dwarfs, 269; red shift, 141, 278; Schwarzschild radius, 497–98; spectra, 281; supergiants, 268; supernova, 148, 585–86; systems, 53; thermal ionization, 489; types, 10, 480; white dwarfs, 84, 269. *See also* Galaxies
Statistics, 114, 221, 435–36, 542
Stefan, Jozef, 515
Stefan's Theory of Black Box Radiation, **515**
Steinberger, Jack, 344, 516
Steinberger's Two-Neutrino Theory, **516**
Stellar. *See* Stars
Steno, Nicolaus, 121, 257, 516–17
Steno's Theory for Fossil Formation, **516–17**
Stensen, Niles. *See* Steno, Nicolaus
Stern, Otto, 459, 517–18
Stern's Theory for the Magnetic Moment of the Proton, **517–18**
Sternberg, United States Surgeon General George Miller, 465
Steroids, 307–8, 468
Stevin, Simon, 210, 429, 471
Stillman, Benjamin, 117
Stillman, Henrietta, 117
Stokes, George, 518
Stokes' Law of Hydrodynamics and Fluorescence, **518**
Stoney, George Johnstone, 518–19
Stoney's Theory of the Electron, **518–19**
Strait of Gibraltar, 7
Strange attractor, 356
Strasburger, Eduard, 519
Strasburger's Law of Cytology, **519**
Strassman, Fritz, 61, 380, 381
Stratification. *See* Geology
Stratigraphy, 6
String/superstring theories, 156, 399, 548, 570–71
Strong force, 235, 560, 578, 580–81
Strutt, John William. *See* Kelvin Rayleigh, Baron
Struve, Frederich, 520
Struve, Otto, 519–20
Struve's Theory of Interstellar Matter, **519–20**
Sturgeon, William, 265
Subatomic particles. *See* Particle physics
Suess, Eduard, 520–21, 559
Suess' Theory of Continental Drift, **520–21**
Sulfa drugs, 140–41
Sullivan, Louis Henri, 109
Sumner, James, 413
Sun. *See* Solar (sun)
Sunspot cycle, 251
Superconducting super collider, 274
Superconductivity theories: BCS, 43–44, 297; Casimir force, 79–80; high temperature, 90–9, 372–73; Josephson junction, 297–98; low temperature, 301–2; Meissner effect, 379–80; two-fluid model for helium, 330;
Superfluidity theory, 330
Swammerdam, Jan, 374, 521
Swammerdam's Theory of Preformation, 374, **521**
Swan, Sir Joseph Wilson, 49–50
Sylvius, Jacobus, 551
Symmetry, 411–12
Synchrocyclotron, 379, 548. *See also* Particle Accelerators
Synge, Richard L. M., 369–70
Synodic period, 97
Synthesization, 275
Syphilis, 140, 198
Szilard, Leo, 521–22, 583
Szilard's Theory of Neutrons Sustaining a Chain Reaction, **521–22**

Talbot, William Henry Fox, 112
Tamm, Igor Yevgenyevich, 523–24
Tamm's Theory of the Cherenkov Effect, **523–24**
Tarski, Alfred, 42
Tartaglia, Niccolo (Fontana), 78, 524–25

Tartaglia's Mathematical Solution to Cubic Equations, **524–25**
Tatum, Edward Lawrie, 343, 525
Tatum's Theory of Gene-Controlling Enzymes, **525**
Taxonomy, 20, 68, 75–76, 109, 120, 353–54, 464, 530
Taylor, Joseph Hooton, Jr., 525–26
Taylor, Richard, 203
Taylor's Theory of Gravitational Waves, **525–26**
Tektites, 9
Telegraph, 266, 307, 368, 506
Telescopes, 212–13, 251–52, 267, 357–58, 419, 482–83
Teller, Edward, 423, 526–28, 542
Teller's Theory for the Hydrogen Bomb, **526–28**
Temin, Howard Martin, 41, 144, 528
Temin's Theory for Transcribing RNA Information into DNA, **528**
Temperature: Celsius (centigrade), 82; definition, 58; Fahrenheit, 174; measurements, 212–13, 306
Tesla, Nikola, 150, 244, 269, 357, 528–29
Tesla's Concept of High-Voltage Alternating Current, **528–29**
Tesselations. *See* Tile theory
Tetanus, 47–48
Thales of Miletus, 12, 529–30
Thales' Theory that Water is the Basis for all Things, **529–30**
Thenard, Louis Jacques, 222
Theophrastus, 530
Theophrastus' Concepts for Plant Classification, **530**
Theorell, Axell Hugo Theodor, 530–31
Theorell's Theory of Enzyme Action, **530–31**
Theorem, definition of, 168
Theory of Everything (TOE), 156, 234, 241, 263, 560. *See also* Grand unification theories
Thermal conductivity, 195
Thermal ionization theory, 489
Thermocouple, 343, 500
Thermodynamics, 92–93; Carnot cycle theory, 78–79; chemical, 228–29, 229–30; conservation of energy, 92; entropy theory, 92; first law, 92, 306; Kelvin's concepts, 305–6; Maxwell's Demon theory, 375–76; negative temperature system theory, 462; second law, 92–93, 306, 331; third law, 404, 506–7; very-large-scale integration, 331. *See also* Heat
Thermoelectricity theory. *See* Electricity
Thermometers, 174, 211–12
Thermometry, 343
Thermonuclear energy: carbon-nitrogen cycle, 54–55; cold fusion, 187–88; energy reaction, 255–56; fusion, 526 27; hydrogen-helium-energy, 420; stellar nucleosynthesis, 195–96. *See also* Hydrogen
Thermoscope, 174
Thermionic. *See* Electricity
Thomson, George P., 122, 127
Thomson, James J., 105, 390, 441, 455, 478, 531–32, 538, 568
Thomson's Electron Theory, **531–32**
Tile theory, 438–39
Timekeeping, 211–12, 282, 284–85
Ting, Samuel Chao Chung, 89, 532–33
Ting's Theory for a New Photon-Like Particle, **532–33**
Tiselius, Arne Wilhelm Kaurin, 533
Tiselius' Hypothesis for Protein Analysis, **533**
Titius, Johann, 59
Tizard, Henry, 93
Todd, Alexander Robertus, 103, 533–34
Todd's Theory for the Structure and Synthesis of Nucleotides, Nucleosides, and Nucleotide Coenzymes, **533–34**
Tomonaga, Sin-Itiro, 145, 183, 498, 534–35
Tomonaga's Theory of Relativistic Quantum Electrodynamics, **534–35**
Tonegawa, Susumu, 295, 535–36
Tonegawa's Theory of Antibodies and the Immune System, **535–36**
Toon, O. B., 487
Torricelli, Evangelista, 428, 536–37
Torricelli's Vacuum and Theorem, **536–37**
Torsion balance, 163
Townes, Charles Hard, 362, 363, 527, 537–38
Townes' Theory for Amplifying Electromagnetic Waves, **537–38**
Townsend, Sir John Sealy, 538
Townsend's Theory of Collision Ionization, **538**
Trade winds. *See* Meteorology
Transistors, 44, 413–14, 503–4
Tricarboxylic acid cycle. *See* Krebs Cycle
Trigonometry. *See* Mathematics
Truman, Harry S., 526
Tsvet, Mikhail Semyonovich, 370

Tuberculosis, 319
Turco, R.P., 487
Turing, Alan Mathison, 394, 538–39
Turing's Theory for Testing Computer Intelligence, **538–39**
Turner, David Warren, 539
Turner, Herbert H., 293
Turner's Theory for Measuring Outer Energy Levels of Molecules, **539**
Twistor theory, 438
Tyndall, John, 463, 540
Tyndall's Theory for the Transmission of Light through Gases, **540**
Typhus, 409–10, 442–43

Uhlenbeck, George Eugene, 541–42
Uhlenbeck's Theory of Electron Spin, **541–42**
Ulam, Francoise, 542
Ulam, Stanislaw Marcin, 526–27, 542
Ulam's "Monte Carlo" System, **542**
Ultrasound, 332
Uncertainty principle, 262. *See also* Bohm; Heisenberg, Werner; Quantum theory
Unified field theory (UFT), 155–56, 262
Uniformitarianism. *See* Evolution
Universal gas equation. *See* Ideal Gas Law
Universe: age, 279, 490–91; changing, 67–68; closed (static), 260, 276–77; dark matter, 30, 39, 479, 520, 572; expanding/contracting, 491; fundamental forces, 235; geocentric, 451–52; heliocentric, 96, 213; horizon paradox, 392; incomplete, 409; inflationary, 149, 204–5, 240–41, 260–61, 508; multiple universe, 169–70; nature and origin, 62, 217–18, 258–59, 347–48; nonhomogeneous, 224; non-static, 347–48; nonuniform, 508; Olbers' paradox, 418–19; size, 97–98; singularities, 260–61, 491, 508; spherical, 85; steady state, 236, 276–77, 483. *See also* Big Bang; Black holes
Upatnieks, Juris, 208
Urea cycle, 320
Urey, Harold Clayton, 197, 250, 542–43
Urey's Gaseous Diffusion and Origin of Life Theories, **542–43**

Vaccination. *See* Inoculation
Vacuum theory, 536–37
Valence. *See* Chemistry
Van Allen, James Alfred, 543–44
Van Allen Radiation Belts, **543–44**
Van de Graaff, Robert Jemison, 546–47
Van de Graaff's Concept of Producing High Voltage, **546–47**
Van der Meer, Simon, 89, 478, 547–48
Van der Meer's Theory of Particles to Confirm the "Weak Force," **547–48**
Van der Waals, Johanne Diderik, 548–49
Van der Waals' Equation for Gas Molecules, **548–49**
Van der Waals force, 79
Van Musschenbrock, Peter, 557
Van Wesel, Andries, 551
Van Wesel, Everhard, 551
Van't, Hoff, Jacobus Henricus, 343, 462, 549–50
Van't Hoff's Theory of Three Dimensional Organic Compounds, **549–50**
Van Vleck, John Hasbrouck, 14, 550–51
Van Vleck's Theory of Paramagnetism, **550–51**
Variometer, 163
Vavilov, Nikolai, 359
Venturi effect, 51
Venturi, Giovanni Battista, 51
Vernalization, 359
Vernov, Sergei N., 545
Vesalius, Andreas, 176, 551–51
Vesalius' Theories of Anatomy and Physiology, **551–52**
Vespucci, Amerigo, 467
Victoria of England, 306
Virchow, Rudolf Carl, 552
Virchow's Cell Pathology Theory, **552**
Virus theories: HIV/AIDS, 214–15, 393–94; influenza, 442–43; interferon, 288; polio, 162–63, 474, 485–86
Vitalism, 571
Vitamins, 468, 531, 534
Volcanoes. *See* Geology
Volta, Alessandro Giuseppe Antonio Anastasio, 118, 124, 157, 217, 417, 552–53
Volta's Concept of an Electric Current, **552–53**
Von Kleist, Ewald Georg, 557
Von Laue, Max, 275, 553
Von Laue's Theory for the Diffraction of X-Rays in Crystals, **553**
Von Neumann, John, 87, 95, 377–78, 553–54
Von Neumann's Theory of Automata, **553–54**
Vulcanization, 506

Waddington, Conrad Hal, 555
Waddington's Theory of Genetic Assimilation, **555**
Waldeyer-Hartz, Henrich Wilhelm Gottfried von, 555–56

Waldeyer-Hartz Neuron Theory, **555–56**
Wallace, Alfred Russel, 76, 121, 556
Wallace's Theory of Evolution by Natural Selection, **556**
Wallach, Otto, 556–57
Wallach's Theory for the Molecular Structure of Organic Compounds, **556–57**
Waller, Augustus, 157
Wallis, John, 77
Walton, Ernest Thomas Sinton, 93, 310, 337, 557
Walton's Concept for Transmuting Atomic Particles, **557**
Water: composition, 81; rate of flow, 537; surface energy with temperature, 163–64
Water screw. *See* Helical pump
Watson, James, 80, 86, 103–5, 160, 188–200, 218, 230, 349, 384, 395, 432, 557, 579
Watson, Sir William, 557
Watson's Theory of Electricity as a Fluid, **557**
Watson-Crick Theory of DNA, **557**. *See* Crick-Watson Theory of DNA
Watson-Watt, Sir Robert Alexander, 557–58
Watson-Watt's Concept of Radar, **557–58**
Wave mechanics. *See* Quantum theory
Wave: amplifying electromagnetic, 537–38; Doppler's principle, 141–42; electric/magnetic field vectors, 306–7; electromagnetic radiation, 152, 376–77, 537–38, 566; fluorescent, 518; gravity, 525–26, 558–59; harmonic analysis, 195; Hertzian, 269–70, 368; light, 94–95, 408–9, 579–80; matter, 127; photoelectric effect, 348; P/S (seismological) 293; Radar, 555–58; Rayleigh, 464; seismographic, 393; sound and light waves, 276; ultrasound, 332; velocity, 270
Weak force: beta decay, 181, 574–75; confirming, 547–48; conservation of parity, 578; definition, 560; example, 567; Higgs bosons, 274; unified field theory, 155–56; unifying theory, 233–35; nuclear interaction, 344–45
Webb, Edward, 5
Weber, Joseph, 558–59
Weber, Wilhelm, 221
Weber's Theory of Gravitational Waves, **558–59**
Wednesday Society, 262
Wegener, Alfred Lothar, 520, 559
Wegener's Theory of Continental Drift, **559**
Weinberg, Steven, 233, 277, 345, 489, 527, 559–60
Weinberg, Wilhelm, 255
Weinberg's Grand Unification Theories, **559–60**
Weismann, Friedrich Leopold August, 560–61
Weismann's Germ Plasm Theory, **560–61**
Weizsäcker, Carl Friedrich von, 561–62
Weizsäcker's Theories of Star and Planet Formation, **561–62**
Weller, Thomas H. 162–63, 474
Werner, Abraham Gottlob, 359, 563
Werner, Alfred, 562–53
Werner's Coordination Theory of Chemistry, **562–63**
Werner's Neptunian Theory (Neptunism), **563**
Weyl, Hermann, 411
Wheeler, John Archibald, 259, 564
Wheeler's "Geon" Theory, **564**
Whipple, Fred Lawrence, 564–65
Whipple's "Dirty Snowball" Theory of Comets, **564–65**
Whitehead, Alfred North, 64, 565–66
Whitehead's "Action-At-A-Distance" Theory of Relativity, **565–66**
Whole number rule, 25
Wickramasinghe, Chandra, 197
Wien, Wilhelm Carl Werner, 566
Wien's Displacement law, **566**
Wigner, Eugene Paul, 527, 566–67, 578
Wigner's Concept of Parity/Symmetry in Nuclear Reactions, **566–67**
Wiles, Andrew, 179
Wilkes, Charles, 117
Wilkins, Maurice, 103–5, 199–200, 349
Wilkinson, Sir Geoffrey, 567
Wilkinson's Concept of "Sandwich Compounds," **568**
Williamson, Alexander William, 567–68
Williamson's Theory of Reversible Chemical Reactions, **567–68**
Wilson, Allan Charles, 492, 568–69
Wilson, Charles T. R., 88, 94, 233, 390, 568
Wilson, Edward Osborne, 248, 569–70
Wilson, Robert Woodrow, 149, 218, 302, 439
Wilson's cloud chamber, 232–33
Wilson's Hypothesis of Cloud Condensation, **568**

Wilson's "Out-of-Africa" Theory, **568–69**
Wilson's Theory of Dynamic Equilibrium of Island Populations, **569–70**
Winchester, Simon, 5
Witten, Edward, 570–71
Witten's Superstring Theory, **570–71**
Wohler, Friedrich, 352, 571
Wohler's Theory for Nonliving Substances Transforming into Living Substances, **571**
Wolf, Maximillian Franz Joseph Cornelius, 572
Wolf's Theory of the Dark Regions of the Milky Way, **572**
Wolfram, Stephen, 571–72
Wolfram's Theory of Complex Systems, **571–72**
Wollaston, William Hyde, 202, 315
Wollman, Elie, 289
Woodward, Robert Burns, 275, 572–73
Woodward's Theory of Organic Molecular Synthesis, **572–73**
Worldwide Web of Internet, 385–86
Wright, Almoth, 442–43
Wright, Sewall Green, 573
Wright's Theory of Genetic Drift (Sewall Wright Effect), **573**
Wrinch, Dorothy Maud, 573–74
Wrinch's Cyclol Theory of Protein Structure, **573–74**
Wu, Chien-shiung, 574–75, 578
Wu's Theory of Beta Decay, **574–75**
Wurtz, Charles Adolphe, 575
Wurtz's Theory for Synthesizing Hydrocarbons, **575**
Wu-Zetian of China, 284
Wynne-Edwards, Vero Cooper, 576
Wynne-Edwards' Theory for Group Selection, **576**

X-Rays, 46, 94–95: astronomy, 290; development of x-ray tubes, 441; diffraction, 274–75, 303–4, 441–42, 553; genetic mutation, 395–96; proton number of chemical elements, 395; Röentgen rays, 475–76; Zeeman effect, 541. *See also* Electromagnetism; Radiation; Radioactivity

Yalow, Rosalynn Sussman, 577–78
Yalow's Theory of Radioimmunoassay, **577–78**
Yang, Chen Ning, 344, 578
Yang's Theory of Nonconservation of Parity in Weak Interactions, **578**
Yanofsky, Charles, 579
Yanofsky's Theory for Colinearity of DNA and Protein, **579**
Yellow fever, 465–66
Young, Thomas, 579–81
Young's Wave Theory of Light, **579–80**
Yucatan Peninsula, 9
Yukawa, Hideki, 580–81
Yukawa's Meson Theory for the "Strong Interactions," **580–81**

Zamenhof, Ludovic Lazarus, 4325
Zeeman, Pieter, 355, 583
Zeeman effect, 251, 355, 515, 541
Zeeman's Theory of the Magnetic Effect on Light, **583**
Zeidler, Othmar, 410
Zeno of Elea, 26, 583–84
Zeno's Paradoxes, 26, **583–84**
Ziegler, Karl Waldeman, 401, 584
Ziegler's Theory of Stereospecific Polymers, **584**
Zihlman, Adrienne, 295–96
Zinn, Walter Henry, 585
Zinn's Concept of a "Breeder" Reactor, **585**
Zuckerandl, Emile, 585
Zuckerandl's Theory for Measuring the Rate of Evolution, **585**
Zwicky, Fritz, 585–86
Zwicky's Theory for Supernovas and Neutron Stars, **585–86**

## About the Author

ROBERT E. KREBS has written seven books for Greenwood Press. He has taught chemistry, biology, and other sciences at several schools and universities. Dr. Krebs has served as a science specialist in the federal government and a research administrator in four universities. He retired as Associate Dean for Research in the Graduate College at the Medical Center of the University of Illinois at Chicago. He continues his lifelong pursuit of fostering scientific literacy through education.